Beyond reductionism

The Alpbach Symposium

Beyond reductionism

New perspectives in the life sciences

EDITED BY ARTHUR KOESTLER &
J. R. SMYTHIES

RADIUS BOOK / HUTCHINSON

HUTCHINSON & CO (*Publishers*) LTD
3 Fitzroy Square, London W1

London Melbourne Sydney Auckland
Wellington Johannesburg Cape Town
and agencies throughout the world

First published September 1969
This edition August 1972

Printed in Great Britain by litho on smooth wove paper
by Anchor Press, and bound by Wm. Brendon,
both of Tiptree, Essex

ISBN 0 09 112410 7 (cased)
    0 09 112411 5 (paper)

LIST OF PARTICIPANTS

LUDWIG VON BERTALANFFY   *Faculty Professor, State University of New York at Buffalo*

JEROME S. BRUNER   *Director of the Centre for Cognitive Studies, Harvard University*

BLANCHE BRUNER   *Centre for Cognitive Studies, Harvard University*

VIKTOR E. FRANKL   *Professor of Psychiatry and Neurology, University of Vienna and Head of the Department of Neurology at the Vienna Poliklinik*

F. A. HAYEK   *Professor of Economics, University of Freiburg, Germany*

HOLGER HYDÉN   *Professor and Head of the Institute of Neurobiology and Histology, Faculty of Medicine, University of Göteborg, Sweden*

BÄRBEL INHELDER   *Professor of Developmental Psychology, University of Geneva*

SEYMOUR S. KETY   *Professor of Psychiatry, Harvard University*

ARTHUR KOESTLER   *Writer, London*

PAUL D. MACLEAN   *Chief, Section on Limbic Integration and Behaviour, Laboratory of Neurophysiology, National Institute of Mental Health, Bethesda, Maryland*

DAVID MCNEILL   *Professor of Psychology, University of Chicago*

JEAN PIAGET   *Professor of Experimental Psychology, University of Geneva*

J. R. SMYTHIES   *Reader in Psychiatry, University of Edinburgh, Scotland*

W. H. THORPE   *Director, Sub-Department of Animal Behaviour, Department of Zoology, University of Cambridge*

C. H. WADDINGTON   *Professor and Chairman, Department of Genetics, University of Edinburgh, Scotland*

PAUL A. WEISS   *Emeritus Member and Professor, Rockefeller University, New York*

Preface

For the last ten years, my main interest has been the history and present state of science, and its impact on our view of the world. The books that were the result of this interest brought in a number of invitations to scientific symposia—inter-disciplinary or technical—and provided the opportunity for informal discussions with representatives of some of the main branches of contemporary research. They made me aware of a certain discontent with the prevailing philosophical bias which—whether explicitly formulated or tacitly implied—seems to linger on as a heritage from the nineteenth century, although the new insights gained by contemporary research have reduced it to an anachronism. This led to the idea of organizing, not yet another symposium, but one with a cutting edge, epitomised in the title.

About half of the presentations were based on prepared papers, the others delivered *ex tempore* and tape-recorded. This presented the editors with the familiar dilemma deriving from the differences in style between the spoken and the written word. We have attempted a compromise by letting some repetitive, over-emphatic or anecdotal passages stand, in order to preserve something of the individuality of the speakers. The same dilemma arose in more pointed form in transcribing the discussions, which all too frequently tend to go off at a tangent. We compromised by letting passages stand which, though not strictly relevant to the main theme, referred to some new or controversial features in the speakers' research, and thus fitted in with the sub-title of the Symposium.

As usual on such occasions, some participants called each other by first names, while others preferred a more formal address or alternated between the two. We left it at that, instead of imposing a stilted uniformity.

Three participants, Jerome S. Bruner, Paul D. MacLean and Jean Piaget, who had prepared their presentations well in advance, were at the last minute unable to attend. Bruner's paper was read by Blanche Bruner, MacLean's by J. R. Smythies and the Piaget-Inhelder paper by Bärbel Inhelder.

W. H. Thorpe acted as Chairman throughout the conference. His influence in guiding and moderating the discussions was greatly appreciated by all.

My warmest thanks are due to J. R. Smythies who, in spite of pressing professional duties, found the time to act as co-editor and honorary treasurer. I am equally indebted to Harold Harris and Dr. Daniel Brostoff of Hutchinson, London, for their patient help with the final editing.

The Symposium was financed by advances against royalties by the English, American and Austrian publishers, and an additional grant of $1,000 from the International Association for Cultural Freedom (Ford Foundation).

There is a name index, but no subject index. As most subjects crop up over and again in the discussions, the latter would have been inordinately long, and more confusing than helpful.

The Symposium was held in Alpbach in the Austrian Tyrol. This peaceful mountain village is the seat of a summer University with facilities for scientific conferences. The terraces and *Weinstuben* of the local inns proved excellent catalysers for informal discussions.

A. K.

London, March 1969

Contents

x *Contents*

THE ALPBACH SYMPOSIUM

Opening remarks
Arthur Koestler

Welcome to Alpbach

I think I owe you a word of explanation concerning the morphogenesis of this somewhat unusual conference. It is unusual in three respects. In the first place, it is not sponsored by a foundation or academic body, but by publishers,[1] who hope that a book of some interest will come out of it. The book is meant to be the end-product, and not merely a by-product of the proceedings.

The second unusual aspect of this conference of scientists is that it was convened by a trespasser from the humanist camp. It was therefore both gratifying and surprising that all those whom I approached, with only one exception, accepted the invitation. Obviously there must exist some common ground to explain this positive response, and I hope I am not wrong in believing that it was more or less the same as outlined in the first paragraph of the original invitation. As that was sent out several months ago, perhaps I may remind you that in outlining the purpose of the conference I quoted Dr. Thorpe to the effect that there is "an undercurrent of thought in the minds of perhaps hundreds of biologists", who are critical of the totalitarian claims of the neo-Darwinian orthodoxy; who refuse to believe that the so-called Synthetic Theory provides all the answers to the problems posed by the phenomena of evolution; and who feel that the theory reflects part of the picture but not the whole picture. Such critical tendencies are also in evidence in the other life sciences, from genetics to psychology. There is, for instance, a

[1] Hutchinson, London; Macmillan, New York; Fritz Molden Verlag, Vienna.

I

growing conviction among psychologists that the behaviourists' S-R schema of chained responses, for all its historical merits, is changing from a once useful tool into an impediment to future progress. The common target of these 'holy discontents'—to quote John Donne— seems to be what von Bertalanffy called the robotomorphic view of man, or more soberly, the insufficient emancipation of the life sciences from the mechanistic concepts of nineteenth-century physics, and the resulting crudely reductionist philosophy.

But up to now, no coherent alternative world-view is in sight. There is a groping for a new synthesis, but also a strong feeling that it should not be a premature, abortive synthesis. Nothing would have been easier than to collect in this room a bunch of amiable cranks to concoct a New Philosophy. In fact, New Philosophies under various names are being cooked up all the time at enthusiastic conferences and seminars from Bombay to California; their motto could be, "Cranks of the world unite, you have nothing to lose but your kinks". As that old cynic said, I can cope with my enemies, but God protect me from my friends. If a new synthesis is to emerge, it will emerge from inside the laboratories.

This brings me to the third unusual aspect of this conference. As you have no doubt noticed, invitations were confined to personalities in academic life with undisputed authority in their respective fields, who nevertheless share that holy discontent. The purpose was to find out whether there is an overlap, a common denominator or pattern underlying these critical attitudes in widely varying fields; a negative agreement which might perhaps one day emerge into something positive. To hope for more would be unrealistic; and even this cautious formulation seems rather a mouthful. Anyway, this is the explanation which I felt I owed you before the proceedings start.

The living system: determinism stratified

Paul A. Weiss[*]

Rockefeller University, New York

Introduction: need for the systems concept

Just like the painter, who steps periodically back from his canvas to gain perspective, so the laboratory scientist emerges above ground occasionally from the deep shaft of his specialized preoccupation to survey the cohesive, meaningful fabric developing from innumerable component tributary threads, spun underground much like his own. Only by such shuttling back and forth between the worm's eye view of detail and the bird's eye view of the total scenery of science can the scientist gain and retain a sense of perspective and proportions. My introductory comments to this symposium constitute one such excursion to higher ground.

Although this has brought me close to the portals of philosophy, I have never consciously trespassed into the disciplinary precincts of that branch of learning, for which I would have felt neither equipped nor qualified. I want to stress that at the outset, lest any of my subsequent remarks be misconstrued as pretentious aspirations at philosophical profundity. They are no more than conclusions and postulates derived from pragmatic insights acquired from the study of living organisms. Considering that in contemporary biology the nexus between inductive experimental fact-finding, on the one hand, and theoretical speculations, on the other, is very tenuous indeed, such examples of convergence from both ends as are attempted by this symposium unquestionably could speed the rapprochement, and this essay is meant to contribute to that goal.

[*] In references to the author's publications in the text, abbreviated to "P.W."

3

My prime object here is to document that certain basic controversies about the nature of organisms and living processes, which have for long failed to melt away in the heat of argument (e.g., reductionism versus holism), readily vanish in the light of realistic studies of the actual phenomena, described in language uncontaminated by preconceptions. In this light (1) the *principle of hierarchic order* in living nature reveals itself as a demonstrable descriptive fact, regardless of the philosophical connotations that it may carry. And further (2) the necessity becomes compelling to accept organic entities as *systems* subject to network dynamics in the sense of modern systems theory, rather than as bundles of micro-precisely programmed linear chain reactions. A strictly mechanistic, machine-like, notion of the nature of living organisms presupposes a high degree of precision in the spatial and chronological programme according to which the innumerable concurrent component chains are composed and arrayed—a conception later amplified, but in no way altered, by letting the programme include equally pre-programmed checking and spare mechanisms to keep the bunch of separate processes from falling apart when faced with the fortuitous fluctuations of the outer world.

Animal behaviour: systems dynamics

The explanation by Jacques Loeb (1918) of animal behaviour in terms of rigidly concatenated reflex sequences, and particularly his proposition of tropisms as paradigms of a precise cause-effect machine principle in organisms, epitomizes that kind of mechanistic preconception. His thesis had, however, two serious flaws. Not only had that particular brand of naïvely mechanistic thinking already become outdated in physics, but studies of the actual behaviour of animals in goal-directed or other forms of directional performances showed none of the presumed stereotypism in the manner in which animals attained their objectives. True, the beginning and end of a behavioural act could often be unequivocally correlated with a vectorial cue from the environment; but the execution of the given act was found to be so variable and indeed unique in detail, from case to case and from instance to instance, that it was gratuitous to maintain that the attainment of essentially the same result regardless of the variety of approaches is simply the blind outcome of a chain of seriated steps appropriately pre-designed by evolution to lead to that end. In other words, organisms are not puppets operated by environmental strings; moreover, the analogy is meaningless, anyhow, if one remembers that the "environment" that pulls the strings of puppets in proper order is as often as not another organism—the puppeteer with his brain or at least some machine contrived by a brain.

In fact, it was precisely the detailed study of the movements and tracks by which some species of butterflies assume resting postures rigidly oriented with regard to gravity and light, that prompted me to disavow the reflex chain theory of animal orientation as unrealistic and to propose in its stead a *general systems theory of animal behaviour* (P.W. 1925). That was forty-five years ago. The editor who prefaced a re-publication of the article in question some thirty-five years later (P.W. 1959) commented that "this paper is one of the earliest examples of system-theoretical thinking in behavioural science from the biologist's point of view. . . . It is remarkable . . . that the basic tenets of the paper seem to have been so largely borne out by later developments". As the following discussion rests on that conceptual framework, I shall set forth the gist of it here.

Analytic thinking—an abstraction

To me, as an observer of nature, the Universe presents itself naïvely as an immense cohesive continuum. However, we usually do not look at it as such. We are used to looking at it as a patchwork of discrete fragments. This habit stems partly from a biological heritage, which makes focusing on "things", such as prey, enemies, or obstacles, a vital necessity; partly from cultural tradition; and partly from sheer curiosity, which draws our attention and interest to limited "objects". These may consist of well-delineated patterns in our visual field; of repetitive arrays of sounds in bird song, melody or human language; of processes of patterned regularity, such as waves. What makes them the focus of our attention, is their reiterative appearance in relatively constant and durable form; at least, they hold together long enough or recur in our experience often enough to deserve a name, in contra-distinction to the more fleeting, far less regular, constellations of their surroundings, which we, for contrast, then call "background". I shall not dwell here on the fact that the distinction between relatively constant and more inconstant patches in the Universe, as we perceive it, is sharpened by peculiarities of our sensory functions (e.g., visual contrast phenomena) and by psychological principles (e.g., those underlying Gestalt psychology). I want to lay stress on the fact that no "part" that we might mentally dissect out because we happen to be especially interested in it or because it has forced itself upon our attention is ever truly isolated or "isolable" from the rest.[1]

The process through which we have come to treat any such conspicuous cluster of properties, called "parts", as ideally isolated, is mostly empirical. We watch the complex move relative to its more variable background, and if we find that it is not perceptibly altered by the translocation, we venture to treat it as independent of its environment. But note the reservations in my phrasing. In the first place, I said "perceptible", which refers to the limited powers of discrimination of the observer and his instruments; and secondly, in speaking of "independence from the environment", one must allow that since "environment" is ubiquitous, we cannot test, hence never discount, "dependencies" upon any of the features of the cosmic environment which are universal. Temperature or radiation, for instance, are cases in point.

Actually, even in that limited definition, independence is not absolute; for all those putatively independent entities are interconnected by the common environmental matrix, in which they lie embedded, so that for every single one of the discrete items, every other item is part and parcel of the former's environment. Our habit of atomizing the Universe mentally into isolated or "insulated" fragments reminds one of the familiar adage according to which "no man is an island", which simply rectifies the optical impression of the naïve observer on the shore by pointing to the submerged connections among the visible peaks. In a similar way, having recognized connections between isolated items, man then goes on to sort those he deems "relevant" from "negligible" ones; which obviously lets the judgement of the describer (or of statistics) intrude into purportedly "objective" descriptions of properties of "objects".

From analysis to synthesis

By raising his sights from single objects to their interrelations with others, man reverses his direction from analysis to synthesis: in doing this, he discovers simple rules which describe the interrelations between such entities to his satisfaction but without shaking his

[1] See the pertinent and elaborate remarks on this point by Whyte (1949).

basic conviction that, to all practical purposes and intents, those entities could be regarded as having primarily an isolated existence of their own, becoming just secondarily coupled depending on "circumstances"; sometimes forgetting that "circumstances"—merely a broader substitute term for "environment"—are ever present. This is of course, as I have said before, a deliberate abstraction, but one to which we must credit the tremendous success of science over the last two millennia.

We have learned that if a finite series of modifications of an entity A is regularly associated with a correlated series of modification in another entity B, a rule can be established from which all future correlations between the two can be extrapolated without further experience. We then proceed to study A in its relation to C, and C again in its relation to B, and so seriatim, to learn how different parts of the Universe, erstwhile mentally dissected and separated, hang actually together. The artificial, but eminently fruitful, method of analysis, adhering to the atomistic concept of Democrit, can thus be partly reversed by putting two and two together—either physically or mentally in our imagination—linking by way of consecutive synthesis such coupled pairs into complex chains and cross braces, constructing compound real or ideal structures, the way a child builds bridges with an erector set.

This brings me to the salient point. The basic streak that runs through practically all of our biological thinking is still that science, given time, will succeed in describing and comprehending, by the consistent application of this synthetic method, all that is within the Universe in entities and properties and processes that are knowable to us, including the phenomena of life. Modern physics has had already to depart from such a micromechanistic, naïve picture of the outer world, but we are not concerned with physics here. We are concerned with *living organisms*, and for those, we can assert definitely, on the basis of empirical investigation, that the mere reversal of our prior analytic dissection of the Universe by *putting the pieces together* again, whether in reality or just in our minds, *can yield no complete explanation of the behaviour of even the most elementary living system.*

The living organism: a system

This sentence contains the key words: "behaviour" and "system". A living system that does not behave is dead; life is process, not substance. A living system is no more adequately characterized by an inventory of its material constituents, such as molecules, than the life of a city is described by the list of names and numbers in a telephone book. Only by virtue of their ordered interactions do molecules become partners in the living process; in other words, through their behaviour. And since this involves vast numbers of disparate compounds, all living phenomena consist of *group behaviour*, which offers aspects not evident in the members of the group when observed singly. Now, this fact is generally disposed of by referring to living systems as "complex"; but the term "complex" need imply no more than a haphazard conglomeration, whereas in the living system we find distinctive orderliness of the complexes. Thus, in contrast to the infinite number of possible interactions and combinations among its constituent units which could take place in a mere complex, in the living system only an extremely restricted selection from that grab-bag of opportunities for chemical processes is being realized at any one moment—a selection which can be understood solely in its bearing on the concerted harmonious performance of a task by the complex as a whole. This is the feature that distinguishes a living system from a dead body, or a functional process from a mere list of parts involved, or a sentence from an alphabet, or in biological terms, ecology from taxonomy.

Although the brilliant progress of biochemistry keeps on increasing the list of opportunities for componental interactions, the rules of order which rigorously restrain them in such co-ordinated fashion as to yield a harmonious group performance of the collective can only be recognized, appreciated and properly described once we have raised our sights from the element to the collective system; and this, as you will see, means passing to a higher level of conceptualization.

Hierarchy of wholes and parts

The mere mention of "levels" brings me to the fundamental distinction between atomistic, micro-mechanistic terms of explanations on the one hand, and hierarchical concepts of organization on the other. The difference is that the latter imply some sort of discontinuity encountered as one crosses interfaces between lower and higher orders of magnitude, while the former, trying to reduce all phenomena to the properties of ultimate elements in their various combinations, are based on the premise of a continuity of gradations all the way up from the single elements to infinite numbers of them. To decide which one of these two contrasting presentations of nature represents the reality of biological phenomena is, of course, not to be left to *a priori* conviction, but is a matter of empirical study. If co-ordinated group performances of a high order of regularity can be proven to be the blind resultant of a multitude of concurrent linear bundles of chain reactions minutely pre-set in spatial distribution and pre-scheduled in duration and sequence, then the former theory could hold sway. If not, then systems theory would have to be granted primacy for the treatment of organized systems; for the systems concept is the embodiment of the experience that there are patterned processes which owe their typical configuration not to a prearranged, absolutely stereotyped, mosaic of single-tracked component performances, but on the contrary, to the fact that the component activities have many degrees of freedom, but submit to the ordering restraints exerted upon them by the integral activity of the "whole" in its patterned systems dynamics.

So, here I have at last put my finger on the sore spot, the touching of which has for ages hurt the protagonists of analytical-reductionist orthodoxy—the concept of *wholeness*. Refusing to look beyond their ultimate and most extreme abstraction, namely, the presumption of truly "isolated" elements in nature, and spurred by the dramatic success of explanations of many complex effects in terms of interactions among such elements, they could not help but ask what there could be then in the universe other than elements and interactions. Well, if this is put as an open question, not just a rhetorical one, then I would answer it as follows: The interaction between a positive and a negative electric charge, or between the earth and a falling stone, can certainly be described, at least in first approximation, without paying attention to what happens in the rest of the universe. And if one watches a multitude of stones falling to earth, the total result can still be represented as the sum of all the individual events.

But there is also another class of interactions, which of necessity escapes the elementarian observer in his preoccupation with the smallest samples, because they pertain to properties peculiar to larger samples only of the universe, ignored in the communitive process which led to the concept of elements in the first place. It is in that latter class that the empirical dichotomy arises between *simple conglomerates* and the type of *ordered complexes* which we designate as *systems*. In other words, systems are products of our experience with nature, and not mental constructs, and whoever without being privy to that primary practical experience would try to abrogate them, could do so only by arrogation.

Reductionism and holism

I shall presently attempt a concise definition of the criteria that mark a complex of parts for designation as a system, but before doing so, let me put in an aside. There is an age-old controversy in biology between the two opposite extremes of "reductionism" and "holism". The former finds concurrently its most outspoken advocates in the field of so-called "molecular biology". If this term is used to imply no more than a deliberate self-limitation of viewpoint and research to molecular interactions in living systems, it is not only pertinent and legitimate, but has to its credit some of the most spectacular advances in modern biology. If on the other hand, flushed by success, it were to assume the attitude of a benevolent absolutism, claiming a monopoly for the explanation of all phenomena in living systems, and indeed were issuing injunctions against the use of other than molecular principles in the description of biological systems, this would obviously show a lack of practical experience with, or disregard of, the evidence for supra-molecular order in living systems. Historically, the term "molecular biology" was coined almost simultaneously, though independently, by Astbury (1951) and myself[2]; it was to indicate, on the scale of orders of magnitude, the lowest level of investigation relevant to the advancement of biological knowledge. But nothing in the nomenclature insinuated that it should assume the role of *pars pro toto*. As I once put it, there is no

[2] On assuming the chairmanship of the Division of Biology and Agriculture of the National Research Council in 1951, I restructured the administrative subcategorization of "biology", previously based on forms of life (botany, zoology, bacteriology, etc.) or on methods of study (anatomy, biochemistry, biometrics, etc.), by a hierarchical system of order according to functional principles in common to living organisms: Molecular, Cellular, Genetic, Developmental, Regulatory and Group and Environmental Biology (see, for instance, P.W. 1952). This scheme of classification, subsequently adopted, with some supplementations, by the National Science Foundation, is now frequently applied in the organization of educational and publication programmes.

phenomenon in a living system that is *not* molecular, but there is none that is *only* molecular, either. It is one thing not to see the forest for the trees, but then to go on to deny the reality of the forest is a more serious matter; for it is not just a case of myopia, but one of self-inflicted blindness.

When people use the phrase "*The whole is more than the sum of its parts*", the term "*more*" is often interpreted as an algebraic term referring to numbers. However, a living cell certainly does not have more content, mass or volume than is constituted by the aggregate mass of molecules which it comprises. As I have tried to illustrate in a recent article (P.W. 1967), the "*more*" (than the sum of parts) in the above tenet does not at all refer to any measurable quantity in the observed systems themselves; it refers solely to the necessity for the observer to supplement the sum of statements that can be made about the separate parts by any such additional statements as will be needed to describe the *collective behaviour* of the parts, when in an organized group. In carrying out this upgrading process, he is in effect doing no more than *restoring information content* that has been lost on the way down in the progressive analysis of the unitary universe into abstracted elements.

In this neutral version appears to lie the reconciliation between reductionism and holism. The reductionist likes to move from the top down, gaining precision of information about fragments as he descends, but losing information content about the larger orders he leaves behind; the other proceeds in the opposite direction, from below, trying to retrieve the lost information content by reconstruction, but recognizes early in the ascent that that information is not forthcoming unless he has already had it on record in the first place. The difference between the two processes, determined partly by personal predilections, but largely also by historical traditions, is not unlike that between two individuals looking at the same object through a telescope from opposite ends.

System, operationally defined

However, whether or not you accept this point of view, has no bearing on the conclusiveness of the epistemologically neutral definition I am about to propose. Pragmatically defined, a system is a rather circumscribed complex of relatively bounded phenomena, which, within those bounds, retains a relatively stationary pattern of structure in space or of sequential configuration in time in spite of a high degree of variability in the details of distribution and inter-relations among its constituent units of lower order. Not only does the system maintain its configuration and integral operation in an essentially constant environment, but it responds to alterations of the environment by an adaptive redirection of its componental pro-cesses in such a manner as to counter the external change in the direction of optimum preservation of its systemic integrity.

A simple formula which I have used to symbolize the systems character of a cell (P.W. 1963) could be applied equally well to systems in general. It sets a system in relation to the sum of its components by an inequality as follows: Let us focus on any par-ticular fractional part (A) of a complex suspected of having systemic properties, and measure all possible excursions and other fluctua-tions about the mean in the physical and chemical parameters of that fraction over a given period of time. Let us designate the cumu-lative record of those deviations as the variance (v_a) of part A. Let us furthermore carry out the same procedure for as many parts of the system as we can identify, and establish their variances v_b, v_c, v_d, $\ldots$ v_n. Let us similarly measure as many features of the total complex (S) as we can identify, and determine their variance (V_s). Then the complex is a system if the variance of the features of the whole collective is significantly less than the sum of variances of its con-stituents; or written in a formula:

$$V_s \ll \sum (v_a + v_b + v_c + \ldots \ldots \ldots v_n)$$

In short, the basic characteristic of a system is its essential *in-variance* beyond the much more variant flux and fluctuations of its elements or constituents. By implication this signifies that the ele-ments, although by no means single-tracked as in a mechanical device, are subject to restraints of their degrees of freedom so as to yield a resultant in the direction of maintaining the optimum stability of the collective. The terms "co-ordination", "control", and the like, are merely synonymous labels for this principle.

To sum up, a major aspect of a system is that while the state and pattern of the whole can be unequivocally defined as known, the detailed states and pathways of the components not only are so erratic as to defy definition, but even if a Laplacean spirit could trace them, would prove to be so unique and non-recurrent that they would be devoid of scientific interest. This is exactly the opposite of a machine, in which the structure of the product depends crucially on strictly predefined operations of the parts. In the system, the structure of the whole determines the operation of the parts; in the machine, the operation of the parts determines the outcome. Of course, even the machine owes the co-ordinated functional arrangement of its parts, in the last analysis, to a systems operation—the brain of its designer.

Hierarchy: a biological necessity

To dramatize the need for viewing living organisms as hierarchically ordered systems, I shall give you the following facts to ponder. The average cell in your body consists of about 80 per cent of water and for the rest contains about 10^5 macromolecules. Your brain alone contains about 10^{10} cells, hence about 10^{15} (1,000,000,000,000,000) macromolecules (these figures may be off by one order of magnitude in either direction). Could you actually believe that such an astronomic number of elements, shuffled around as we have demonstrated in our cell studies (see below), could ever guarantee to you your sense of identity and constancy in life without this constancy being insured by a superordinated principle of integration? Well, if you could, for instance by invoking a micro-precisely predetermined universe according to Leibniz' "prestabilized harmony", the following consideration should dispel that notion. Each nerve cell in the brain receives an average of 10^4 connections from other brain cells, and in addition, recent studies on the turnover of the molecular population within a given nerve cell have indicated that, although the cells themselves retain their individuality, their macromolecular contingent is renewed about 10^4 times in a lifetime (P.W. 1969a). In short, every cell of your brain actually harbours and has to deal with approximately 10^9 macromolecules during its life. But even that is not all. It is reported that the brain loses, on the average, about 10^3 cells per day irretrievably rather at random, so that the brain cell population is decimated during the life span by about 10^7 cells, expunging 10^{11} conducting cross linkages. And yet, despite that ceaseless change of detail in that vast population of elements, our basic patterns of behaviour, our memories, our sense of integral existence as an individual, have retained throughout their unitary continuity of pattern.

Those looking at biology exclusively from the molecular end might feel satisfied by calculating that a contingent at any one time of 10^{15}

brain molecules in intercommunication could numerically account for any conceivable number of resultant functional manifestations by their mass. However, this misses the real problem. It is redundant for science to confirm as conceivable that which from experience we already know to happen; what it has to explain, is not that it happens, but why it happens just the way it does. And this is exactly where the above molecular computation fails abysmally, for it ignores the crucial fact that contrary to that "conceivable" infinite number and variety of possible kaleidoscopic constellations and combinations, the real brain processes, taken as a whole, retain their overall patterns.

This particular example has taken us right up to one of the highest levels of organismic organization—the brain. It brings back to me the *spiritus loci* of this delightful village, site of our symposium. Here, one of the great physicists of our age, Erwin Schrödinger, spent periods of time in work and contemplation. In his lecture series on "What Is Life?" (1945), he grappled with the very problem I have tried to dramatize here, namely, the contrast between the degrees of potential freedom among trillions of molecules making up the brain on the one hand (or for that matter, on an even larger scale, the whole body), and on the other hand, the perseverance in an essentially invariant pattern of the functions of our nervous system, our thoughts, our ideas, our memories (and as for the whole body, of our structure and the harmonious physiological co-operation of all our parts). He was forced to the conclusion that, as he put it, "I . . . that is to say, every conscious mind that has ever said or felt 'I' . . . am the person, if any, who controls the 'motion of the atoms' according to the laws of nature."

Let us disregard the implied allusion in this statement to a brain-mind dualism, for the main emphasis lies on the word "control"— the subordination of the blind play of atoms and molecules to an overall regulatory control system with features of continuity and relative invariability of pattern; in short, the postulation of a systems principle. What the theoretical physicist, however, did not seem to have appreciated—and given the lack of detailed empirical familiarity with living objects, could not possibly have apprehended—is that the integral systems operation, whether of the body as a whole, or of the brain within it, deals with the molecules not directly, but only through the agency of intermediate subordinate sub-systems, ranged in a hierarchical scale of orders of magnitude (see the description of hierarchical order in cells further below). Each sub-system dominates its own subordinate smaller parts within its own orbit or domain, as it were, restraining their degrees of freedom according to its own integral portion of the overall pattern, much as its own degrees of freedom have been restrained by the pattern of activities of the higher system of which it is a part and participant.[3]

This is the gist of all the lessons learned from biology, as one descends stepwise from the organism, through its constituent cells, on through their organelles, themselves composed of macromolecular complexes, down to the macromolecules and smaller molecules, which are the link to inorganic nature. The principle is valid for the single cell as much as for the multicellular community of the higher animal, and for the latter's development as much as for its homeostatic maintenance of physiological equilibrium in later life. On each one of the mentioned planes or levels of this systemic hierarchy, we encounter the same type of descriptive rule summarized in the inequality formula outlined earlier; namely, that any one of the particular complexes that show that high degree of constancy and unity that marks them as systems loses that aspect of invariance the more we concentrate our attention on smaller samples of its content. So, at each level of descent, we recognize entities comparable to relay stations sufficiently well defined to be described in their own terms (e.g., organs, cells, organelles, macromolecules; or brain functions, as expressed in concepts, thoughts, sentences, words, symbols), but whose methodical behaviour on that level cannot be ascribed to any fixity of regularities in the behaviour of the units of next lower order; just as knowing purely the properties of those intermediary "relay" entities would not permit us to describe by sheer additive reconstruction the behavioural features of their next superordinate level in precise and specific terms.

You are aware that I have tried to translate the formula, "the whole is more than the sum of its parts" into a mandate for action: a call for

[3] Some authors have endowed systems and subsystems of this description with symbolic names, such as "orgs" (Gerard, 1958) or "holons" (Koestler, 1967). If I do not follow suit, it is from fear that such terms might again be naïvely misconstrued for labels of disembodied superagencies conceived as something that might after all somehow some day *materialize*, distilled off and separable from the conservative *dynamics*, whose special rules those terms aim at categorizing. The history of science has amply documented the conceptual hazards inherent in raising adjectives to the rank of nouns; particularly, in the description of living phenomena, where the temptation to personify nouns is ever present.

spelling out the irreducible minimum of supplementary information that is required beyond the information derivable from the knowledge of the ideally separated parts in order to yield a complete and meaningful description of the ordered behaviour of the collective. Our adoption of the traditional reference to hierarchically ordered systems in terms of "levels" is simply a concession to our habit of thinking in spatial imagery. In our imagination, we visualize the system as a whole on one plane; we then dissect it mentally or physically into its components, which we display on another, a lower, plane, the way we teach anatomy to students. Yet, what we must bear in mind is that, in reality, the system and its parts are co-extensive and congruous, that nothing need be presumed to have been disrupted or lost in the dissection process except the pattern or orderly relations among the parts, and that what the "level" we are speaking of signifies, is really the level of attention of an observer whose interest has been attracted by certain regularities of pattern prevailing at that level, as he scans across the range of orders of magnitude. He scans as he would turn a microscope from lower to higher levels of magnification, gaining detail at the expense of restricting the visual field, and he finds noteworthy constancies on every level. As long as we remain conscious of the fact that any geometric image (or verbiage) that we might choose as a visual (or verbal) model of hierarchic structure is a simplified artifact reflecting the inadequacy of our faculty for visualizing abstract concepts, it becomes rather immaterial which one we use. In this sense, they all become equivalent, whether one prefers the laminated structure intimated by the term "level" or Arthur Koestler's tree scheme of reanastomosing arborizations or my own preference for "inscribed domains" (see figure 6).

Open systems

One further qualification of the preceding characterization of systems as phenomena of our experience should be mentioned. In a purist view, if we deny the primacy of the atomistic notion of truly isolated entities in the universe, then evidently we cannot admit the existence of wholly autonomous, tightly bounded, systems of any order of magnitude and complexity, either. It was in forward reference to this point that I have used repeatedly such notorious escape clauses as "to all intents and purposes", "relatively bounded", "relatively constant", "essential", etc., and this applies similarly to the distinction between "relevant" and "negligible", both of which must be graded on a scale of "more or less", in answer to such questions as: "Relevant to what? Negligible in what context?"

This reminder is called for because of the justified emphasis placed by Bertalanffy (1945) on "open systems". True to my concept of the primacy of continuity and interrelatedness throughout the Universe, I must, of course, consider *all* systems as "open"—ideally and theoretically. But turning practical, I recognize that I can circumscribe many systems sufficiently broadly to let me deal with them empirically as if they were truly autonomous. In other words, on practical considerations, we accept their putative deviations from absolute autonomy as "negligible" (negligible; not non-existent), treat them as "essentially" autonomous, and call them "closed" systems. If later we discover that we have erred by drawing the boundary too narrow, that is, by leaving out some "essential" interrelation formerly ignored as "negligible", e.g., interaction with the environment— we then just go and correct our error by extending the borders of our definition. Logically, this does not change the system, but merely rectifies our earlier mistake. Basically, all systems must be expected to be open somewhere somehow. But leaving pedantry aside, we might just as well be practical and close them by empirical boundaries, subject to amendment.

Systems—theoretically founded

Having now at length presented the case for the hierarchical organization of living systems in rather assertive form, the time has come to document those assertions. The documentary evidence will have to be cursory, confined to a few illustrative examples. I shall present it in two parts: a brief theoretical one and a more elaborate concrete one dealing with the living cell.

On the theoretical side, there is a strictly logical test for the identification of a system. It rests on the nature of the interrelations between the units conceived of atomistically, through primary abstraction, as isolated, separate and autonomous. As pointed out earlier, when we reverse our steps from analysis to synthesis, we can identify unequivocal correlations between the behaviours of two such units (A, B) once we have recognized them as mutually dependent. If we then test a third unit (C), whose properties we know, in its dependence on the state of the two others, we might arrive at an even higher synthetic insight, explaining A+B+C, and so forth, by stepwise additions (see the erector set analogy, p. 7). You note, however, that this would apply only for those particular cases in which our original primary abstraction has been empirically validated, that is, on the premise that the abstracted entities have actually been proved to be relatively autonomous. The fundamental distinction of a system is that this premise definitely does not apply as far as the relations among its constituents are concerned. Let us assume, for instance, a triplet of units, A, B, and C, each of which depends for its very existence upon interactions with, or contributions from, the other two. Then, obviously, we could not achieve a step-wise assembly of this triplet, the way we did before by first joining A to B and then adding C; for in the absence of C, neither A nor B could have been formed, existed or survived. In short, the *coexistence and co-operation of all three* is indispensable for the existence and operation of any one of them. This theorem reminds one of the many-body problem in physics.

In empirical study, processes in living systems reveal themselves as just such networks of mutually interdependent tributaries to the integral operation of the whole group. It is impossible here to elaborate this summary statement any further, but a few simple examples might help to clarify its meaning. Systems of this type of "physical wholeness" can be represented by inorganic analogies. A self-supporting arch is one example. One could never close an arch by piling loose stones upon one another because they start to slip off at the curvature. In other words, an arch as a self-supporting structure can only exist in its entirety or not at all. Though static, it is a system. Of course, human imagination has found ways of building arches piece by piece, by attaching every unit to its neighbour by cements, or by building a scaffolding, which holds the individual pieces in place until the keystone has been inserted to join the two halves and give the total structure its static equilibrium. But those are contrivances of a living system, the resourceful human brain, enabling a system to be synthesized from parts, a feat which could never have been accomplished without such help from another system: System begets system.

This conclusion leads right to an example in living systems, namely, the reproduction of the macromolecules in the living cell. Even though this process is commonly referred to as "synthesis", it is radically different from what goes under the same name in inorganic chemistry. If chlorine and hydrogen are brought together, they will combine to hydrochloric acid, even if none of the end product has been present before. By contrast, the assembly of simple constituents into complex macromolecules in organic systems always requires the presence of a *ready-made model* of the product or, at any rate, a *template* of the same high degree of specificity, to guide the proper order of assemblage. The best studied case is, of course, the transcription of genes, segments of a string of deoxyribonucleic acid (DNA), into a corresponding sequence of ribonucleic acid (RNA), the orderly array of which is then translated into a corresponding serial pattern of amino acids in the formation of a protein.

Although this copying process of patterns and its various derivative manifestations, such as the highly specific catalysis of further macromolecular species through the enzymatic action of proteins, is often referred to by verbs with the anthropomorphic prefix "self-", these processes are no more "self"-engendered than an arch can be "self"-building; for in order to occur at all, they require the specific co-operation of their own terminal products—the enzyme systems which, being indispensable prerequisites for all the links in the metabolic chains, including those for their own formation, thus close the circle of interdependent component processes to a coherent integrated system. Only the integral totality of such a system could with some justification be called "self-contained", "self-perpetuating", and "self-sustaining".

The living cell—a system

These theoretical reflections have a direct bearing on the consideration of a living cell.[4] It is impossible to convey a reasonably accurate conception of a living cell by illustrations on a printed page or by museum models, even when supplemented by verbal description. As a matter of fact, the frozen immobility of those text book illustrations has led to such abstruse misconceptions of a living cell that any portrayal would have to dwell more on what the actual cell is not, than on what it is. Yesterday, I took the opportunity at this symposium to show you motion pictures of living cells in action under a variety of controlled experimental conditions. The purpose was to

---

[4] Much of the specific documentation of examples and conclusions presented in the following sketchy survey of order in the living cell may be found in P.W. (1968).

free the mental concept of the cell from the strait-jacket to which static textbook illustrations have committed it. These experimental methods have opened to our view a microcosm of microscopic and sub-microscopic structures in various arrays of great regularity, specific for each kind and state of cell. However, what we perceive as static form is comparable to a single still frame taken out of a motion picture film. Taken all by itself, a static picture fails to reveal whether it portrays a momentary state of an on-going process or a permanent terminal condition. Unless this ambiguity is constantly borne in mind, one runs the risk of mistaking the static picture of the cell for evidence of a mosaic of well-consolidated structures. The following examples will help to clarify this.

Organelles—subsystems

Figure 1 (*pp.* 22–3) depicts a small fraction of a section through a cell as seen under the electron microscope. The two conspicuous sausage-like organelles are mitochondria, which are the "power plants" of the cell. Parallel to them you see a series of collapsed sacs ("cisternae") spaced out at regular intervals. These and the tubules and vesicles, seen in cross-sections, are dotted on the outside with particles, so-called ribosomes, also seen scattered throughout the field. Each mitochondrion shows in its interior transverse folds of surface membrane, spaced out again with some semblance of rhythmicity. What cannot be seen directly, but has been revealed by painstaking further research, is that the walls of these structures are populated by complexes of enzyme systems, not scattered at random, but arrayed in sequential order in accordance with the consecutive metabolic steps which they are to subserve.

Now, to break down in your minds the illusion of fixity evoked by this fixed specimen, let me point out that practically all you see in this picture is fleeting; and that goes even more so for what you do not see. A cell works like a big industry, which manufactures different products at different sites, ships them around to assembly plants, where they are combined into half-finished or finished products, to be eventually, with or without storage in intermediate facilities, either used up in the household of that particular cell or else extruded for export to other cells or as waste disposal. Modern research in molecular and cellular biology has succeeded in assigning to the various structures seen in the picture specific functional tasks in this intricate, but integrated, industrial operation. There is a major flaw, however, in the analogy between a cell and a man-made factory. While in the latter, both building and machinery are permanent fixtures, established once and for all, many of the corresponding sub-units in the system of the cell are of ephemeral existence in the sense that they are continuously or periodically disassembled and rebuilt, yet always each according to its kind and standard pattern. In contrast to a machine, the cell interior is heaving and churning all the time; the positions of granules or other details in the picture, therefore, denote just momentary way stations, and the different shapes of sacs or tubules signify only the degree of their filling at the moment. The only thing that remains predictable amidst the erratic stirring of the molecular population of the cytoplasm and its substructures is the overall pattern of dynamics which keeps the component activities in definable bounds of orderly restraints. These bounds again are not to be viewed as mechanically fixed structures, but as "boundary conditions"[5] set by the dynamics of the system as a whole. At the present, any more specific phrasing would be gratuitous if account is taken of the fact that a system not only maintains its unitary and integral dynamics beyond internal disturbances, but that systemic unity and typical pattern can be restored even after thorough disruptions of continuity. The case is best illustrated by the readily observable transformations of that remarkable structure in our picture, the mitochondrion. Not only is it highly mobile, squirming worm-like back and forth across the cell space to places where energy is needed for special work, but it frequently breaks into pieces, which then can fuse again with other pieces. In fact, by placing a cell into a slightly acid medium, all its mitochondria can be made to break up into small spherical beads which, upon return of the cell to normal medium, merge again into strings, eventually resuming the appearance and internal structure of a normal mitochondrion, as represented in our picture. On the next lower level, even mitochondrial enzyme systems, when dispersed, can reaggregate into their typical space order, and it has been demonstrated that such clusters can carry out their specific enzymatic function only in those particular ordered constellations.

What we are learning from such lessons is that the features of

---

[5] The concept of "boundary conditions" in living systems presumably owes its origin, and certainly its articulation, to Polanyi (1958). It is crucial to the position taken in this paper.

order, manifested in the particular form of a structure and the regular array and distribution of its substructures, is no more than the visible index of regularities of the underlying dynamics operating in its domain. The near-constant interval, for instance, between the collapsed cisternae or between the internal cross folds of a mitochondrion, reveals simply a rhythmicity in the interactions within such a group, which results in characteristically spaced crests and valleys of conditions favouring the aggregation and assembly of higher order arrays. We encounter here the phenomenon of the emergence of singularities in a dynamic system—"unique points", or lines or planes—comparable, for instance, to nodal points in a vibrating string. This may serve as an example for the "emergence" of sub-patterns within a system with defined boundaries by the free inter-active dynamics. It is the converse of attaining a given form by the consecutive stacking of modules on top of one another, or of the turning out of a machine product in precisely programmed steps.

Significantly, living systems contain models of *both* types, that is, of free integral systems operations, as well as of serial machine procedures. Much argument in biological theorizing could have been avoided if this fact had been more generally recognized. Even so, the more general and primary type is the systemic one, for when we look at the components through which a very machine-like operation in an organism is carried out—for instance, at the individual nerve cells which compose a stereotyped reflex arc—we find that these elements themselves operate, within their own active domain, according to the systems principle, so that the "causal chain" reveals itself as a series of systems operations.

Let us remember, then, in summary, that as we move down from the cell as a whole to progressively smaller samples of its entity, we encounter (a) rather well-defined and relatively stable complexes of functional and structural properties which are embedded in, and mutually related through, (b) matrices of much less well-defined, more fleeting configurations, allowing their constituent parts of elements a much higher range of freedom than could be reconciled with a micromechanical concept of a cell.

Ascent to suprasystems

Let us now proceed in the opposite direction and look at those higher-ordered cell patterns, of which the organelles, dealt with in

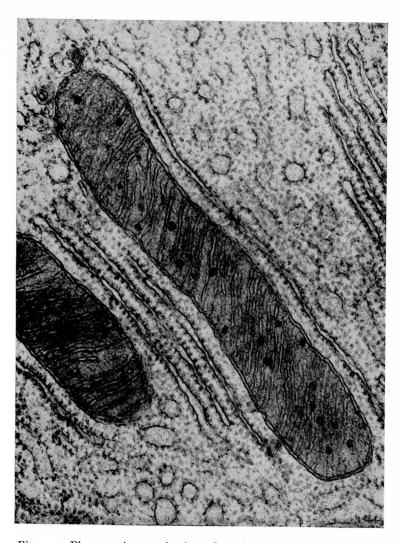

Figure 1 Electronmicroscopic view of an ultrathin section (thickness: 0·00005 mm) through part of a mammalian cell, magnified about 60,000 times. Explanation in text. (From D. W. Fawcett, *An Atlas of Fine Structure: The Cell—Its Organelles and Inclusions*. W. B. Saunders Co., Philadelphia and London, 1966.)

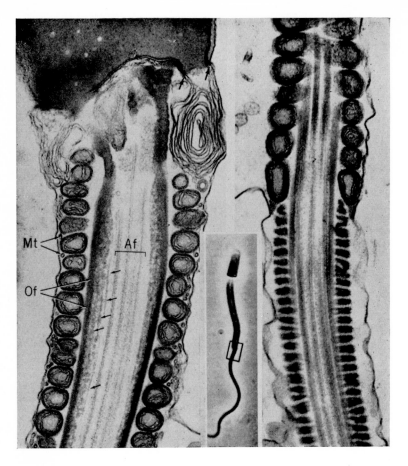

Figure 2 Spermium, as seen (centre) at microscopic magnification and (left and right) in electronmicroscopic views of anterior and posterior regions. (From Bloom and Fawcett, *A Textbook of Histology*, Ninth Edition. W. B. Saunders Co., Philadelphia and London, 1968.)

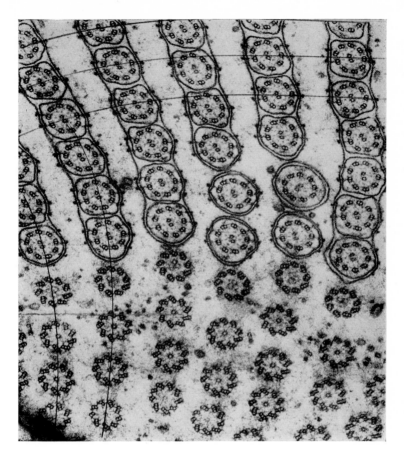

Figure 3 Electronmicroscopic view of an ultrathin slice grazing the surface of a ciliate infusorium. Magnified about 50,000 times. For details, see text. (Courtesy of I. Gibbons.)

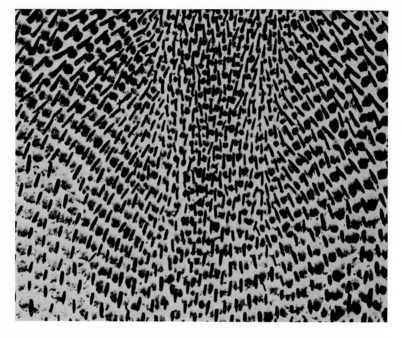

Figure 4 Low-power photograph of the surface of a leopard's tongue. (Courtesy of the late Professor Zeiger, Hamburg.)

the foregoing as relatively sovereign units, appear now as subaltern components, I choose as my first example the spermium. It is shown in Figure 2, in low magnification in the inset, flanked by higher-power magnifications (under the electron microscope), the left one taken near the head, the right one at the base of the tail. You realize that this whole complicated and beautifully organized structure is the content of a single cell. The two rows of roundish bodies in the sheath are mitochondria (Mt), arrayed in file, while the cylindrical core contains long whip-like cilia (Af) of a remarkable fine structure to be described presently. The detailed architecture of sperm is characteristically different for each species, hence is referred to as "genetically determined" (see later). Unfortunately, we are completely ignorant as to how the germinal sperm cell ("spermatocyte") transforms itself into such a remarkably complex, typically structured architecture. Yet, what we can infer from Figure 2 is that it is not formed by stacking up of appropriately pre-hewn pieces, like building a smokestack from bricks, though even such a human construction would not end up straight unless kept in alignment by over-all controls of the assembly process. The individual mitochondria, for instance, are capriciously different from one another, and yet, the composite structure as a whole is of impressive regularity of over-all design. In short, what we have illustrated above for a single mitochondrion, namely, the far greater regularity of the total configuration of the individual organelle as compared to the capriciousness of detail among its component features, repeats itself now on the higher level, at which those same organelles that we had met as systems in their own right appear as sub-units of the higher system of the cell.

These few examples may suffice to illustrate, first, that in the cell certain definite rules of order apply to the dynamics of the *whole* system, in the present case reflected in the orderliness of the over-all architectural design, which cannot be explained in terms of any underlying orderliness of the constituents; and second, that the over-all order of the cell as a whole does not impose itself upon the molecular population directly, but becomes effective through intermediate ordering steps, delegated to sub-systems, each of which operates within its own, more limited authority. What I have tried to summarize here for the internal operation of the cell as a system, is then re-encountered on a still higher level as the basic principle in the functional organization of cells into tissues and organs, up to the body as a whole. To give just a cursory illustration of this ascent

B

on the hierarchical scale, I have selected the following two examples.

Figure 3 (*pp.* 22–3) shows a tangential section grazing the surface of a ciliate protozoan. Each one of the round structures is the cross-section through a single cilium. Being slightly oblique, the section takes us through layers of different depths,. from naked cores, through the appearance of a sheath around each single cilium, to the presence of a common envelope around each row. Each core shows a circumferential array of nine doublets of fibrils with a little hook, arranged pin-wheel fashion, with radial connections to a central pair of fibres. All cilia, from those of protozoans and algae through the whole plant and animal kingdom all the way up to man, are built with only minor modifications according to this standard fine-structural pattern. As were the mitochondrial rows in the sperm tail, so here again the individual cilia are aligned in rows into higher-order assemblies, which as you note, combine a remarkable geometric regularity in the gross with capricious deviations in detail. You also note, besides the equidistant parallel tracts of rows, a further lateral alignment among the cilia, which runs crosswise from row to row, as indicated by the lines in the top part of the picture. The fact that the actual positions of individual cilia deviate from the exact nodal points, at which the ideal line systems of this planar grid intersect, gives further evidence of the systemic nature of this architecture with its well-defined over-all design, but with departures from it in the details of the component units. It is not too far-fetched to see in such grids on the sub-microscopic level an analogy to the lattice structure of crystals—a concept which I have documented under the name of "macrocrystallinity" in several of my earlier publications. To dwell on it here would take us too far afield.

The next hierarchic step takes us now from this highly structured single-celled organism to Figure 4 (*pp.* 22–3), which illustrates the occurrence of similar geometric patterns in the tissues of multi-cellular animals. The picture gives a low-power enlarged view of the surface of a leopard's tongue. Two types of sub-structures are visible, representing tastebuds and papillae. As in the previous case, they are again arrayed in a regular grid. In this case, however, each unit consists itself of a large number of cells. Microscopic study shows that even though each unit has the shape, size and orientation typical of its kind, there is not only no corresponding regularity in the detailed configuration of the population of component cells, but there is actually continual change in the composition of the cell population, cell death being compensated by cell proliferations. Each unit behaves as a multicellular system.

Yet, these multicellular units are themselves subordinated to a pattern of still higher order, which defines the positions of the units as the intercepts of a grid of two essentially perpendicular line systems. As before, one notes a sufficiently wide range of departures from mathematical accuracy to rule out any notion that the total pattern could be merely the net resultant of a neat serial stacking of the individual units. It is important to stress that although some irregularities of such adult patterns reflect secondary distortions during the growth process, there is plenty of evidence at hand to prove that even the earliest stages in their development fail to display the rigorous microprecision of arrangement among the elements as to position, alignment and interstices, which one would have to postulate in order to ascribe the total pattern to a stepwise compilation of micro-patterned building blocks. In fact, if the assurance of an orderly architecture at the end had to rest on nothing but an orderly initial pattern, no such final orderliness could ever be expected unless systemic regulation on the way would hold the pattern together; for otherwise the fortuitous local differences in growth activities would lead not only to the usual minor distortions, but to wellnigh complete obliteration, of the initial pattern.

I do not mean to labour the point further. I have set forth a series of examples to document that the study of the behaviour of cells and cell groups, whether in development or physiological functioning, makes it imperative to postulate systemic order on *supramolecular* levels; that is, dynamics within collectives which restrains the degrees of freedom of the components in such a manner as to insure concert in attaining, maintaining, and if necessary, restoring, the integral pattern of the whole; and further, that each constituent subsystem in the hierarchy has its own, more limited, degree of systemic sovereignty to deal with its subordinate component units in a similar manner as it has been dealt with as a member of the superior collective.

This statement is, in essence, a purely descriptive one. It is noncommittal in regard to preferences for couching systems properties in either "holistic" or "reductionist" terms; for as I indicated at the beginning, and have discussed more fully elsewhere (P.W. 1967), both are in a demonstrable complementarity relation in the sense that either one conveys information which the other cannot supply. Consequently, the acknowledgement of *field continua* as ordering principles in systems on the *integral* level is as valid and indispensable as is the practical acceptance, on the *differential* level, of *discrete singularities* within those continua, whether sub-atomic particles, atoms, molecules, molecular assemblies, organelles, cells, or cell assemblies, as in our last picture. I have yet to encounter any phenomena in the living system which could be adequately described without reference to such a dualistic scheme.

Let me also point out once more that my sole aim is the purely pragmatic task of listing, in the spirit of Kirchhoff, the irreducible minimum of descriptive statements necessary for the representation of natural phenomena not only truthfully, but *completely*. This precept must also pervade the terminology we use to express our observations and inferences. Even though such terminology will differ according to whether we move from the undismembered whole down to the parts or back up from the atomized elements, eventually the two versions must become unequivocally related and mutually consistent. This implies that, as I said earlier, nothing need be assumed to exist in a whole separately and separably, than what exists in the totality of its constituent parts. What must be added on the way up is only the restoration of information content of which the system has been deprived in its physical or mental atomization.

The Brain—a system

It might have been more pertinent to the topic of this symposium to deal specifically with the functions of the *brain* as a hierarchically organized system rather than with the general cellular system. The advantage of the latter, however, is that its morphological aspects provide us with directly perceptible indicators of what remains invariant during a given period and what does not; for as indicated before, living form must be regarded as essentially an overt indicator of, or clue to dynamics of the underlying formative processes. Thus, cell structure serves in much the same way as the height of a mercury column in a thermometer or the excursions of a needle in a galvanometer serve to signal changes of thermal and electric states. Yet, for the systems dynamics of the brain we possess no equally reliable recorders. True, our introspective self-knowledge confirms to the private satisfaction of every one of us that all those conclusions about systemic order and its hierarchic structure that I have outlined for the body and its cells are equally valid for our brain activity. But when we look for more "objective" testimony, we find ourselves up against the limits of our methods of detection and assay. Microanatomy gives us only numerical, geometric and topographic data; using electric instruments, we get electric answers; using chemical techniques, we get chemical information; and so forth. However, we still do not have any inkling of how those fragmentary items of information, obtained analytically, could be combined into a faithful image of the unitary and orderly behaviour of our central nervous system, of which we are privately conscious, and the expressions of which we can observe in the overt behaviour of others.

Nonetheless, the outlook is not quite so dim. Critical studies of brain alterations are increasingly providing us with clues, if not to how the systemic behaviour of the brain does arise and operate, at least to how it does not. Pertinent evidence comes mostly from neuropathology and the study of experimental brain lesions, but partly also from the poorly exploited field of neuroembryology, which presents us with embryonic brains displaying typically patterned activity prior to any experience of the outer world, and above all, prior to the emergence of structures and functional properties, the significance of which would otherwise have to be tested laboriously and traumatically by artifical elimination in later life. In a monograph, to which I must refer here for further information (P.W. 1941, reprinted in P.W. 1968), I have summarized some of the crucial evidence for the systemic hierarchical principle of operation of the lower parts of the central nervous system. The conclusions reached there can be plausibly extended to the brain as a whole.

Of all the features of our subjective knowledge of our brain activities for which we want science to produce an objective record, one of the most hotly debated ones has been the aspect of "freedom of decision" or "free will". The issue has been argued almost entirely on philosophic grounds. That is a domain which, as I stated in the introduction, I feel reluctant to enter. However, since some of the philosophical discussions have hinged on the interpretation of certain unfavourable verdicts pronounced by science, it seems indicated in this place to re-examine briefly the tenability of the respective positions. The way I see it, looking from the outside in, the problem of free will has been treated in general as a corollary of the problem of *determinism*, and the problem of determinism, conversely, has been laid at the doorstep of science for an opinion. As long as science keeps on presenting nature as a micromechanical precision machinery run by strict causality, the concession of any degree of freedom of choice to any natural phenomenon would be inadmissible by the code of that brand of science, and hence, would have to be denied to all processes of nature, including human brain functions. One would then be forced to adopt the alternative of ascribing "free will" to the intervention of extra- or supernatural powers.

Determinism, stratified

To me both of these extreme positions seem to be untenable. The flaw lies in equating science with the doctrine of micro-precise causality, or as I shall call it in the following, "*micro-determinism*". This brings me to the major lesson to which I have been building up in this paper and which I have anticipated in the title as "Determinism Stratified", which is precisely what the study of living nature teaches us.

There may be philosophers or theologians who derive comfort from the idea of a Laplacian universe made up of a mosaic of discrete particles, operating by laws of micro-causality. I submit, however, that modern science cannot deliver such a picture in good faith, least of all life science; and since all science is the product of human brains and brains are living systems, it is quite likely that this abrogation of scientific rationale for micro-causality applies to science in general.

Scientifically, the term "determined" can only mean "determinable"; and similarly, "indeterminacy", whether in the sense of Heisenberg or in the way I shall use the term, can only mean "indeterminability". The scientific concept of "determinability" is of decidedly empirical origin. As we observe a given macrosample of the Universe over a given stretch of time, we note certain unequivocal correlations between configurations of its content at the beginning and at the end of that period of change. If then we find those correlations consistently verified, we set them up as "laws", from which to extrapolate future changes with a sense of certainty. As our primary experience in this operation has only correlated macro-samples with macro-samples, predictability based on it can likewise be no finer than *macroscopic*. So, legitimately, we could only speak of "macrodeterminacy".

The concept of "microdeterminacy" is then derived secondarily by a hypothetical downward extension—"atomization", as it were—of empirical "macrodeterminacy". That hypothesis submits that one would observe the same high degree of consistency of correlation from beginning to end that had been ascertained for the macrosample to hold true for every one of those fractional samples. In other words, the structure of the well-defined macro-change would be simply a composite of the mosaic of micro-changes assumed to be equally well-defined, even if not necessarily determinable.

This view is demonstrably untenable in its application to living systems. We have recognized the state and changes of such systems as being conservatively invariant over a given period, and hence predictable, without a correspondingly invariant micro-mosaic of the component processes. We had to conclude, therefore, that the patterned structure of the dynamics of the system as a whole "co-ordinates" the activities of the constituents. In atomistic micro-deterministic terms, this "co-ordination" would have to be expressed as follows: Since any movement or other change of any part of the system deforms the structure of the whole complex, the fact that the system as a whole tends to retain its integral configuration implies that every change of any one part affects the interactions among the rest of the population in such a way as to yield a net countervailing resultant; and this for every single part. Couched in anthropomorphic language, this would signify that at all times every part "knows" the stations and activities of every other part and "responds" to any excursions and disturbances of the collective equilibrium as if it also "knew" just precisely how best to maintain the integrity of the whole system in concert with the other constituents. Although rarely expressed so bluntly, much of this imagery lurks behind such equally anthropomorphic terms as "organizers", regulators", "control mechanisms" and the like, which particularists have had to invoke in order to fill the information gap between what one can learn from isolated elements and a valid description of group behaviour.

The Boltzmann theorem and thermodynamics have realistically by-passed that gap by confining safe statements about macrorelations to macrosamples only. They relate unequivocally the average state of a system at time t_1 to its average state at time t_2, but realize that tracing an individual molecule through that course is not only not feasible but would be scientifically totally uninteresting and inconsequential; for it would in each individual instance and instant be of nonrecurrent uniqueness, hence valueless for any detailed predictability of future micro-events. If physics has had the sense of realism to divorce itself from microdeterminism on the molecular level, there seems to be no reason why the life sciences, faced with the fundamental similitude between the arguments for the renunciation of molecular microdeterminacy in both thermodynamics and systems dynamics, should not follow suit and adopt "macrodeterminacy" regardless of whether or not the behaviour of a system as a whole is reducible to a stereotyped performance by a fixed array of pre-programmed micro-robots. Since experience has positively shown such unequivocal macrorelation to exist on various supramolecular levels of the hierarchy of living processes in the documented absence of componental microdeterminacy, we evidently must let such positive scientific insights prevail over sheer conjectures and preconceptions, however cherished and ingrained in our traditional thinking they may be.

Macrodeterminacy

In order to drive home this lesson, I am adding in Figure 5 (*pp*. 38–9) a diagrammatic model of macrodeterminacy from a recent book (P.W. 1969 b), in which I have dealt with the problem more fully. The diagram shows the transition of a living system from a state S′ to a state S″. As is indicated in figure 5A, the system S′, comprising subsystems A′, B′, C′, D′ E′, develops between times t_1 and t_2 into the modified system, S″, each sub-system of which at t_2 can be traced back to a corresponding sub-system at t_1; this makes the pattern of t_2 explicable, that is, predictable, hence, determinable or "determined", as a direct transform, piece by piece, of the macroconstellation of component sub-systems at t_1, the sub-systems having kept their relative positions and relations.

This kind of mosaic correlation between two stages permits an embryologist, for instance, to identify specific regions of an early embryo as being the predictably earmarked forerunners for the formation of heart, liver, kidney, brain, etc., respectively. Yet, looking at the diagram 5B, we note that such clear-cut correlations no longer hold for smaller samples of each sub-system, represented by the various symbols. In other words, if we were to follow individual cells of these various prospective organ areas from the earlier into the later period, we would find them to take far more fortuitous courses, differing individually from case to case; this is indicated by the lack of correspondence between the upper and the lower sets of lines connecting symbols from S' to S" for two embryos of the same species. This fact is so well established in embryology that one has even got to referring to the process which changes a cell from its originally "indeterminate" and multivalent condition into one definitely committed for a given fate, as the process of "determination".

The same principle as illustrated here for development repeats itself at all orders of magnitude; let us review, for instance, the establishment of organelle structures within the cell, or of the finer sub-structures within organelles.

A mitochondrion, in being reconstituted after fragmentation, will order lipid molecules into the characteristic lamellar configuration seen in Figure 1, but the new position will not be a precise replica of the former one, nor do the lipid molecules in the pool know which ones will be recruited. Likewise on the next lower level, the enzymes in regularly seriated clusters, which are to dot the new mitochondrial lamellae, do not know their final arrangements until they are in place; and so forth.

I could go on to confirm the validity of this principle of *determinacy in the gross despite demonstrable indeterminacy in the small* for practically any level and area of the life sciences. In order to take account of this hierarchical repetitiveness, I have suggested the simile of "grain size" of determinacy as an empirical measurement for the degree of definition and predictability at any given level. The mosaic of organ rudiments mapped out in the early embryo, for instance, is very "coarse-grained", whereas the mosaic of genes in the chromosome is far more "fine-grained" (see below). It is immaterial at this juncture whether or not the principle is rigorous in a philosophical sense. What counts is that, scientifically speaking, it is as realistic and logical a proposition as we can deduce from the accessible facts.

As you will note, one could turn this renunciation of the primacy of microdeterminacy into a positive scientific declaration in favour of the existence of "free will". I prefer to give it a more restrained interpretation, for it really implies no positive commitment. What it does, is simply to remove the spurious objections and injunctions against the scientific legitimacy of the concept of freedom of decision that have been raised from within the scientific sector, or from other camps leaning on supposedly scientific verdicts. I cannot see that science can prove free will but, on the other hand, I can see nothing in what we know in the life sciences that would contradict it on scientific grounds. To go beyond this statement would be a matter solely of private belief.

A canon for determinacy

Lest there be misunderstandings about my thesis of macrodeterminacy which is not explainable in terms of aggregates of microdeterminate events, let me review briefly the major way stations to this conclusion.

1 Nature presents itself to us primarily as a continuum.
2 In scanning this continuum, we recognize complexes of phenomena which retain identity and show a high degree of stability and persistence of pattern, in contrast to other samples with less cohesive features.
3 Success of science over the ages has validated the abstraction involved in our dealing with such reasonably constant entities as if they had an autonomous existence of their own.
4 Some phenomena of nature can be reconstructed in practice, or at least in our minds, from the analytical knowledge of properties, interrelations and interactions of such conceptually isolated entities.
5 Some of the sciences, particularly the physical sciences, have confined themselves mostly to the consideration of phenomena amenable to the treatment according to Point 4.
6 In the life sciences, there are likewise many questions which can be answered satisfactorily by the recombinatory method according to Point 4.
7 Understanding of the integrality of a living system, however, has proved, on logical grounds, refractory to the same methods, and the empirical study of life processes has discounted, on factual grounds, the probability of future success.
8 The preceding point implies that it is logically and factually gratuitous to postulate that the methods of Point 4, successful in Points 5 and 6, must necessarily also be sufficient to restore completely the loss of information about the dynamics of *systems* suffered in their analytical atomization. By the same token, restrictive injunctions against descriptions of living phenomena in terms other than those compatible with Points 5 and 6 can no longer be upheld.

9 *The systems concept* proves applicable to the description of those phenomena in living systems which defy description purely in terms of micro-mechanical cause-effect chain reactions; it thus lends substance to the principle of systemic organization.

10 Applying the systems concept, an organism as a system reveals itself as encompassing and operating through the agency of sub-systems, each of which, in turn, contains and operates through groups of systems of still lower order, and so on down through molecules into the atomic and sub-atomic range.

11 The fact that the top level operations of the organism thus are neither structurally nor functionally referable to direct liaison with the processes on the molecular level in a steady continuous gradation, but are relayed step-wise from higher levels of determinacy (or "certainty of determinability") through inter-mediate layers of greater freedom or variance (or "uncertainty of determinability") to next lower levels of again more rigorously ascertainable determinacy, constitutes the *principle of hierarchical organization.*

12 Although I have emphasized for didactic reasons the relatively conservative features of systems, the uni-directional change of systems must not be overlooked. We find it expressed, for instance, in the mutability of systemic patterns in evolution, ontogeny, maturation, learning, etc., as well as in the capacity of recom-bination of systems into what then appear as super-systems with the emerging properties of *novelty* and *creativity.*

This set of twelve points represents a sort of conceptual canon, based on empirical studies, against which theoretical pronouncements and formulations in the life sciences ought to be checked. As you will recognize, some statements in the current literature would fail the test of validation in terms of these criteria, while some others would even seem totally irreconcilable with the principle of *stratified* determinism. One in this latter class that comes immediately to mind is the prevailing notion of *"genetic determinism"*. A few comments on this issue are therefore called for. A less cursory discussion of the subject may be found in the book referred to above (P.W. 1969b).

Genetic determinism, scrutinized

The term "genetically determined" means three different things to three different groups of people: (1) the broad-gauged student of genetics, who is thoroughly familiar with the underlying facts and uses the term simply as a shorthand label to designate unequivocal relations between certain genes and certain "characters" of an organism; (2) scientists in various other branches who are not familiar with the actual content of the term and accept it literally in its verbal symbolism; and (3) the public at large, to whom the term frequently imparts a fatalistic outlook on life, frustrating in its hopelessness, of an inexorably pre-set existence and fixed course towards a pre-ordained destiny. In the present context, I shall concern myself only with the first of these three groups, because many of its practitioners seem addicted to a doctrinal orthodoxy that is clearly at variance with the picture of living processes I have tried to present in this article. The source of the discrepancy is easy to spot. It lies not in factual inconsistence, but merely in the phraseology used, which, if one stops to think, turns out to be a queer hybrid between brilliantly established analytic facts and scientifically spurious anthropomorphic lingo. Like the mule, the hybrid has proved to be viable but not fertile. Very briefly, here is an account of its origin.

Basically, genetics proceeds by the same analytical technique that I have described in the early part of this paper as the progressive dissection and subsequent mental isolation of features of the world around us that have aroused our interest by their constancy. As we sharpen our view, our attention is drawn to specific *differences* between entities whose generic aspects are indistinguishable. Eyes, as the organs of vision, are in essence the same throughout a species, but the iris of some appears black, in others blue, and in albinos transparent. And similarly, the same object, hair, occurs in distinctive varieties of dark or blond or red as well as straight or wavy or crinkled. In our analytic mood, we then abstract those various differential criteria from their generic carrier subjects and confer upon them a measure of autonomy, which they may or may not merit. Since "genetically determined" black pigment, for instance, can be bodily extracted from black eyes and black hair as a proteinaceous substance and since the "genetically determined" absence of it in blue eyes and blond hair is traceable to a gene defect, we feel entitled to deal with it without regard of the eyes or hair that carry it, that is, to concede to the gene-colour correlation substantive identity. But does this also entitle us to think that an eye can really be resolved simply into a bundle of such "gene-determined" attributes, so that if we just kept on stripping them off one by one, we would end up with no substantive entity left over of which those attributes were merely qualifying variables? In other words, what makes for "eyeness" in the integral formative dynamics of an eye? Evidently, by consistently promoting in this manner *adjectives* to the rank of *substantives*, we have, by definition, vested genes with the exclusive "responsibility" for organization and order in an organism. (It is worth noting, incidentally, that the term "responsibility" so commonly found in reductionist "explanations", is again an anthropomorphic reference to a system—the human brain.)

It has been one of the most spectacular triumphs of analytical science to have demonstrated that many of the observed *differentials* among features within a given species of organisms can be unequivocally correlated with corresponding *differentials* between the chromosomes of the respective varieties and that these chromosomal differences can be further resolved into differences in the array of sequences of nucleic acid residues along the backbone of the helical macromolecule DNA. Segments of those gigantic macromolecules for which such differences of composition can be demonstrated or compellingly inferred are, therefore, the molecular counterparts of what with a purely symbolic term for units of inheritance of characters used to be called "genes". So far, so good. Overt differences between "characters", that is, specific attributes of a body, have been linked indisputably with specific differences in the structure of a macromolecule. This is all beautiful and straightforward.

One then was faced, however, with the major problem of how it would be possible, true to the spirit of scientific parsimony, to reconstruct from the knowledge of those elementary units the higher entity of an organism functioning in harmonious co-ordination. Let us recall that genetic methodology can only test *differences* between organisms. As I just pointed out, it can deal with differential properties of eyes or of hair, but cannot elucidate the nature of "eyeness" or "hairness", as such. In other words, we know of no "genes for eye" or "genes for hair" that could explain the basic formative dynamics by which those systems attain and retain their generic configurations, even though we know plenty of genes referring to differences in their specific styles of architecture and properties of building stuffs.

Lest there be further misinterpretations, let me remark that such phrases as "genes for eye*less*ness" (anophthalmia) or "genes for hair*less*ness" do not refer to properties of eyes or hair at all, but signify differentials among other conditions of the body, either permissive or repressive, as the case may be, for the processes of eye formation or hair formation, without being in the least concerned with the organizational pattern of those processes themselves. In conclusion, there is neither logical or factual support for the supposition that organization can be explained in reference to gene interactions *alone*.

Nevertheless, the claim of the gene for recognition as the sole ordering principle in organism persists with undiminished tenacity: the rigorously and conservatively ordered array of components in exploring how far one could effectively synthetize any system of *all* order in an organic "system"; the term "system", in that event, being decidedly a misnomer. As far as I can discover, this claim rests on sheer assertion, based on blind faith and unqualified reductionistic preconceptions. But if one examines the phraseology in which the claim has been presented, one recognizes that its proponents have by no means been unmindful of the problem of the organizational wholeness of the organism. Their error lay in the manner in which they tried to overcome, or rather, gloss over, the difficulty of the problem.

Instead of reversing the rigorous objective methodology, which has marked the successful descent from the organism to the gene, and exploring how far one could effectively synthesize any system of higher order from nothing but genes in free interaction in an environment devoid of order, they had recourse to pretentious anthropomorphic terms, which evaded or obscured the issue. One simply bestowed upon the gene the faculty of spontaneity, the power of "dictating", "informing", "regulating", "controlling", etc., the orderless processes in its unorganized milieu, so as to mould the latter into the co-ordinated teamwork that is to culminate in an accomplished organism. It is at once evident that all those terms are borrowed from the vocabulary of *human systems behaviour*, especially the brain; in short, terms reserved for precisely the kind of complex group dynamics that cannot be pieced together by sheer summation of componental contributions. Therefore, to contend that genes can "determine" the systemic wholeness of the organism, while at the same time being driven to invoking the systems analogy of brain action in order to endow gene action with the required integrating power, is logically such a circular argument that we can readily dismiss it from serious scientific consideration.

Besides this *logical* flaw, the speculation about gene monopoly in organization meets also with *factual* contradictions. They pertain to the very premise that genes operate in a random, disorganized matrix. In the first place, unless we want to vest genes with animistic powers, we cannot grant to them the faculty for spontaneous "action". They can only *inter*act; but interact with what? With a disordered molecular population, to which they, as the expression goes, "transmit the needed ordering information"? The transfer of order from DNA through RNA to protein is well attested, comparable to the translation of *words* from one language into another. But how to get from words to the meaningful *syntax* of language? One might concede that the change from the coded polypeptide chains to proteins and tertiary conformation of the latter into highly specific configurations might be regarded as steps up to a next higher level of organization, like forming words from phonemes. But then, where do we go from there? Even considering that proteins with enzymatic functions are instrumental in the synthesis of other species of macromolecules, this still would leave us only with a bag of molecular units milling around in thermal agitation. And if one were to point to the further fact that such compounds can link up chemically into still more complex composite units with solid structure, one would get even further away from the realistic picture of a living cell, which, as I have indicated, rules out any portrayal of organization that intimates a piece-meal origin through the consecutive assemblage of building stones.

Development—a hierarchic systems operation

The conflict vanishes with the realization that genes, highly organized in themselves, do not impart higher order upon an orderless milieu by ordainment, but that they themselves are part and parcel of an ordered system, in which they are enclosed and with the patterned dynamics of which they interact. The organization of this supra-genic system, the organism, does not ever originate in our time by "spontaneous generation"; it has been ever present since the primordial living systems, passed down in uninterrupted continuity from generation to generation through the organic matrix in which the genome is encased. The organization of this continuum is a paradigm of hierarchical order. I have schematized it in Figure 6 (*pp.*38–9) by concentric shells which, in this case, coincide with physical enclosures. The diagram is self-explanatory. The profusion of arrows indicates pathways of all possible interactions that must be taken into account in studying the dynamics of this system, organism.

Now, let us start our consideration of the time continuum of living generations with a particular egg, containing in its nucleus the inner sanctum, the chromosomes with their genes. An egg is not a nutrient solution for the genome to feed on, but it is a full-fledged organism, equivalent in its systematic features with other single-celled organisms (e.g., the protozoans illustrated in Figure 3), having derived these features directly from its prior existence as a germ cell in the maternal body. We know that in its crust, the egg cytoplasm consists of an orderly mosaic pattern of fields (see P.W. 1938), the typically different properties of which map out the dynamics of specific organ districts. As the egg nucleus divides, the derivative daughter-nuclei, therefore, come to lie in specifically disparate domains, which later, in their ensuing interactions with the nuclear genome, become further modified; and as this interplay between plasma and nucleus in the steadily increasing cell population continues, the genic environment becomes progressively diversified. In all this, the genes themselves do not significantly change; their pattern is replicated stereotypically in practically all cells of a given individual throughout development. Yet, each cell's genome is, and always has been, a captive of an *ordered* environment. While the genome contributes to the specific properties of that environment in mutual interactions with it during the whole course of embryogenesis, it is only by virtue of the primordial frame of organization of the *cytoplasm* of the egg that an individual can maintain from the very start the unity of overall design, to which the masses of freely mobile scalar entities of lower order, including the cells and nuclei with their chromosomes and genes, owe the "microdeterminacy" of their eventual fates.

In this incessant interplay, the latitude for epigenetic vagaries of the component elements on all levels, indicated by the arrows in Figure 6, is immense[6]. And yet, the end products turn out to be far more similar than one could expect if there were no conservative systems dynamics in operation to guard the overall design. Hierarchic

[6] For a penetrating discussion of the epigenetic course of development in relation to genic participation, consult Waddington (1962).

(Unfortunately, Dr Waddington was unable to attend the first meeting of the Symposium, at which this paper was presented. *Eds.*).

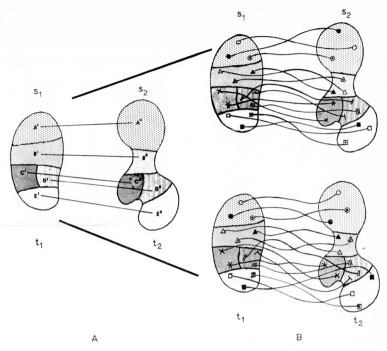

Figure 5 Diagram symbolising the changes in the development of two individual germs from the stage at time t_1 to that at time t_2; the upper and lower courses correspond in 'macroview' of overall pattern, but differ in 'microview' of detail, as explained in text.

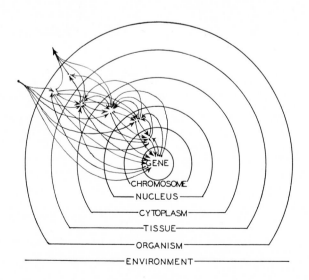

Figure 6 Interactive relations among the hierarchically ordered subsystems of an organism. (From P. W., 1968.)

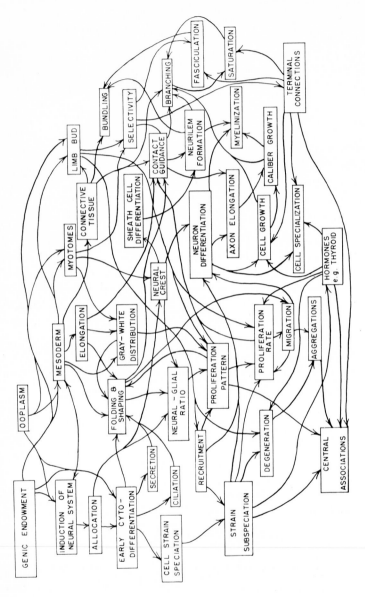

Figure 7 Some demonstrated interdependencies and interactions in the development of a mature nervous system (bottom level) from a fertilised egg (top level). (From P. W., 1968.)

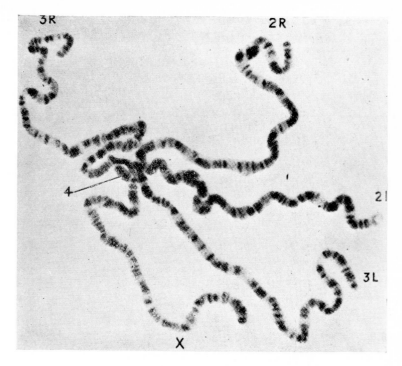

Figure 8 Giant chromosomes from the salivary gland of an insect. The aperiodic 'banding', expressing the linear sequential order of genes, is as strictly identical among the chromosome sets of all cells of an individual as are all the telegraphic tapes on which the same message is printed in Morse code. (From Sinnot, Dunn and Dobzhansky, *Principles of Genetics,* 4th ed. McGraw Hill Book Co., 1950.)

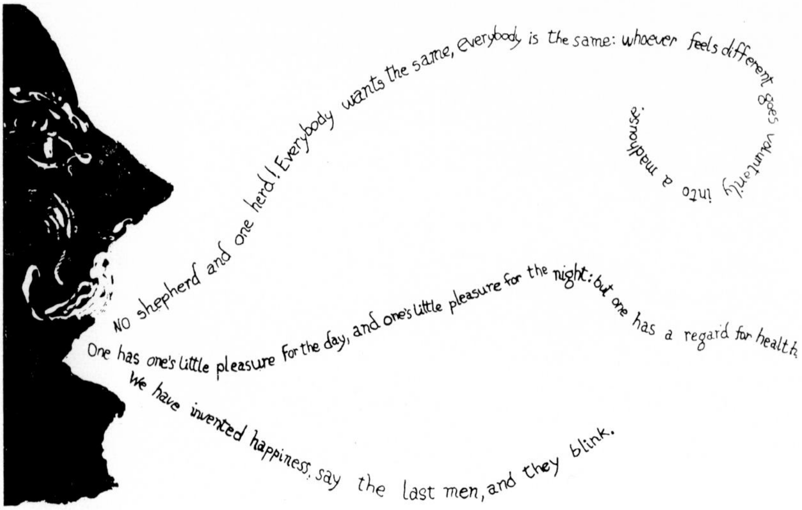
No shepherd and one herd! Everybody wants the same, everybody is the same: whoever feels different goes voluntarily into a madhouse.
One has one's little pleasure for the day, and one's little pleasure for the night: but one has a regard for health.
We have invented happiness, say the last men, and they blink.

Figure 9 Signature for an English edition of a book by Nietzsche with quotations. (Courtesy of Pratt Institute, New York.)

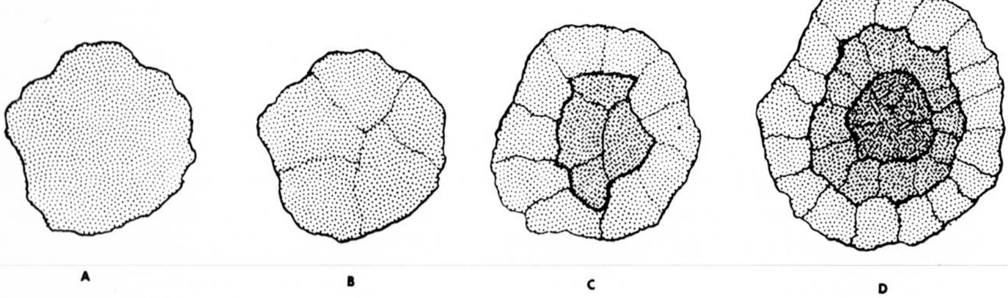
A
B
C
D

Figure 10 Diagram, illustrating the emergence of differentials in collectives of erstwhile equivalent units as a result of numerical increase. (From P. W., 1968.)

stepwise delegation of tasks to sub-systems is ostensibly nature's efficient device to let an organism keep order without having to deal with all its trillions of molecules directly.

In conclusion, genetics alone can account rather precisely for the differences between attributes of systems, such as cells or organs or organisms, but for a complete definition of the integral subjects which carry and display those attributes, genetics must call on *systems dynamics* for supplementation; and reductionism, similarly, on holistic succour, often disguised by anthropomorphic phraseology. The dynamism of organization is *dualistic*. The coarse-grain macro-determinacy of the systemic plasma domains, in which the genomes reside, is, through the interactions with the fine-grain microdeterminacy of the rigorously structured genes, progressively translated and consolidated into more detailed and specific definition—"from the general to the specific"—without the organism or its parts ever losing their free systemic dynamics entirely while they are alive. The loss of those integrative dynamics is the mark of death.

In order to make more concrete this necessarily general presentation, let me give just one specific illustration for the systemic network operation through which the macro-pattern of an egg is progressively converted into the detailed structural and functional pattern of the mature individual. Figure 7 (*pp.* 38–9) is a flow chart of some of the major tributory processes that lead from the egg with its dual endowment—genes, embedded in ooplasm (top of figure)—to the finished functioning brain and spinal cord (bottom of figure). The boxed-in titles are labels for some of the major component processes that have been singled out from the total developmental fabric for analytical study; the arrows represent actually identified dependencies between those processes, that is, pairwise dynamic connections between boxes, each of which in isolated consideration, appears as a linear cause-effect chain. But as soon as we raise our sights back from the artificially detached single threads to the total context, we recognize the true intricacy of the systemic fabric, in which "everything depends on everything else", just as on the lower level of the molecular ecology of the cell.

If the total network were not so coherent and conservatively unitary in its overall dynamics, no two nervous systems could ever resort from the developmental cauldron with even the faintest mutual resemblance; for it has been precisely the analytical study of the single "cause-effect" nexuses which has impressively demonstrated the enormous unpredictable variability of each tributary component from case to case, which, mountingly compounded, as ontogeny proceeds, would blur and finally obliterate all initial order. So, what holds them all together in co-ordination under an overall design? The usual answer is: "control processes". If these are visualized as coequal to arrows, the answer is partly correct, partly misleading. As I pointed out earlier on, there are examples of programmed linear "feedback" pathways to be found in organisms. But they are decidedly *not* fair samples of the general method of organismic systems operation. Paraphrasing the dictum about "the whole and its parts", the development of the nervous system is "more" than the sum of arrows in the diagram, even if additional "feedback control" arrows were included; incidentally, to satisfy the facts, the number of the latter would have to be infinitely large.

The dualism of "interaction"

Since much of this last comment was called forth by the unduly broad implications of the term "genetic determinism", especially in its articulated version of "genetic code", I may be pardoned for adding the following, slightly facetious, pictorial analogy to language. Figure 8 (*pp.* 38–9) shows a set of fixed and stained giant chromosomes of an insect; their aperiodic band structure corresponds to the underlying stereotyped aperiodic ("coded") seriation of the genes along the chromosoma laxis. Figure 9 (*pp.* 38–9) is from the frontispiece of a translation of Nietzsche's "Thus spake Zarathustra", which I discovered at the Pratt Institute in New York. It shows sentences exuding from the poet's mouth (though hardly simultaneously). The resemblance between the two pictures is not entirely fortuitous, for the meaningful seriation of letters in a word has been rightfully used as model for the meaningful serial linkage of molecular elements into genetic subunits to serve as code for the transcription of "information" from DNA to RNA and subsequent translation to protein. Quantitatively, the alphabet with its twenty-six letters is richer than the genetic code with only three-letter "words", but otherwise the analogy is of intriguing pertinence. In fact, the visual counterpart to the analogy, shown in our two pictures, goes even one step further; it does so by what it fails to show, for both figures depict senders without receivers, implying "action" without "reaction". Unless heard and understood, the poet's utterance would be that of a "voice crying in the wilderness": and unless faced with an appropriately "interlocutory" field structure around it, the gene string, likewise, cannot make sense.

This analogy is another reminder of the inevitability of a dualistic reference scheme for the exhaustive description (= explanation) of living systems, as exemplified by the whole-part, field-particle, cytoplasm-gene, etc. relationships. Due to the understandable prevalence of the analytical methodology and reductionist phraseology of our scientific past, we are faced with a conspicuous asymmetry of knowledge in favour of the second member of these conjugated pairs. Some signs of redress of that imbalance, however, have appeared, and it would have been profitable to accentuate them further on this occasion; for instance, by presenting examples of *field effects* as concrete demonstrations of the macrodeterminacy of systems. I have steered clear of this, because of limitations of time and listeners' patience. But as a parting bid for the acceptance and exploration of field concepts of group behaviour in biological systems, I want to give at least a glimpse of a simple model, which I have found helpful in the past both for making "field principles" more tangible and for keeping mysticism out-of-bounds.

The model, diagrammed (*pp*. 38–9) in Figure 10, starts from a single unit, let us say, a cell (A), assumed for simplicity to be internally homogeneous, but in steady equilibrated interaction (e.g., exchange of substance and energy) with the ambient medium across its free surface. Let us then increase the number of such units, either by cell division or aggregation (B). At first, all the members of the group are equal and share in the free surface, hence continue to exist as before. But as their number increases, when a critical number is reached, a new situation suddenly emerges (C) which splits the mass of previously equivalent units into two radically disparate groups— an *outer* group, still in direct contact and exchange with the original medium, and an *inner* one, shut off from access to the old medium and encircled instead completely by its former equals. As the inner units (cells), in response to this drastic change of their conditions, switch into a different (metabolic) course, the formerly uniform group acquires a core-crust (inner-outer) differentiation. If then the outer layer interacts with the newly modified inner group, a third type of unit will arise between them (D), and so forth, in ever mounting complexity, but obviously conforming to an orderly overall architectural pattern. This inner-outer differentiation is one of the most elementary instances of a group pattern being definitely predictable as a whole, while at the same time the fates of its components are still indefinite, thus illustrating macrodeterminacy in the absence of microdeterminacy, as in Figure 5. Yet, the main point is not just this early indeterminacy of the members of the group, but the fact that their future fate is determined by their place within the dynamic configuration of the whole group and can be predicted and understood only in reference to the "whole". For example, if such a group were cut in two, then future "inner" units would become "outer" ones, and the whole, though of reduced size, would turn out as of harmonious structure and proportions. In essence, this is the story behind the phenomenon of "twinning".

Epilogue

I herewith rest my case. Biology has made spectacular advances by adopting the disciplined methods of the inorganic sciences and mathematics, but it has not widened its conceptual framework in equal measure. My aim was to illustrate not only the need, but the feasibility of such an adaptive move. It may entail attempts in physics to enlarge its conceptual structure so as to be able not only to encompass living nature, but to fulfil the postulates raised by the realities of phenomena germane to living systems. But pending a successful achievement of this task, and even if it were doomed to failure, biology must retain the courage of its own insights into living nature; for, after all, organisms are not just heaps of molecules. At least, I cannot bring myself to feel like one. Can you? If not, my essay may, at any rate, have given you some food for thought.

DISCUSSION

THORPE   When you objected to the statement that "the whole is more than the sum of parts" was your objection simply and solely based on the view that the word "more" implies quantity only?

WEISS   As far as I am concerned, "more" states more identifying data or so.

THORPE   How then would you rephrase it?

WEISS   The information about the whole, about the collective, is larger than the sum of information about the parts.

THORPE   I see. And this is of course shown by your formula?

WEISS   Yes, essentially. It shows what I said at the beginning, that molecular biology, for instance, is not going to give us *all* the information we need. Some of the most vital information, yes.

HAYEK   Do you remember the Woodger discussion on this problem of parts and wholes. I thought he gave a perfectly satisfactory explanation.

WEISS   It started in the "Woodger days" at Cambridge, years and years ago. Why didn't it take? Because there was what I called yesterday a confidence gap. The hard-line, workbench biologists spoke contemptuously of "generalists". This is why I put a lot of hardware into my presentation here. You have to come up with hard facts to be listened to. Logically he has been absolutely correct.

THORPE   How do you relate this to Elsasser's thought, as expressed in his last book *Atom and Organism*? If I remember correctly, he there points out that the study of organisms reveals regularities in biological processes which are intrinsically impossible to deduce mathematically and in their entirety from the laws of physics.

WEISS   Well, it depends on how you define physics.

THORPE   I think that Elsasser's argument carries the implication that was in fact stated a long time ago, I think by Clerk Maxwell, to the effect that the name of physical science is often applied in a more or less restricted manner to those branches of science in which the phenomena considered are of the simplest and most abstract kind. That is to say, one excludes the consideration of the more complex phenomena such as those observed in living beings. To put it another way, physics is the most restricted of the sciences, because it is the most restricted in the field of phenomena to which it is devoted. On the contrary, the investigator in an unrestricted science such as biology, must be prepared to follow his problems into any other science

43

whatever, provided, of course, that he thinks that methods and techniques of that science will help him to solve his problems. As Carl Pantin recently pointed out (*The Relations Between the Sciences*, C. P. A. Pantin, Cambridge University Press 1968) "physics and chemistry have been able to become exact and mature just because so much of the wealth of natural phenomena is excluded from their study".

WEISS We had at Christmas a rather unsatisfactory discussion among outstanding people, with Gerald Holton as chairman and organizer of the meeting on the question "Does life transcend the physical sciences?". Well, how can you answer that question, unless you have two symposia, first, one on "what is life?" and a second one on "what is physics?". For instance, the laws of physics as I used to learn them right after the first World War, could not possibly have predicted solid state physics.

BERTALANFFY This reminds me of a story illustrating the point. When 35 years ago, I had to undergo the habilitation colloquium as Dozent at the University of Vienna, the late Professor Schlick, founder of neo-positivism, was on the Committee. He asked me: "*Herr Kollege*, you contended in your books that biology cannot be reduced to physics. How do you define 'physics'?" My answer, considered rather "fresh" at the time, was: "I am very sorry, *Herr Professor*, that I cannot readily give such definition, but I would be delighted to hear yours." As Schlick was a grand man, he let me pass but never answered the question.

There is no immutable corpus of laws of physics. In our own time, we have seen the emergence of quite novel concepts and laws in quantum, atomic and particle physics. It is quite possible and to some extent verified, that the inclusion of biological phenomena leads to an expansion of physics when defined as the science of inanimate nature. The self-duplicating DNA helix, the genetic code, also the expansion of the classical entropy principle with the inclusion of open systems, are essentially novel compared with "physics" understood as the content of current textbooks or as the science of inanimate phenomena.

WEISS Physics has, by self-confinement, carved out, quite legitimately, a partial aspect of the universe in which to take an interest. Physics decided not to concern itself for instance with individual phenomena, but only the overall rules. In geology, there is also a *general* geology concerned with the principles of erosion and vulcanism and so on, but one cannot predict from that, for instance, the exact form of Pike's Peak. Nor does the geologist care to.

To take up another point, the so-called step-by-step hypothesis indicates complete ignorance of the basic requirements of the living system. It is a system to begin with, because the essential component processes are in the circular interdependence of a fabric that you cannot reconstruct, piece by piece, from the properties of the pieces, which must hang together or else the fabric cannot exist at all. One component missing, all the others fall apart. You can synthesize DNA if you have the enzymes and other prerequisites made by somebody else; you borrow things and then you cheat by not mentioning the fact that you borrowed something from another system which was already the kind of system you are trying to explain.

BERTALANFFY   Not to mention the fact that the spontaneous formation of open systems maintaining themselves in a non-equilibrium state, by random events, is thermodynamically un-understandable.

THORPE   What was your last remark?

BERTALANFFY   Thermodynamically it would not work because if you have molecules, of, let's say, nucleic acids plus A, plus B, plus enzymes, for instance, you have the famous primeval soup. But they would never arrive at the idea of forming an open system, remaining in a state of high improbability; what they would do instead would be breaking down into nucleic acids again.

WEISS   That's right . . .

THORPE   Another point that Elsasser makes in the book I referred to just now, is that physics is the study of homogeneous systems and biology the study of inhomogeneous system. That is to say, that all atoms or all molecules of given species are all exactly alike. By contrast, no two organisms are ever exactly alike. Thus physical classes are homogeneous; biological classes are inhomogeneous. He then comes to the conclusion, which I think very closely concerns our discussion today, that the hypothesis that system events can always be adequately predicted from previous system events, using nothing but the laws of physics, constitutes the mechanistic or reductionist hypothesis. I would like to know how far you feel yourself in agreement with Elsasser's ideas on these matters.

WEISS   What he says is quite true even in physics, because if you take, for instance, meteorology, you cannot derive the morphology of a thundercloud; all you can do is homogenize it and all you get is H_2O. This self-restraint of physics is fine, because it gave us some of our overall rules and laws of nature. In the living system, physics spells out merely a set of injunctions of what must *not* be assumed to happen in nature. It is a restrictive frame and tells you only when you go wrong. It sets a boundary condition for expectations, but leaves the content of the frame undefined.

THORPE   Yes, Carl Pantin whom I referred to just now, makes this point very nicely, that physics is a restrictive science in that it restricts itself intentionally in various ways. Biology, by contrast, is an unrestrictive science. Thus biology calls in all the possible explanatory principles from whatever science may be able to aid in its classification, whereas physics is basically a restrictive science.

HAYEK   I would not really go so far as to say that biology is *not* a restrictive science. Physics has probably narrowed down things more and allows only one outcome. Biology restricts less, leaves a wider range of possibilities still compatible with its theories.

THORPE   Yes, I think that's fair enough.

WEISS   In our educational system we are acting very much like newspaper editors, who highlight the spectacular, and neglect the far more constant phenomena. For instance, we are making much of the marvels of evolution without ever telling our kids that the basic implements of all living systems are exactly the same and have remained the same, from the simplest living system that we know, all the way up to man. They should at least be told about this tremendous conservatism in the biochemical performances and structural elements of all organisms. This includes all the biochemical mechanisms of macromolecular synthesis, energy utilization, respiration, storage, proliferation, cell division, membrane structure and function, contractility, excitability, fibre-formation, pigmentation, and so forth. All of these processes have remained unaltered in essence through the ages. There has been a reshuffling, re-sorting and recombination with emergent novelty and progressive improvements—yes; but it all had to start with the full complement of the minimum vital implements to begin with because it's still the same assortment in the simplest amoeba and the highest metazoan. And this fact has not been publicized, even though it has been stressed for instance by Florkin and Baldwin, the latter in Cambridge having been the first to make that point.

I go even further; I am rather radical in approaching this problem. I may be wrong, but I discussed it with astronomers and cosmologists, and asked them by what right we can contend that the universe has been chaos to begin with; why don't we rather assume since we see everything in ordered constellations, that order is *primary* and that our concept of randomness is merely a construct—a simplifying abstraction—of the human mind that finds it more convenient to deal with this simplified concept? I was surprised that most of those I consulted—save for extreme "big bang" believers—answered: Yes, we couldn't contradict this, because we never find chaos; we have to have at least an inhomogeneity of two to begin with. But as soon as we are getting to have three, we are coming face to face with the analytical insolubility of the many-body problem. You cannot have three mutually indispensable components by building up one plus one. And in accepting such systems, physicists now are far beyond the physicists of even fifty years ago.

MCNEILL   I was fascinated during this talk by an impression of *déjà vu*. The structure of the argument we have just heard has an analogue in psycholinguistics. Some have believed that sentences are collections of elementary "particles"—words—and that the characteristics of sentences can therefore be understood by adding together the characteristics of the elementary "particles". This is a reductionist position in Dr. Weiss's sense. We study words to understand sentences. It has been attacked along lines much like Dr. Weiss's—essentially by showing that sentences have properties not possessed by words, just as he showed that higher structures have properties not possessed by lower ones. There are all kinds of examples. Except for the adjectives *eager* and *hard, John is eager to please* and *John is hard to please* have the same words in the same order, but are basically different structurally. *John* in one case is the subject of a sentence and in the other the object of a verb—the difference is not trivial. The relations occupied by words are properties that emerge only in sentences, and cannot be inferred from any fact about the words themselves. All of modern linguistics is a renunciation of the kind of reductionism Dr. Weiss attacks.

BERTALANFFY   This is one example that essentially parallel developments occur in various fields independently of each other. Incidentally I should emphasize that Paul was one of the first to introduce the concept of "system" in biology in 1924.

WEISS   I would like to add a further statement to Thorpe's question about physics. I said it is up to the physicist to decide whether he wants to raise his sights in order to encompass what we present as biological phenomena in disciplined documentations. The biologist can say: I wash my hands, I deal with the phenomena of living nature rigorously in their own right, and if the physicist cannot come along with me by expanding his own vision accordingly, there is no reason for us to downgrade nature to meet his inadequacy.

THORPE   I suppose our idea of the primacy of physics is really due to the idea that it all started in a nebula the make up of which was entirely random. Given that, you have to explain all subsequent developments in terms of the random motions of particles in your hypothetical nebula. That, I think, is really the point isn't it? Has the primacy of physics gone or has it not?

WEISS   Demokrit has been a remarkably useful appearance in the history of science, but a monopoly to atomism is no longer thinkable, because in pure physics it hasn't worked either. There they have the field and particle dualism and complementarity principles and whatnot. Nothing that I am saying about molecular biology should be construed as a lack of appreciation of the tremendous advances made in that field. It's only a warning against the monopolistic position often taken there, which starves out some of the equally important but sorely neglected problem areas in biology.

HYDEN   I would like to make a comment on your determinism stratified and the micro-freedom, as you called it, or the freedom of the micro-system. You brought in the range of the numbers and I think that was a very good point, because the high numbers have to do with freedom, but the freedom in the micro-system depends not only on the immediate environment of other micro-systems, it depends also on the outer environment. And if you have factors which will activate this cell system, then you will have more freedom, of course, because the number of molecules produced will be truly enormous. If you have activation of the genes, you have more freedom; it can be restricted if the genes will never be activated in the micro-system. Thus there is interaction and inter-activation, either with an outer system, or the system of another cell, of a neighbouring cell or something like that. If you go back in evolution, there must have been one step, certainly a saltatory step, when the interaction produced a redundancy of molecules, of very similar molecules, yet slightly different from each other, and that created in themselves a greater freedom. We have data to support this.

WEISS   Yet the question remains, that even the largest amount of numbers of units, in *numerical algebraic* terms, cannot logically lead to *geometric* patterns of a high degree of ordered regularity. This would be creation *de novo* every time it occurred and the terminology that is used in discussion of these questions is indeed veiled animism.

THORPE Can you say that the terms "coding", "readout", etc. are veiled animism?

WEISS No, they are tactical devices which are very expedient and quite legitimate. Animism is introduced only in terms like "dictate", "inform", "regulate", "organize", etc., as actions—actions by whom? Spirits? Chemicals endowed with spirits?

KOESTLER What worries me is, once you start on this game of semantic puritanism, it becomes a sort of boomerang. You used, for instance the word "freedom", but how do you define freedom? And if you define freedom by unpredictability, then the question arises, unpredictable by whom or by what?

WEISS No, I can put that into operational terms. The concept of measurement of "freedom", as in physics, by identifying "degrees of freedom" give the terms a rather definite meaning.

KOESTLER But it opens more questions than it answers.

WEISS It perhaps shows the finite limits of how far we can go in scientific "explanation". I agree that it leaves a twilight zone of indefiniteness. But my point is: I am not against anthropomorphism, but I am against the practice of conceding to anthropomorphic terms explanatory value. That is what I mean.

BERTALANFFY Yes, exactly.

THORPE To raise a different point: how far do you think it is generally useful or possible to say that, in an enzyme-substrate system, for example, the greater amount of the information is in the enzyme as opposed to the substrate. Is this a useful way of looking at the problem of biological organization?

WEISS Oh, certainly, there is no question about it.

THORPE In other words, if you have an enzyme and the substrate —most of the information is in the enzyme, but the substrate is just as important?

WEISS That is right. Even more, because if you have an interface, and you have enzyme systems, for instance, that only operate at a pH that you find at an interface, the interface contributes again an additional step of relevance. The enzyme interaction in a test tube will then not be a model for what you will actually get in the organized living system. Even a scramble of the mitochondrial enzymes in isolation does not do the same things as does their ordered sequential arrays in the intact mitochondria. Chemists have shown that, if you break mitochondrial enzymes into fragments, the fragments do not operate enzymatically; they are substructures; only after they have been recombined in the proper kind of seriation is the enzymatic capacity restored.

THORPE Can one take this one step further? Would you say that the DNA is not a self-reproducing molecule?

WEISS The term self-reproducing is to me an anthropomorphism— because nothing is "self"-reproducing, but there are "self-perpetuating" patterns.

THORPE Could you then re-phrase this in a non-anthropomorphic way?

WEISS Well, yes, it has a pattern which it can enlarge or transfer on less ordered elements in the surroundings. I mean that's an expression of facts.

KOESTLER If you want to avoid anthropomorphism at any price, you can say the DNA can Xerox itself. (Laughter). But I want to go back for a minute to the conundrums of freedom by taking up a point which Hydén raised. I do not know whether this analogy holds. Holger talked about the redundancy arising from molecules which are nearly identical, but not quite. Now, thinking of cloning. If you produce a clone from a soma cell, you produce a number of near-perfect copies of that individual organism, yet they will not be identical because of their interaction with environment, and so on. Is there an analogy with Holger's argument?

HYDEN Yes, if you have a number of molecules, which have weight in this environment for the system and can be used at a very limited frequency. If you think of the slight changes in the molecules, they differ only very little from each other. But taken as a group, these differences could be useful in that environment, in the interplay, and therefore the number of molecules would give more freedom in that environment.

WEISS Right, but how do molecules form a triangle?

HYDEN Well, you showed some examples here on the screen.

WEISS Molecules forming triangles without any informational data telling them how to form triangles?

HYDEN No, of course not, not even if there are great numbers, unless they were restricted by the space.

WEISS It is this kind of problem we have to come to grips with.

HYDEN Yes, but Arthur took up the clone, and although the clone is quite another phenomenon in a way, its members may slightly differ from each other, depending on the pH, depending on the oxygen, and so on. If in a clone of very simple organisms there were suddenly an outburst of nearly identical molecules, that would give in that environment quite another freedom.

WEISS   At the same time if you take different cells from different organs in higher organism, and you culture them, they do not lose their type-specificities. If you have a differential among these types A, B, C, D; A may change to A^1, B may change to B^1, C to C^1, but the differential A:B:C:D is retained, in their respective offsprings despite the fact that they all have the same common genes. This is the problem of differentiation; not cyto-differentiation of the individual cell, such as, for instance a myoblast arranging its actomyosin in such a way that it lines up like recruits into cross-striated columns, but "strain-differentiation", in which a myoblastic cell strain goes on for ever and ever after to produce only that particular type of cell, and for this there has yet been no adequate explanation. For cyto-differentiation one can assume that certain parts of the genome are inactivated or blocked or derepressed so that only a certain part of the tremendously rich repertory of the genome is ready for action. That can remain so as long as that particular cell individual lasts, and as a result, the cell can manifest a single-tracked expression of characteristics or criteria of its cytoplasm or of its specific products. But how, in cell proliferation does that kind of specific change come about, which is a "product", then is reflected back upon the genome in such a way that the genic interaction will henceforth constantly turn out exclusively that particular kind of copy? This is the true problem of differentiation. It's not generally faced. I said the genome is a captive and it can of course respond only to what its inner sanctum hears from its immediate surroundings. So, since the output of the cell strain must continue to be of the kind that has once been initiated, one must conclude that the genome—the animal in the inside cage—moulds its cage and from then on the cage moulds the animal to reproduce again the same kind of cage, and so on *ad infinitum*. This is a very basic problem.

SMYTHIES   May I take up again your analysis of the phrase that the whole is more than the sum of the parts. You made an interesting point here that "the more" refers to the number of *statements* that can be made. Now, undoubtedly this requires a certain amount of qualification; but surely it refers not to the *number* of statements (whichever way you measure the number of statements) that can be made, but to the *quality* of the statements?

WEISS   Yes, I said "information", I did not say statements. You need additional information about the state of the higher complex. The statement about this relation here is a statement which you cannot derive from knowing only the situation inside.

SMYTHIES   Another way of looking at it would be to say—Arthur in his book* has made this point and it has been made before—that statements referring to lower levels are contained as units in the higher-order statements. The fundamental logical difference is really that the higher-order statements contain as units the totality of statements referring to lower-order systems.

WEISS   Yes. We are not dealing with mere increase in number. We take a simple case as a model, and the reality of course is infinitely more complex, meaning *more* of the same. But each one of the co-operating units has its variants, has its lack of micro-precise, pre-ordained mechanical "one-way trackedness", let's call it. And yet the overall pattern of the collective is preserved. And this kind of thing is what I tried to express here, because complexity is only one part of it; as you go up you find increasing constancy and regularity, and it is this "rule of order", this regularity which we must accept in order to comprehend what is going on in the larger complex. Sheer numerical complexity becomes ordered complexity.

SMYTHIES   That is one way of differentiating between the higher levels in the hierarchy and the lower levels, but I think there are other logical differences between the high-order statements and the low-order statements. One of those that were mentioned is that the higher-order statement contains the lower-order ones as units.

WEISS   That's right. I mean a system operates through sub-systems. And when I speak of components I take whatever is of interest at the level of consideration at the time. When I am going from astronomy down through all forty orders of magnitude in the universe, I have to define the one level I am considering. And then, depending on that level, I can empirically identify how much of the regularities I observe can be derived by coincidence, as it were, by blind repetitiveness from the information I have about the sub-systems at the next lower level ("mechanisms"). And if I cannot then I have to make an extra statement for that level in order to give me complete information.

* Koestler (1967).

KOESTLER   I am afraid we are rehashing here principles which date back to Lloyd Morgan about new relationships that emerge on higher levels, and to Needham in the Twenties—you remember his *Time the Refreshing River*, where he emphasizes the emergence on higher levels of new types of laws, of new properties, emergent relationships; so we are not saying very much that is new. But I think if one goes half the hog, one might just as well try to go the whole hog. You mentioned Schrödinger, our genius loci, who is buried here in Alpbach. Schrödinger—the passage which you read by him reflects his mystic side, his gentle, playful side. But the essential thing in that book *What is Life?* is of course his statement that living organisms feed on negentropy. The organism lives on negative entropy. It goes against the Second Law of Thermodynamics—it does not break down, it builds up—and so does evolution, building up to higher levels, emergent levels. I am perhaps anticipating what Ludwig will say this afternoon—that the alternative to reductionism is a theory of "build-upism", a theory which accounts for this building-up that goes on, somehow, all the time, to higher levels of complexity. The universe did it and does it all the time, and our theories must account for it.

WEISS   But wait a minute . . .

KOESTLER   What I am saying is not new either; it goes back, for instance, to Woltereck's principle of "anamorphism", the emergence of higher forms, as a primacy principle, as a primary postulate.

SMYTHIES   It is a logical necessity deriving from your basic scheme of innate order. If you order things this way, then it is logically necessary that you obtain statements of the higher order.

KOESTLER   Exactly.

SMYTHIES   It is not any property of inanimate matter or anything else.

WEISS   What I said was that, for instance, innerness and outerness are emergent properties of a multi-unit system. This is an example of the emergence of new properties. But as for the whole philosophical attitude: we first break down to the micro-micro-level and deal with the units which we isolate by artificial abstraction—that is the primary act, it is a dissection. And then we are talking not about emergence, but about re-emergence: we come back up again, to where we have been originally in our primary experience.

BERTALANFFY   In my opinion, the terms "reduction" and "emergence" are both obsolete, and do not correspond to fact. Except for trivial cases, such as the reduction of acoustics to mechanics or the application of physical and chemical laws to biological phenomena, there never was "reduction" to pre-existent concepts and laws when new fields of phenomena were incorporated. Electrodynamics was not "reduced" to classical mechanics, even less, of course, quantum and atomic physics to classical physics, nor molecular biology to bio-chemistry of twenty years ago. What did happen was an expansion of concepts, models and laws, through which the relatively lower levels themselves were enriched and modified. Thus, for example, the atom developed from a little solid corpuscle into Bohr's model with fundamentally novel quantum conditions, etc. This then led to further syntheses.

"Reduction" and "emergence" are both metaphysical notions and hypostatizations of concepts into supposed reality. Reductionism makes the metaphysical claim that "atoms" with properties known (or assumed) at a certain time, are the ultimate reality into which everything should be resolved; "emergentism" claims that this cannot be done. What we actually see is progress toward ever more refined and synthetic concepts. Hence higher levels can not be reduced to a pre-given set of statements about the entities on lower levels; on the other hand, they become explainable within the system concerned and a wider theoretical framework (e.g. explanation of chemical properties in terms of atomic and quantum physics).

REFERENCES

ASTBURY, W. T. (1951) *Adventures in Molecular Biology*, Harvey Lectures, Series 46, pp. 3–44.

BERTALANFFY, L. v. (1952) *Problems of Life. An Evaluation of Modern Biological Thought*, J. Wiley & Sons, New York.

GERARD, R. W. (1958) *Concepts of Biology*, National Academy of Sciences, National Research Council. Publication 560, Washington D.C.

KOESTLER, ARTHUR (1967) *The Ghost in the Machine*, Hutchinson & Co. London, and Macmillan, New York.

LOEB, J. (1918) *Forced Movements, Tropisms, and Animal Conduct, Monographs on Experimental Biology*, Lippincott, Philadelphia and London.

POLANYI, M. (1958) *Personal Knowledge*, Routledge, London and University of Chicago Press, pp. 328–35.

SCHRÖDINGER, E. (1945) *What is Life? The Physical Aspect of the Living Cell*, Cambridge University Press.

WADDINGTON, D. C. (1962) *New Patterns in Genetics and Development*, Columbia University Press, New York & London.

WEISS, PAUL (1925) *'Tierisches Verhalten als "Systemreaktion". Die Orientierung der Ruhestellungen von Schmetterlingen (Vanessa) gegen Licht und Schwerkraft'*, Biologia Gen., 1, pp. 168–248.

———— (1939) *Principles of Development. A Text in Experimental Embryology*, H. Holt & Co., New York.

———— (1941) 'Self-differentiation of the basic patterns of coordination', *Comp. Psychol. Monogr.*, 17, pp. 1–96.

———— (1953) 'Medicine and society: the biological foundations', *Jour. Mt. Sinai Hospital*, 19, pp. 716–733.

———— (1959) 'Animal behaviour as system reaction: orientation toward light and gravity in the resting postures of butterflies (Vanessa)', *General Systems: Yearbook of the Society for General Systems Research*, 4, pp. 1–44.

———— (1963) 'The cell as unit'. *Jour. Theoret. Biol.*, 5.

———— (1967) '$1+1\pm2$. (When One Plus One Does Not Equal Two)', in: G. C. Quarton, T. Melnechuk, F. O. Schmitt (eds.) *The Neurosciences: A Study Program*, Rockefeller University Press, New York, pp. 801–821.

———— (1968) *Dynamics of Development: Experiments and Inferences*, Academic Press, New York.

———— (In press) *Life, Order and Understanding*.

———— (1969) ' "Panta Rhei", and so flow our nerves'. *Proc. Am. Philos. Soc.*

WHYTE, L. L. (1949) *The Unitary Principle in Physics and Biology*. H. Holt & Co., New York, Cresset Press, London.

Chance or law
Ludwig von Bertalanffy

State University of New York at Buffalo

After Weiss' splendid presentation, I feel in a somewhat disadvantageous position. The problems we are concerned with are largely similar. But while Paul could draw on his rich personal experience to present a vast and lively perspective of biological organization, the considerations to be offered by me will be largely in the way of abstractions and generalities. This presentation will be in two parts. Following Koestler's request, first a brief statement of General System Theory will be given; secondly, I shall attempt to apply the views so obtained to the Synthetic Theory of evolution as a focal point of the present meeting.

1 General systems

As we are trying to give local colour to the present session, I also may be permitted to evoke a personal memory. It was twenty-one years ago when, in the second year of the Alpbach College, I first presented the idea of general system theory to a wider forum (1948). So it is a pleasure to discuss the topic again, after it has developed into a well-cultivated field with a large literature, textbooks, journals, working groups and organizations.[1]

General system theory was first proposed as an abstract field of

[1] Among them *General Systems; Journal of Mathematical Systems Theory;* and the *Society for General Systems Research*, an associate of the American Association for the Advancement of Science with local chapters in the U.S. and Canada. Some works which may serve as an introduction to General System Theory are indicated in the References.

study. Since then, the "systems approach" has become widely used in industry, commerce, defence, and society in general. "Systems" has become a sometimes overworked catch-word. Systems analysis, systems engineering and similar denominations emerged as professions which did not exist a comparatively short time ago. The systems approach became necessary in view of the complexities of modern technology and society in general. We shall not dwell on the darker aspects of this development which as any in science, presents a Janus face of possibly beneficial and menacing consequences. The systems approach may, as Boguslaw (1965) emphasized, lead to further dehumanization, making human beings ever more into replaceable wheels of what Lewis Mumford (1967) so pungently called the megamachine of society. It is fair to say that, apart from its uses by the industrial-military complex, the trend may also be utilized for more desirable purposes. Coming from Canada, I may quote the premier of Alberta, Mr. Manning (1967), who wrote the systems approach into his political platform saying, "an inter-relationship exists between all elements and constituents of society. The essential factors in public problems, issues, policies, and programmes must always be considered and evaluated as interdependent components of the total system."

This is a fair statement of the problem. However, in the present discussion we are not concerned with practical aspects of the system approach, nor with its close connections with computer science, automation, cybernetics, and related technological fields.

The point of interest to the present Symposium is different. We seem to witness, in contemporary science, a change in *basic categories* of knowledge, of which the complexities of modern technology represent only one, and possibly not even the most important manifestation. As Paul has so vividly demonstrated in his presentation, we are forced, in all fields of science, to deal with complexities, wholes, systems, organizations or whatever terms you choose to express the problem. This implies a shift or re-orientation in scientific thinking which, at present, is far from being accomplished, but where promising beginnings have been made.

This development is of course related to the revision of the so-called mechanistic view and approach in science, which needs no detailed survey in this circle. Suffice it to recall that not even physics is "mechanistic" any more since matter de-materialized in modern theory, determinism reached its limits in quantum theory, and ultimate particles are entities defined only by highly abstract mathematical expressions. The life sciences have experienced the phenomenal upsurge of molecular biology; but notwithstanding or perhaps just because of this progress, new problems have appeared or old problems have reappeared with enhanced emphasis. The ultimate "reduction" of the phenomena of life to the molecular properties of DNA and related substances as promised in popular accounts of molecular biology, appears somewhat less than convincing. If anything, organisms are organized things, with respect to both structure and function, exhibiting hierarchical order, differentiation, interaction of innumerable processes, goal-directed behaviour, negentropic trends, and related criteria. About these, the mechanistic approach—not excluding molecular biology—is silent. The reason is not simply imperfect knowledge, so that the discovery of some new enzyme, or a new electromicroscopic structure would close the gap. The trouble is rather that the conventional categories, concepts and models of physics and chemistry do not deal with the organismic aspects that I have mentioned. They seem to leave out just what is specific to living things and life processes; and new categories appear to be required.

A similar discontent is manifest in psychology. Certain developments in contemporary society are, as it were, an experimental refutation of the mechanistic scheme. Consciously and purposely, psychology in the first half of this century set out to model itself after the paradigms of physics. The result was the "robot model" of man, dominated by the stimulus-response (S-R) and conditioning scheme of behaviour derived from animal experimentation, and bypassing or denying any specificity of human as compared to animal behaviour. This scheme was germane to otherwise opposed systems, such as classical and neo-behaviourism, psychoanalytic theory, the brain as a computer, etc; and it was applied in a large-scale experiment to modern society. According to the theory, proper conditioning, gratification of needs and drives, relaxation of tensions, etc., should result in psychosocial equilibrium.

In an affluent society where material needs are satisfied, sexual impulses easily gratified, and the human animal is properly conditioned by the techniques of the mass media, a state approaching universal bliss and happiness should be in view. Unfortunately, the experiment led to results contrary to expectation. Affluent society found itself beset by dissatisfaction and social unrest, juvenile delinquency, new forms of mental disorder such as existential neurosis, and other well-known phenomena. Again, a revision of basic categories appears to be due, and is manifest in recent develop ments in psychology. As Piaget and Inhelder represent one of the most important modern currents in that field, I am happy to report that Piaget has expressly related his conceptions to the general system theory of the present speaker (quoted in Hahn, 1967).

Again, the same problem appears in the social sciences. We may consider individuals as robots, and even transform them more and more into robots of consumption, of politics and of the industrial-military complex. But we pay for this dearly by moving nearer to *Brave New World* or *1984*; by neuroses, hippies, drug addiction, riots, wars and other symptoms of a sick society.

Thus we see, and this is by no means a new discovery, that similar problems arose in very different fields, or perhaps one and the same problem appeared in different disguises. Hence the question, what is the common denominator of these developments?

One oft-quoted answer was given by Warren Weaver (1948), co-founder of information theory. Up to recent times, Weaver said, science was concerned with *linear causality*—cause and effect, two-variable problems, stimulus-response, and the like. This was the prototype of thinking in classical physics. Somewhat later, the problem of *unorganized complexity* appeared which is essentially answered by statistical laws. Think, for example, of the motion of molecules in a gas, expressed by the gas laws. The paradigm of laws of unorganized complexity is the second principle of thermo-dynamics with its many formations and derivations. Now, however, we are confronted with *problems of organized complexity* at all levels of the hierarchic structure of the universe.

We can put this also in a slightly different way. The prototype of physical thinking were *process laws* such as Newton's law or the laws of electrodynamics, whose paradigmatic expression is in differential equations. The question of organization apparently demands other types of laws of which, at present, we have only a vague conception. To arrive at such organizational laws, *Gefügegesetz-lichkeiten* (cf. Bertalanffy, 1969) or whatever expression we use, we need, on the one hand, empirical investigation and definition in each case and at each particular level. On the other hand we need a general conceptual framework which transcends that of traditional science.

It perhaps deserves mentioning that this is a new approach essentially different from the classical. The latter was epitomized in the Third Axiom of Descartes' *Discours de la méthode*, which states that one should "start with the simplest objects most easy of discernment, in order to rise gradually step by step to an understanding of the most complex"—a statement which, of course, summarizes the analytical or reductionist approach. In contrast to this, we are now confronted with problems loosely indicated by terms like system, organization, and the like.

In summary, the classical form of process law is the differential equation. The laws of unorganized complexity are founded in the theory of probability. The laws of organized complexity are essentially system laws.

We may therefore epitomize the quest for a general system theory as follows:

First, up to recent times, "exact" science was held to coincide in essence with theoretical physics. With the new advances in the biological, behavioural, and social sciences, new problems have arisen which are not amenable to a "physicalistic" approach and require new conceptual tools.

Secondly, among these problems are multivariable interaction, organization, hierarchic order, differentiation, negentropic trends, goal-directed processes, and so forth. These concepts did not occur in physics; for this very reason they were bypassed, denied, or declared unscientific and metaphysical by mechanistic science. However, we cannot discuss a living organism without considering beyond isolable physico-chemical processes, the organization of processes in the way of maintenance, reproduction, development, etc., of the system as a whole. Similar considerations apply to psychology and social science. In fact, conceptual and in part even material models can be developed to account for such aspects which are programmatically excluded in mechanistic science. Hence an expansion of scientific categories, concepts and models became necessary, and is taking place at present.

Thirdly, for the reasons indicated above, there is a need for the development of "general system theory", that is a discipline (or disciplines) concerned with properties and principles of "wholes" or systems in general, irrespective of their particular nature and the nature of their components. The models and principles thus developed are intended to be interdisciplinary, that is, applicable to systems of various kinds and encountered in different scientific disciplines. To give a simple well-known example: the feedback model originally came from technology (exemplified by the thermostat, feedback control in radio receivers, and automation in general); but the same feedback scheme can be applied to many physiological regulations (homeostasis), to certain psychological and social phenomena, etc. Such models are therefore *isomorphic* in different branches of knowledge and can often be transferred from one realm to another. Thus there exist isomorphisms between systems different in their nature, components, acting forces, etc., with respect to certain general and abstract characteristics. Or, in still different words, there appear to be general system laws applying to systems of various kinds, irrespective of what their nature, components and interactions may be.

These ideas are hardly older than some twenty or thirty years. No wonder, then, that these developments are in their beginnings. There has been some justified criticism of unsatisfactory attempts in this new field, but one should remember that the establishment of classical physics, the "mechanization of the world picture" (Dijksterhuis 1961) took some 300 years from Nicholas d'Oresme and the Occamists to Galileo and Newton; and that the range of phenomena initially covered by physics was very spotty indeed. Considering the facts of history, the considerable number of approaches to systems theory is rather remarkable, although the gap between mathematical formalism and empirical facts is often large, and system-theoretical explanations are at present only possible in a small range of phenomena. Some of these approaches may be briefly mentioned:

"Classical" general system theory, so called, because it uses "classical" mathematics, i.e., calculus. It is capable of establishing certain general system principles and to apply them to concrete cases, for example in the theory of open and closed systems.

This leads to "non-classical", recent approaches. Sets of simultaneous differential equations as a model become unmanageable if the number of variables increases and especially if the equations are non-linear. *Computerization* and *simulation* have opened an essentially new approach. For instance, Hess (1969) has calculated the 14-step reaction chain of glycolysis in the cell in a computer model of more than 100 non-linear differential equations; similar analyses are routine in economics, market research, etc.

Related to this (and especially to the theory of open systems), is *compartment theory*, i.e., investigation of systems consisting of sub-units between which transport processes take place. Mathematical difficulties become great in the case of three- and multi-compartment systems; but Laplace transforms and the introduction of net and graph theory make analysis possible and lead to a theory of high sophistication.

The most general formal properties of systems can be axiomatized in terms of *set theory*. This provides mathematical elegance and generality, although the connections of axiomatized system theory with concrete system problems are tenuous.

Many system problems concern structural or topological properties of systems, rather than quantitative relations. *Graph theory* and *net theory* are available for such approach. Waddington's "epigenetic landscape" also belongs to these topological approaches.

The best known of the various systems approaches is *cybernetics*, the theory of "communication and control", with its basic concepts of information transfer and feedback. The cybernetic model is capable of describing the formal structure of regulatory phenomena even when the actual mechanisms are unknown and the system is a "black box" defined only by input and output. For this reason, the same cybernetic scheme may apply to hydraulic, electric, physiological, etc. systems. The highly sophisticated theory of servo-mechanisms in technology has been applied to natural systems only to a limited extent.

Although the feedback scheme is widely applicable, especially to homeostatic phenomena in biology, the frequent identification of "cybernetics" with "systems theory" is incorrect. Cybernetic systems represent one highly important, but rather special case of "general systems".

Other systems approaches include *information theory* (information defined by an expression isomorphic to negative entropy offering a possible measure of organization), *the theory of* (abstract) *automata*, *game theory* (concerned with the behaviour of supposedly rational players and their interactions in a "system"), *queuing theory*, and others.

These are mathematical tools in the process of development which, hopefully, may be applied to problems of wholeness, organization, teleological behaviour and so forth.

Of course, there are many system problems which at present elude mathematical theory. Perhaps the most important one is that of hierarchical order. Its relational aspects can be described by the "tree" of topology, but obviously hierarchical order in living nature is intimately connected with problems of differentiation, open systems, the negentropic trend and others which transcend present mathematical and physical theory. Considerations such as those by Koestler (1967) on "open hierarchical systems" may eventually lead to a more rigorous theory.

The essential point is that new conceptions have, as it were, removed blinds confining vision in the former, physicalistic, mechanistic and reductionist approach which obscured essential aspects of reality. One may be aware of the shortcomings of these new conceptions, but they nevertheless signify a basic re-orientation.

A "scientific revolution" in strictly technical terms, such as those elaborated by Kuhn (1962), means the appearance of new conceptual schemes of "paradigms", bringing to the fore aspects which previously were not perceived or were even suppressed in "normal" science. There is a shift of focus in the investigation of problems; a new emphasis on philosophical analysis which is not felt necessary in periods of growth of "normal" science; new paradigms emerge, and there is lively competition between rival theories. These criteria Kuhn derived from a study of the classical revolution in physics and chemistry, but they also fit recent developments in our own field.

The scientific revolution of the 16th–17th century put an end to the ancient Aristotelian conception of the universe. But this was not a sudden shift from a "primitive" world view to the "correct" world picture of post-Renaissance science. It was a long-drawn process in which Aristotelian science, which, in the words of Santillana (1955), was "common sense itself", was gradually supplanted by new paradigms such as analysis into elementary events, the conception of "blind natural forces" expressed in deterministic laws, and so forth. Presently another re-orientation appears to be taking place. While this development is far from being completed, it heralds new paradigms of system and organization. The world begins to look not like a chaotic "game of dice", to use Einstein's expression; but rather as a great organization in the sense of Nicholas of Cusa, Leibniz and Goethe.

2 Chance and law in evolution

The present meeting was asked to discuss, in Thorpe's and

Koestler's words, "an undercurrent of thought in the minds of scores, perhaps hundreds of biologists over the past twenty-five years who are critical of the neo-Darwinian theory's claim that random mutations *cum* natural selection provide all the answers to the phenomena of evolution"; while similar critical tendencies are in evidence in the other life sciences. My discussion of current thinking about evolution will be general, partly speculative and, hopefully, heretical. The views to be offered are not new; but as they are hardly expressed in current literature, they seem to deserve explicit statement.

Neo-Darwinism or the "Synthetic Theory" of evolution has incorporated genetics, cytology, molecular biology, physiological and population genetics, into its framework. For the present purpose it will suffice to outline briefly the main differences between the Neo-Darwinist and Darwin's original theory.

First, Darwin's rather vague concept of hereditary variation has been replaced by the concept of mutation in modern genetics and molecular biology. Mutation (in its principal form) is defined as a change in a chromosomal gene, ultimately in a DNA molecule. Secondly, the Darwinian conception of selection as a bloody "struggle for survival" was replaced by that of differential reproduction, that is, those mutants gradually prevail that produce the largest number of offspring. Thirdly, instead of the survival of the fittest or best adapted individual, what really matters according to the modern theory is rather a continuous change in the gene pool, the genetic constitution of populations.

What has *not* changed compared to Darwin's original conception is the essentially accidental nature of evolution. Mutations are errors sometimes happening in the reduplication of the genetic DNA code, comparable, according to a well-known simile (Beadle, 1963), to typing errors made by a not-too-careful secretary when copying an original script. Selection, acting by way of differential reproduction, is the directive agent in evolution. This implies that it is ultimately the environment, and its changes, that determine the course of evolution. Environmental changes, however, are accidental with respect to the organism concerned. Therefore evolution is essentially "outer-directed", to adopt Riesmann's sociological term.

In the words of a recent book (Williams, 1966), the theory of selection is "based on the assumption that the laws of physical science plus natural selection can furnish a complete explanation for any biological phenomenon, and that these principles can explain adaptation in general and in abstract and any particular example of an adaptation".

This, as the author correctly states, is the common belief among biologists, expressed both in current teaching and in research. The objective of the present discussion is, of course, not to "refute" neo-Darwinism as a scientific theory; its achievements are implicitly taken for granted, and specific questions under dispute are beside the present argument. The question raised here refers to the "nothing-but" claim of synthetic theory, i.e. the statement that the theory is *in principle* capable of furnishing a complete explanation of evolution.[2]

From the appearance of Darwin's work in 1859 to the 1920s, there has been an enormous amount of discussion about the theory of selection, which is today forgotten (Eduard von Hartmann's (1875) critique was one of the earliest and is still the best). The main arguments were reviewed by me in 1949 (1960), but evolutionists paid little attention. This is a pity; for many pertinent questions were left unanswered and swept under the carpet. I, for one, in spite of all the benefits drawn from genetics and the mathematical theory of selection, am still at a loss to understand why it is of selective advantage for the eels of Comacchio to travel perilously to the Sargasso sea, or why *Ascaris* has to migrate all around the host's body instead of comfortably settling in the intestine where it belongs; or what was the survival value of a multiple stomach for a cow when a horse, also vegetarian and of comparable size, does very well with a simple stomach; or why certain insects had to develop those admirable mimicries and protective colorations when the common cabbage butterfly is far more abundant with its conspicuous white wings. One cannot reject these and innumerable similar questions as incompetent; if the selectionist explanation works well in some cases, a selectionist explanation cannot be refused in others.[3]

In current theory, a speculative "may have been" or "must have

[2] I would resent being labelled as an "anti-selectionist"—a reflection of the simple mentality of Western movies, dividing the world into good guys and bad guys. Calling attention to problems and limitations is not being "anti"— unless "pro" means swallowing a pseudo-religious or political dogma hook, line and sinker.

[3] For some botanical examples illustrating similar questions, cf. Gavaudan (in Moorhead and Kaplan, 1967).

been" (expressions occurring innumerable times in selectionist literature) is accepted in lieu of an explanation which cannot be provided. For example, paleontologists give ingenious reasons why such and such a shape of tooth or bone or reconstructed muscle was advantageous to a carnivore, tree-climber, or runner, but it must remain subjective opinion what selective value to attribute to a sabre tooth or the rudiments of an eye millions of years ago.

This is at the bottom of the question, recently (Moorhead and Kaplan, 1967) discussed anew—namely, whether the principle of selection is a tautology. It is in the sense that the selectionist explanation is always a construction *a posteriori*. Every surviving form, structure or behaviour—however bizarre, unnecessarily complex or outright crazy it may appear—must, *ipso facto*, have been viable or of some "selective advantage", for otherwise it would not have survived. But this is no proof that it was a product of selection.

Furthermore, selection *presupposes* self-maintenance, adaptability, reproduction, etc. of the living system. These therefore cannot be the *effect* of selection. This is the oft-discussed circularity of the selectionist argument. Proto-organisms would arise, and organisms further evolve, by chance mutations and subsequent selection. But in order to do so, they must already *have* the essential attributes of life.

I think the fact that a theory so vague, so insufficiently verifiable and so far from the criteria otherwise applied in "hard" science, has become a dogma, can only be explained on sociological grounds. Society and science have been so steeped in the ideas of mechanism, utilitarianism and the economic concept of free competition, that instead of God, Selection was enthroned as ultimate reality. On the other hand, it seems symptomatic that the present discontent with the state of the world is also felt in evolutionary theory. I believe this is the explanation why leading evolutionists like J. Huxley and Dobzhansky (1967) discover sympathy with the somewhat muddy mysticism of Teilhard de Chardin. Riddles that cannot be dealt with in the accepted scientific frame of reference, pop up as mythology and metaphysics.

In the following, I would like to enumerate, in a rather unsystematic way, some of the problems which seem not to be answered by conventional theory, and which deserve empirical research and conceptual re-evaluation.

(A) A general progression of evolution towards higher organization (comparable to a similar trend in individual development) is a pheno-menological fact, that is, a matter of the paleontological record. This transition toward higher organization is not an expression of a sub-jective value judgement, nor connected with vitalism, a drive toward perfection or other metaphysical supposition; it is a statement of fact describable at any length in anatomical, physiological, behaviour-al, psychological, etc. terms. This ascent or "anamorphosis" in evolution is not refuted by cases where evolution was arrested or reversed under conditions which, as a rule, can well be circumscribed, such as life in the deep sea (*Lingula*), in caverns (blind cavernicolous animals), or parasitism. The enormously varying "tempo of evolu-tion" (cf. Mayr in Moorhead and Kaplan, 1967) obviously requires explanation but is no disproof of the fact that, by and large, a transi-tion towards higher forms and functions has taken place, nor a proof that this was completely outer-directed.

The conventional theory of evolution considers *adaptation* and *evolution* under the same terms of reference, both to be explained by random mutation, selective advantage, differential reproduction, etc. However, in my opinion, there is no scintilla of scientific proof that evolution in the sense of progression from less to more compli-cated organisms had anything to do with better adaptation, selective advantage or production of larger offspring. Adaptation is possible at any level of organization. An amoeba, a worm, an insect or non-placental mammal are as well adapted as placentals; if they were not, they would have become extinct long ago.

Nor is there an apparent adaptive difference between specializa-tion and non-specialization. It may be advantageous to the specialist to have a monopoly for some environmental niche (within limits, as over-specialization may lead to extinction); but it is also advan-tageous not to be specialized and being able to roam over the wild blue spaces; in fact, we find both categories in the past and present faunas and floras. If differential reproduction and selective advantage are the only directive factors of evolution, it is hard to see why evolution has ever progressed beyond the rabbit, the herring, or even the bacterium which are unsurpassed in their reproductive capacities.

Neither is it evident that the conquest of new environmental niches is connected with an up-grading of the organizational level; as already said, such conquest was achieved at quite different levels of organization. In contrast to the theory of selection, L. L. Whyte

(1965) has pointed out that the decisive evolutionary steps, the branching-out into many different types, seem to have occurred just when the ecological niches were relatively empty, as in the conquest of land by vertebrates. That is, these explosive phases seem to have happened when competition was at a minimum rather than maximum. On the other hand, vertebrates never superseded lower forms of life, which indeed are irreplaceable in the ecological cycle as a whole.

The identification of evolution with adaptation is therefore by no means proved. It is a debatable point, not an *a priori* principle of evolution.

(B) According to the orthodox hypothesis, evolution (and biological phenomena in general) are to be explained by "the laws of physical science" plus "natural selection".

However, what we actually see is a formidable range of organization from the various levels of systems in physics to electron-microscopic and macroscopic systems in biology (mostly of highly dynamic nature) to still higher entities of sociological nature as Weiss (this volume, cf. also 1962) has admirably surveyed. The entities of physics are but a small sector of this spectrum.

"Organizing forces" and "laws of organization"—in whatever way defined—are evident at the various levels, from protein and nucleic acid helices to fibrils and fibrillar structures, to organelles like mitochondria reconstituting themselves after severe disturbances, to tissues like hydra or embryonic organs re-organizing themselves after dissociation. The nature of these "organizing forces" is at present insufficiently known, but their effects can be directly "seen" or demonstrated by experiment. We have no reason to doubt that they are ultimately "physical" in nature, that is, rooted in the elementary forces (nuclear, electric, gravitational) known to physics. But the "laws of organization" require formulation in each particular case and at each level. At each level, properties appear that are "non-summative", i.e., belong to the respective system and cannot be obtained from the isolated parts. For example, it has been pointed out (Platt, 1961) that the mere increase in molecular size beyond some 500 atoms leads to "new" or "emergent" properties and "new" laws of physics which, combined with the open-system nature of organismic systems, go some way in accounting for "strange" processes in biology. According to Pattee (1961), the order in biological macromolecules is not adequately explained as an accumulation of genetic restrictions via selection, but replication presupposes well-ordered rather than random sequences.[4]

Thus there are principles of "self-organization" at various levels which require no genetic control. Immanent laws run through the gamut of biological organizations. But then, these obviously are not the outcome of random mutations *cum* selection, nor is evolution completely outer-directed.

(C) "Organizational laws" will also have to be pursued at the macrolevel of evolution. Several approaches come to mind.

Evolution is said to be opportunistic, that is, a viable adaptation can be reached in any number of ways. This is plausible for structures such as the horns of antelopes, cited by Simpson, or the floating devices cited by Mayr; that is, structures that have few technological constraints. But human technology shows that, as a rule, its products are not opportunistic in the sense that a given purpose has any number or even a large number of solutions. A watch, an automobile or a computer can be construed only in certain ways (and certainly is not constructed by tampering with materials at random which, of course, forms the ancient argument against the chance origin of "living machines"). Similarly, if you—or nature—wish to get an eye, you have to go over the steps of pigment spot, cup eye and camera eye. If a respiratory pigment is needed, there are apparently a limited number of ways available, leading to the relatively small number of such pigments. A circulatory system, kidney, or neural computer can be constructed only along certain lines, and they are not opportunistic in the sense that any construction will do. Hence the similarities of eyes, circulatory or nervous systems in independent phylogenetic lines. This, of course, is the well-known fact of convergent evolution. However, there seem to be *technological constraints* in evolution which deserve investigation in modern ways, possibly in the line of optimality principles in biology (Rosen, 1967).

"Organizational laws" in evolution appear to be expressed by *parallelisms* at the genetic, developmental and organizational levels (for further discussion, cf. Bertalanffy, 1960, p. 92 ff.). Only certain

[4] "We find in none of the present theories of replication and protein synthesis any interpretation of the origin of the genetic text which is being replicated, translated and expressed in functional proteins, nor do they lead to any understanding of the relation between particular linear sequences or distributions of subunits in nucleic acid and proteins, and the specific structural and functional properties which are assumed to result entirely from these linear sequences" (*Pattee, l.c.*)

pathways appear to be open at the level of possible gene mutations; of viable developmental processes; and of organizational configurations. In contrast to the study of homologies which predominate in evolutionary considerations, this lends emphasis to the investigation of *analogies* at various levels.

Regularities in evolution, as proposed by Rensch (1961), deserve further investigation. Principles like allometric increase very probably approach "laws in evolution".[5]

(D) The organismic conception has definite implications with respect to the genetic code. The code, as we presently know it, is represented by the chain of nucleotides in the chromosomal DNA molecule, base triplets "spelling out" the amino acids for the synthesis of specific proteins in the well-known way via messenger and transfer RNA, ribosomes, etc. But the code, obviously, cannot be a fortuitous string of "words" (nucleotide triplets) comparable to the "word salad" produced by a schizophrenic. In some way, there must be organization—otherwise the DNA code would produce, at best, a heap of proteins, not an organized system, a bacterium, fly or human being. We presently know the *vocabulary* of the genetic code, the triplets standing for amino acids; but we do not know its *grammar*, the meaning[6] of the message as a whole. This "grammar" (or programme, or algorithm for producing an organism, contrasted to bits of information provided by individual codons) is unknown, but must be postulated. If speculation is permitted, we may venture to think that whereas the "genetic code" as presently known is a structural order, a linear arrangement of codons, the "grammar" would rather be a dynamic order (as may be expected, considering the generally dynamic nature of organismic entities). Instead of a linear graph representing the arrangement of codons, the "grammar" may rather be expressed by some fantastically complex set of reaction equations.

Such a dynamic programme (rather than spatial arrangement) would alleviate the contradiction between the particulate features of genes (as demonstrated by translocations, inversions, even simple

[5] I think Waddington's "principle of archetypes" is closely related to "evolutionary constraints" as outlined above (and by Rensch).

[6] It may be noted that the question of "meaning" comes in here just as it does in the psychology of learning (*vs.* "atomistic" conditioning), "semantic" information (*vs.* information measured in bits), linguistics (structure of language *vs.* a series of signs), etc. It is always essentially the same problem, i.e., significance with respect to a whole.

recombination) and the holistic features they also exhibit (in position effects, mutual interference, regulator genes, etc.).

This admittedly is speculation. The point to be maintained is that we should beware of simplistic notions, of believing that present molecular biology is able to "reduce" life to the properties of DNA molecules, and of naïve catch-words like mutation and selection as ultimate answers.

(E) There is another aspect which is rarely envisaged. If we are unwilling to accept preformation, i.e., the assumption that all human genes were present in the primeval amoeba and that evolution consisted merely in genetic loss or deletion (a notion even more fantastic than that of the 17th century preformationists who assumed that all future humans were preformed in Eve's ovary), the information carried in the chromosomes must, in some way, have accumulated during evolution. Little is known, however, about the origin of new genes, excepting rare cases of duplication when, according to current opinion, genes so formed became available for new functions. This seems a rather poor mechanism for the great sweep of evolution. One would expect, for instance, some proportionality between number of codons and taxonomic level. However, there is no clear relationship between chromosome number, DNA content, etc., and taxonomic position. These, for a variety of reasons (especially the unknown redundancy of the code), are poor measures of information content; but there certainly is an unanswered question.

(F) A seemingly fundamental contrast between inanimate and living nature has been known for a long time. According to the second principle of thermodynamics, the general trend of (macro-)physical events is toward maximum entropy, i.e., more probable states, disappearance of differentiation, and states of maximum disorder. In contrast to the continual degradation of energy by irreversible processes, and in apparent violation of the second principle, living systems maintain themselves at a high level of order and in states of fantastic improbability, in spite of irreversible processes continuously going on. Even more, both in individual development and in evolution, living systems tend toward states of increasing order, organization, and improbability. In contrast to the principle of increasing entropy in physics, there appears to be a negentropic trend in living nature; and this naturally has often served as a vitalistic argument.

A partial answer to this well-known riddle is simple. The second principle of thermodynamics, in its classical form, applies by definition only to closed systems. In open systems, an expansion of thermodynamics is necessary which is known as irreversible thermodynamics. In the generalized entropy function according to Prigogine, we find therefore not only a term expressing the production of entropy due to irreversible processes, but also one for entropy transport, which can be due to the importation of material rich in free energy into the system. Because the second term may overcompensate the first, the entropy balance in an open system may be negative. As an open system the organism may retain a constant level of negentropy, i.e., of high order and organization, or may even evolve towards states of decreasing entropy and increasing order.

The next question is the definition of the state toward which systems, closed or open, are tending. Closed systems approach equilibrium, defined by maximum entropy, probability and disorder. Open systems, under certain conditions, may approach a steady state. In a steady state, the system remains constant over time, although irreversible processes are going on, and remains at a distance from maximum entropy. We do not at present have a general thermodynamic definition of the steady state. One attempted definition, known as Prigogine's theorem, states that open systems tend toward a state of minimum entropy production. However, Prigogine's theorem applies only under severely restrictive conditions (and in chemical reaction systems only to near-equilibrium states). More recently, Prigogine (1965) has combined thermodynamic with kinetic considerations; the result of their application to living organisms remains to be seen. Another, more revolutionary answer was proposed by the Russian biophysicist Trincher (1965). He proposed to supplement the physical principle of increasing entropy by biological principles of adaptation and evolution which essentially amount to the introduction of the quantity "information" into the respective equations. This, however, seems to create a dualism between inanimate and living nature; while the entropy principle has a microphysical basis in Boltzmann's derivation, i.e. the definition of entropy as probability, such a basis is not given in Trincher's theory.

Thermodynamic considerations introduce another dimension into the problem of evolution. Apparently, when looking at living systems, we encounter two *Urphänomene*, as B. Hess (1969) called them. Any organismic system comprises on the one hand, a thermodynamically open system; on the other, an information-carrying system represented by nucleoprotein molecules. If one of these is lacking, we cannot speak of a "living" system. Contrary to widespread opinion, a virus or nucleoprotein molecule capable of self-duplication is, by itself, merely a chemical in a flask, but not a living system; it has no metabolism and displays no "life" phenomena; it exhibits replication only in a living cell (or under suitable experimental conditions), i.e., when a metabolizing system is provided. On the other hand, an open chemical system as used in industrial chemistry (or a flame, or "open" enzyme system), is not "living" because life phenomena such as heredity, morphogenesis, etc., are bound to nucleic acid molecules.

Considered thermodynamically, the problem of neo-Darwinism is the production of order by random events. Order and local decrease in entropy can, of course, take place if "organizing forces" are present (e.g., the formation of a crystal in a solution due to lattice forces). In the absence of such forces, the second principle of thermodynamics and its equivalent in information theory (Shannon's Tenth Theorem stating that information can be converted into noise but not vice versa) will prevail.[7] Considered in terms of the theory of automata, a machine (closed system) can change its structure or "learn" owing to input of information, but it cannot progressively increase its structure because this requires input of energy and is possible only in an open system. We may expect further insights from a future synthesis of irreversible thermodynamics, information theory, and the laws of supermolecular organization.

Such considerations particularly apply to the problem of the origin of life, which is usually considered in terms corresponding to the synthetic theory of evolution, i.e., chance appearance of organic substances and subsequent selection. According to well-known experiments, the formation of amino acids and other organic compounds is well possible in an artificial "primeval atmosphere", energy being provided by electric discharges. It may also be that, given the necessary time, eventually macromolecules, including self-duplicating DNA molecules have formed by chance in the organic "soup" of the primeval ocean. Organic substances may form structures of the nature of coacervates. What is at present quite inexplicable is why and how organic substances, nucleoproteins, or

---

[7] Cf. Eden (in Moorhead and Kaplan, 1967, p. 11): "Every attempt to provide for computer learning by random variation in some aspect of the programme and by selection has been spectacularly unsuccessful."

coacervates should have formed, against the second principle—systems not tending toward thermodynamic equilibrium but "open systems" maintaining themselves at a distance from equilibrium in a most improbable state. This would be possible only in the presence of "organizing forces" leading to the formation of such systems. Before such systems had emerged, selection could not even start to act.[8]

(G) Lastly, I submit that considerations and theories more general than usually envisaged may be required in the theory of evolution.

General phenomenological aspects of organismic systems are hierarchical order, a trend toward higher organization, differentiation of originally unitary systems into subsystems, progressive centralization, etc. These, however, are not specific of the biological realm; they are also found in psychological, sociological, etc. systems. Of course, the nature of the systems concerned, of their component entities, of the "forces" in the system are quite different. Nevertheless, there are close similarities or isomorphisms, that is, formal correspondences between these widely different systems and phenomena. If so, general system theory comes in.

A general system theory of dynamic hierarchic order (Koestler, 1967), then, is a pressing problem. One semi-mathematical approach is that of Simon (1965) mentioned by Koestler in his last book (1967).[9]

One important difference should be noted. Hierarchical order in physical systems, such as the space lattice of a crystal (or the watch in Simon's parable), results from the union of originally separate systems of lower order, in this case, of atoms or radicals. In the biological realm, *differentiation hierarchies* are predominant, which are rarely encountered in inanimate systems; although one may see a trace in the "differentiation" of a flame into different regions. Differentiation or self-organization is impossible within closed systems or ordinary automata because increasing order presupposes import of energy or "negative entropy" and is possible only in open systems. Hence a system theory of dynamic hierarchic order will,

---

[8] Cf. the considerations by Schulz (1950) amounting to the conclusion that the formation of macromolecular compounds (nucleo-proteins) depends on the pre-existence of metabolizing systems ("pre-vital metabolism"), with catalyzers, cyclic processes, etc., whose origin seems not to be explicable by presently known physico-chemical laws.

[9] When Simon in his watchmaker's simile concludes that evolution is feasible, in reasonable times, only "by stable intermediate forms"—that is, that the works don't break down again at each step—it is but another expression for what has been termed "organizing forces" in the present context.

in a way, have to combine general system theory with thermo-dynamic considerations. To use Koestler's brilliant simile, it will be a unified theory of "The Tree and the Candle".

Conclusion

It appears, therefore, that the synthetic theory of evolution is not a complete explanation and that the problem needs a new approach. This ought to include exploration of organismic systems beyond the molecular level; regularities in evolutionary processes; the "grammar" of the genetic code; thermodynamic and information-theoretical considerations; a theory of dynamic hierarchic order; generally speaking, the consideration of evolution as not completely "outer-directed" but co-determined by laws at the organismic levels. These are researchable problems deserving further exploration.

Oswald Spengler, in his *Decline of the West* fifty years ago, wrote that the concept of entropy is the most profound in western physics, expressing, as it does, the direction of time, the trend of the universe toward an ultimate heat death, and in general the historical sense of our civilization projected into the universe. Although not in-fluenced by Spengler, Eddington (1958, p. 103 f.) in a little-known passage, arrived at a similar conclusion:

> From the point of view of the philosophy of science the conception associated with entropy must, I think, be ranked as the great con-tribution of the nineteenth century to scientific thought. It marked a reaction from the view that everything to which science need pay attention is discovered by microscopic dissection of objects. It provided an alternative standpoint in which the centre of interest is shifted from the entities reached by customary analysis (atoms, electric potential, etc.) to qualities possessed by the system as a whole, which cannot be split up and located—a little bit here, and a little bit there. We often think that when we have completed our study of *one* we know all about *two*, because "two" is "one and one". We forget that we have still to make a study of "and". Secondary physics is the study of "and"—that is to say, of organization.

Eddington's reflections from the physicist's viewpoint, are rein-forced by those of the biologist. The general characteristics of living systems, such as organization, hierarchic order, self-maintenance, differentiation, anamorphosis, also revolve around Eddington's "and" (which we call "system"); and around the entropy principle (in our terms, the expansion of thermodynamics to include open systems). Certainly, the problems of biological research are much richer than the stereotype of "laws of physics and chemistry" plus "selection". They surely will require not less but probably much more scientific ingenuity and sophistication than theoretical physics. In its philosophical aspects, this amounts, as already said, to the replacement, or rather the generalization of the Newtonian universe of blind forces and isolable causal trains, and of Darwinian living nature as a product of chance, by an organismic universe of many levels, the laws of which are a challenge to future research.

DISCUSSION

THORPE It seems to me that there is an outstanding problem raised by our discussion—namely the problem of fixity in evolution. What is it that holds so many groups of animals to an astonishingly constant form over millions of years? This seems to me to be *the* problem now—the problem of constancy; rather than of change. And here one must remember that the genetic systems which govern homologous structures are continually changing. Thus the control system is continually changing but the system controlled is constant, and constant over millions of years. This problem seems to me to stick out like a sore thumb in modern evolution theory.

Another point which I would be glad if Professor Bertalanffy would discuss, is the difference between animals and plants. I think it is very striking that as he said, the *obvious* feature of evolution is the progress in organization amongst animals, a progress which does not seem to have anything to do with the survival value of the new forms. This seems not to be so in plants. I do not think one can find anything like the same "progress" in organization in the plant world. One feels that this must have something to do with mobility; and that an animal is characteristically an organism which has to move about, has to change and develop its organization so, presumably, as to be able to adopt new environments where the selective pressure against it is reduced. So my question here is—how far would you think this to be an adequate basis for explaining this striking difference between animals and plants?

BERTALANFFY I see no difference, in principle, between animals and plants. Certainly the paragon of evolution is the phylogeny of vertebrates, and already invertebrate phylogeny is problematic in many respects; mainly because the main phyla were already present at the beginning of the paleontological record, and because the probability of preservation is slimmer than with vertebrate bones and teeth. But the evolution from blue algae to phanerogams seems well comparable to that of animals, although the factor of mobility may have contributed to the greater variety of animals. Certainly, synthetic theory is presumed to apply to both animal and plant evolution; and also the problems appearing are similar.

WEISS I would like to say something about a bad habit of thought which tempts us to make a direct jump from a gene to a property. To the embryologist who knows the complications of development from

77

genes to phenotype, this kind of shortcut means completely faulty reasoning. Let's assume, for the sake of argument, that a mutation is really just a simple shift in the nucleic acid sequence. This shift in the DNA is transcribed to the RNA and so you get a shift in the amino-acid sequence of the protein—a tertiary shift. But your protein is not a simple linear chain of polypeptides. The tertiary configuration, the protein chain, is tied together by cross-linkages, by links or bridges which form an organized, three dimensional structure. You can't shift linear sequences in such a configuration at random without upsetting its cohesion and balance. All right. Then what does your "mutated" protein do—it operates, then, in a context which has not been directly affected by the mutation, so now there are discrepancies between this newly-formed protein and its environment, and the basic requirements for even the first kind of reaction are missing—in other words, this isn't going to work. The mismatch is going to be there from the beginning, and your mutated sequence is never going to form even a cell. There must be innumerable such errors occurring that never have a chance to develop, let alone to create, an organism that will outbreed other organisms. In other words, what sense does it make to try to infer the number of mutations from the number of detectable changes we find in the terminal product? This kind of thinking leads to astronomical improbabilities. One of the few people who has really come to grips with the frightful complications of this problem is L. L. Whyte in his *Internal Factors in Evolution*. He didn't solve the problem, the key is still missing, but he put his finger on it, and he had to take a frightful beating from the geneticists for it.

There is an analogy between the methods we use to interpret brain lesions and our ways of interpreting mutations. We look at something which comprises a large number of stable entities, plus one which has been changed, but which is only a small fraction of the whole complex. So we are relying on the identification of differential features between the complex X, which is not accessible to testing, and which is the bulk, the mass of the complex; we can only test for differences, for single factors like the colour of the hair or colour of the eye. You can test for the factors which suppress a digit in malformations, but you cannot test for the factors which produce digits. This morning we talked about the constancy of basic forms in living systems, but we only have a way of getting at the differentials. This is a methodological limitation, this putting the spotlight on the differentials, personifying them and isolating them from the context. What we do then is to compare X minus A with X minus B; we have abstracted the A and the B and then jump to the conclusion that A is the seat of a particular colour of an eye or curvature of an eye. This is a basic error in method.

HAYEK I am not fully convinced of Bertalanffy's statement about adaptability—this question of best adaptation seems to me ambiguous. We may say that a fish is perfectly adapted to its normal environment, but that does not exclude that some change in its organization would make it better adapted to some new, Martian environment. It is a question of adaptation to new niches. You can argue that higher organization does not create better adaptation, but then adaptation becomes the wrong word, because I am talking of the possibility of surviving better in another environment—other than that to which the organism was so far adapted. If you include this factor I wonder whether then your statement is still true?

BERTALANFFY This, I am afraid, is just the sort of argument I criticized, and which reminds me of the saying, if Grandma had wheels, she would make a lovely autobus. It can, of course, be said that higher organization "may" or "must" have been of selective advantage in some way, environment or niche. If selection is taken as an axiomatic and *a priori* principle, it is always possible to imagine auxiliary hypotheses—unproved and by nature unprovable—to make it work in any special case. But this procedure corresponds exactly to that of epicycles in the Ptolemaic system: If planetary motion is *a priori* cyclic, then any orbit, however seemingly irregular, can be explained by introducing more epicycles. Similarly, some adaptive value (and consequent selective advantage, survival due to differential reproduction, etc.) can always be construed or imagined. But the procedure of arbitrarily introduced auxiliary hypotheses proves neither the Aristotelian-Ptolemaic, nor the neo-Darwinian system.

There is a similar logical flaw in oft-heard arguments on the spontaneous origin of DNA and other organic compounds. Stanley Miller *et al.* experimentally simulated the conditions which presumably prevailed in the early history of the earth and showed that amino acids and other organic compounds may be formed. But this is something different from the assertion that, given only sufficient time, any chemical configuration which is possible, may naturally have been formed by random events. A pair of nylon stockings is, in its chemical structure, much simpler than a structure of nucleic acid chains, but we do not expect to find any nylons formed by "spontaneous generation" and "reaction at random".

KETY   It is hard to defend the thesis that increased complexity makes for wider adaptability. The mammal is not more adaptable than the fish. The mammal can survive in water, but he has not retained the adaptation of the fish—he has lost that and acquired another kind of adaptation. I wonder, however, whether one could say that it is necessary for a new organization to be *more* adaptable than another, or simply that it should be moderately adaptable. And then one could say that the reason there has been evolution is that higher complexity is less probable than lower complexity; and so it would take longer for higher complexity to arise, not by competition with less complex forms, but simply because of the elaborate process by which it has to be organized.

HAYEK   This seems to suggest another ambiguity of the term adaptability. It may well be true, as you suggest, that in general the more highly organized organism has a smaller range of adaptabilities than a lower organized one, and yet it may be quite possible that only the more highly organized one has the capability of adaptiveness to some one particular circumstance. It may have a small range of adaptability, and yet the possibility of adapting itself to one particular circumstance may be much higher than others. One must be very careful how one uses the word adaptability.

WEISS   Particularly if one takes the naïve approach to look at those characters which are particularly conspicuous. In human physiology the most adaptive features—if you deviate from them you get sick and you die—are really the adjustment, say, between the pulmonary circulation and the heart and the breathing mechanism, that includes the nervous system and its control of the vascular muscles. Here we have a tremendously complicated network of basically unrelated things. How these things can mesh and come out as a better functional machine, so to speak, is a problem one has to face; but so long as we look at unit characters like the colour of hair and the colour of eyes, we are sort of missing the essentials.

THORPE   Here again we must remember that because of their higher development, the vertebrates show a very great increase in the ability to use their own environment. Such animals can go to other environments so that the selective forces may thereby be changed or redirected. But on the other hand, by achieving that complexity these animals have lost one of the very easy ways of getting over difficulties which the lower animals have to face—namely hibernation, aestivation, etc. Lower animals, and still more plants, can form cysts, resistant eggs, seeds, spores, etc. which can in extreme cases survive almost all hazards.

WEISS   All I am saying is that I would not even start on the question of better adapted or less adapted, because we do not know the balance sheet. We buy something which is very spectacular and looks like a great advance, and we pay the price for it in some other processes in the organism which makes it more susceptible, less viable in the long run, for the species.

KOESTLER   To make confusion worse confounded I would like to refer to one of my hobby horses, the puzzle of the marsupials, which just do not fit into the orthodox theory. You have in Australia a whole lot of these marsupials, which look to a surprising degree like less efficient copies of the placentals. If you look at the Tasmanian wolf's skull and the Siberian wolf's skull—as drawn for instance in Hardy's Gifford lectures—you find that these two creatures which have developed independently over eight hundred million years have skulls which at first sight are indistinguishable. In those eight hundred million years, since Australia became an isolated island, not a single unorthodox animal emerged there, except for the moderate unorthodoxy of the kangaroo and the wallaby. What bothers me is that this problem does not bother the naturalist. It bothers me that these fantastic resemblances have arisen independently from each other, but no science-fiction creature, no bug-eyed monster, was produced by random mutations. Now, Hardy in his last book, and als￢ Waddington, have dug out Goethe's archetypes and revived the idea that evolution can only run in archetypal grooves, gradually actualizing potentialities which are already present in the amoeba. This is a terrifying perspective, if one tries to be logical about it.

BERTALANFFY   I believe what Waddington means by archetypes are evolutionary constraints. I have indicated a few approaches to this problem.

KOESTLER   That is not an answer, if I may say so, Ludwig. If you look at orthodox textbooks like G. G. Simpson's, you will see that he explains this striking similarity between the marsupial wolf and our wolf by simply saying that both are predators who prey on animals of roughly the same size and speed, so the selective pressure is on both the same, and therefore the appearance of both is the same. This seems to me question-begging on a heroic scale.

WEISS   I would say, much more dramatic are for instance the similarities between, let's say, the visual apparatus of squids and mammals. There is a lack of realism in people who work with mathematical symbols to explain evolution through mutation, because their idea is that "anything goes". Only an almost infinitesimally small number of component changes are sufficiently mutually conformant to produce even the first step towards anything.

KOESTLER   Do you realize that what you say leads to the conclusion that evolution is the fulfilment of a preordained programme, pre-ordained by way of constraints?

WEISS   You cannot play more melodies on the piano than you have keys. I mean it is this kind of thing.

KOESTLER   Sherrington, if you remember, once asked the question: why haven't we got six limbs like the insects, one of them a pair of wings?

THORPE   Well, our time is up and we must now conclude this discussion.

REFERENCES

BEADLE, G. W. (1963), *Genetics and Modern Biology*, American Philosophical Society, Philadelphia.

VON BERTALANFFY, L. (1947), *"Das Weltbild der Biologie"*, *"Arbeitskreis Biologie"*, *Alpbacher Hochschulwochen;* in: (1948) S. Moser (ed.) *Weltbild und Menschenbild*, Tyrolia Verlag, Salzburg.

———— (1960), *Problems of Life*, Harper Torchbooks, New York (In German: 1949.)

———— (1969), *"Gefügegesetzlichkeit"*, in: J. Ritter (ed.) *Historisches Wörterbuch der Philosophie*, I, Schwabe, Basel.

BOGUSLAW, W. (1965), *The New Utopians*, Prentice Hall, Englewood Cliffs.

DIJKSTERHUIS, E. J. (1961), *The Mechanization of the World Picture*, Clarendon Press, Oxford.

DOBZHANSKY, TH. (1967), *The Biology of Ultimate Concern*, The American Library, New York.

EDDINGTON, A. (1958), *The Nature of Physics*, University of Michigan Press, Ann Arbor.

HAHN, ERICH (1967), *"Aktuelle Entwicklungstendenzen der soziologischen Theorie"*, *Duetsche Zeitschrift für Philosophie*, 15, pp. 178–191.

VON HARTMANN, E. (1875), *Wahrheit und Irrtum im Darwinismus*. C. Duncker, Berlin.

HESS, B. (1969), *"Modelle enzymatischer Prozesse"*, *Nova Acta Leopoldina.*

KOESTLER, ARTHUR (1967), *The Ghost in the Machine*, Hutchinson, London.

KUHN, T. S. (1962), *The Structure of Scientific Revolutions*, University of Chicago Press, Chicago.

MANNING, HON. E. C. (1967), *Political Realignment—A Challenge to Thoughtful Canadians.* McClelland and Stewart, Ltd., Toronto/ Montreal.

MOORHEAD, P. S. and M. M. KAPLAN (eds.) (1967), *Mathematical Challenges to the Neo-Darwinian Interpretation of Evolution*, Symposium Monograph No. 5, Wistar Institute Press, Philadelphia.

MUMFORD, L. (1967), *The Myth of the Machine*, Harcourt, Brace, New York.

PATTEE, H. H. (1961), "On the origin of macromolecular sequences". *Biophysical Journal*, *I*, pp. 683–710.

PLATT, J. R. (1961), "Properties of large molecules that go beyond the properties of their chemical sub-groups". *Journal of Theoretical Biology*, *I*, pp 342–358.

PRIGOGINE, I. (1965), "Steady states and entropy production", *Physica*, 31, pp. 719–724.

RENSCH, B. (1961), *"Die Evolutionsgesetze der Organismen in naturphilosophischer Sicht"*, *Philosophia Naturalis*, 6, pp. 289–326.

ROSEN, E. (1967), *Optimality Principles in Biology*, Butterworth, London.

DE SANTILLANA (1955), *The Crime of Galileo*, University of Chicago Press, Chicago.

SCHULZ, G. V. (1950), *"Uber den makromolekularen Stoffwechsel der Organismen"*. *Die Naturwissenschaften*, 37, pp. 196–200, 223–229.

SIMON, H. A. (1965), "The architecture of complexity", *General Systems*, 10, pp. 63–76.

TRINCHER, K. S. (1965), *Biology and Information: Elements of Biological Thermodynamics*, Consultants Bureau, New York.

WEAVER, W. (1948), "Science and complexity", *American Scientist*, 36, pp. 536–544.

WEISS, PAUL (1962), "From cell to molecule", in: J. M. Allen (Ed.), *The Molecular Control of Cellular Activity*, McGraw-Hill, New York.

WILLIAMS, G. C. (1966), *Adaptation and Natural Selection*, Princeton University Press.

WHYTE, L. L. (1965), *Internal Factors in Evolution*, Braziller, New York.

Works relating to General System Theory
VON BERTALANFFY, L. (1967), *Robots, Men and Minds*, Braziller, New York.

VON BERTALANFFY, L. (1968), *General System Theory. Foundations, Development, Applications*, Braziller, New York.

———— and A. RAPOPORT (eds.) Society for General Systems Research, 12 vols. since 1956, Bedford (Mass.).

BUCKLEY, W. (ed.) (1968), *Modern Systems Research for the Behavioral Scientist. A Sourcebook*, Aldine Publishing Co., Chicago.

DEMERATH III, N. J. and R. PETERSON (eds.) (1967), *System, Change, and Conflict. A Reader on Contemporary Sociological Theory and the Debate over Functionalism*, Free Press, New York.

GRAY, W., N. D. RIZZO and F. D. DUHL (eds.) (1969), *General Systems Theory and Psychiatry*, Little Brown, Boston.

GRINKER, ROY R. (ed.) (1967), *Toward a Unified Theory of Human Behavior*, 2nd ed., Basic Books, New York.

MESAROVIC, M. D. (1961, 1964) *Systems Research and Design, Views on General Systems Theory*, Wiley, New York; (1968), *Systems Theory and Biology*, Springer-Verlag, New York.

Biochemical approaches to learning and memory
Holger Hydén

University of Göteborg, Sweden

I would like to sketch a molecular mechanism for learning, storage and retrieval of information which could operate at the system level as well as at the level of the single neuron. This theory is based on biochemical data obtained through learning experiments in animals. Later, I would like to discuss changes in brain with age which, methodologically, may be considered a different problem, but is clearly linked with the memory problem. In this context, the uptake of injected macromolecules—DNA in this case—by brain cells and its possible effect will also be briefly discussed.

Introduction

Does the engram exist in the brain in the sense of a memory trace which is engraved by experience and, furthermore, has it biochemical correlates available for analysis? It seems appropriate to cite from Lashley's paper (1950): "In Search of the Engram". On the basis of animal experiments carried out over many years, Lashley came to the conclusion that there are no special nerve cells for special memories. According to Lashley, it is not possible to demonstrate the isolated localization of a memory trace anywhere within the brain. Particular regions may be essential for learning a particular activity. Within such regions the parts are functionally equivalent. The engram is represented throughout the region.

This seems to be a good starting point for a discussion of biochemical aspects of learning and memory. But the discussion must also include another problem, namely, the individual's chances of actualizing his brain potentialities during his life cycle.

One of the definitions of learning refers to the capacity of a system to react in a new or modified way as a result of experience. Memory is the capacity to store information which can later be retrieved with high distinctiveness to steer the function in accordance with the new information.

An old theory states that when something is learned and stored, there occurs in millions of nerve cells a facilitation of the message at the synapses. This is an attractive view, but only part of the truth, because the activity flow can be made very specific and one big neuron may have 10,000 synaptic knobs on its surface. Figure 1 is a photograph of a surviving nerve cell, isolated by micro-dissection and stained to show the synaptic knobs, which are in focus. They appear as small dots on the cell surface. The facilitation theory only scrapes the surface of the problem.

In learning, a short-term memory is first established. This lasts for seconds to hours and is liable and susceptible to interference of various kinds. This storage or fixation process takes place during training to learn and immediately after. Short-term memory differs in nature from long-term memory. Long-term memory may last for the better part of a life-cycle, although, of course, it may undergo modifications. It is remarkably resistant to poisoning and shocks, and can be retrieved in a fraction of a second.

It is important to bear in mind that there are three principal components of the brain. Neurons, which differ biochemically, e.g. with respect to transmittor substances (Dahlström and Fuxe, 1964) and RNA composition (Hydén, 1962). They do not divide and only a small part of their genome is active. Glia, the second type of brain cells, also seldom divide and are not electrically excitable. They are rich in lipoproteins and in rapidly turning over RNA (Egyhazi and Hydén, 1966). Figure 2 depicts schematically the relationship between neurons and the surrounding glia. To the right is seen a capillary, at the top and to the left parts of two neurons. In between are various types of glial cells which cover the surface of the neurons except at the sites of the synapses, and whose delicate membraneous processes interlace.

The third component of the brain consists of extra-cellular tissues which seem to amount to around 20 % (Harreveld *et al.*, 1965). They presumably contain mucopolysaccharides and mucoproteins.

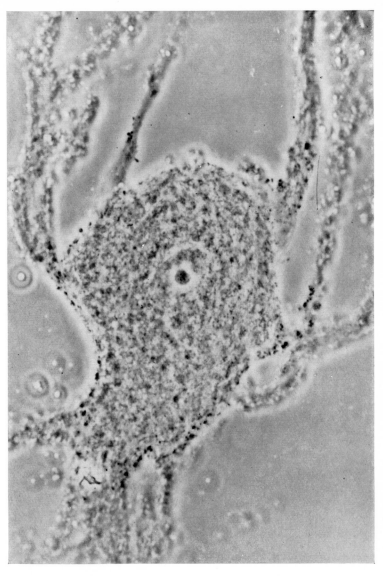

Figure 1   A single nerve cell stained to show the synaptic knobs.

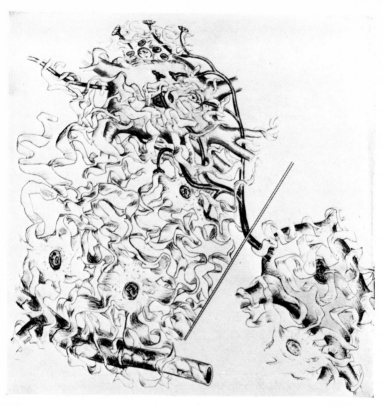

Figure 2 Schematic drawing of the relationship between neurons and the surrounding glia.

The genetic roots of learning mechanisms

If the outer part of the neuron is the site where the pulse-coded nerve impulses impinge, the inner part handles the energy demands, but also the synthesis of specific substances, and regulates the expression of the gene activities. Synaptic facilitation, therefore, can be expected to be regulated by genetically controlled mechanisms.

Any substance which can serve a memory mechanism for both innate behaviour and acquired learning for a lifetime must be assumed to have an easy access to the genes.

It may be argued that learning involves an additional activation of the genome of the neuron which, therefore, becomes more richly patterned biochemically and structurally, and can respond over a wider range with more selectivity. This type of reasoning may seem to lead to a pessimistic view, and some people find the idea hard to accept. It would set limitations on the capacity of the brain. It would mean that only genetically pre-programmed factors in our brain cells could be selected for a certain purpose and utilized (Plato, ed. 1961). On the other hand, one may ask whether there exists an instructional mechanism in the brain for learning at the highest level, i.e. problem solving, and which operates at the system level in the brain hierarchy, and in the individual brain cell. This idea goes back to John Locke (1931) who maintained that the brain at birth is a tabula rasa on which experience chisels the writing. Phrased in modern language, both selectional and instructional mechanisms may be supposed to operate.

Some neurons are clearly pre-programmed genetically to handle certain types of stimuli. Examples are neurons in the visual cortex. Experience can modify these: for example when a tone and a light flash are given *in pair* to an animal, gradually the visual and hearing centres both learn to respond to only one stimulus (Morrell 1967).

We may now ask if a brain mechanism really could exist in which the genes and the DNA-code not only permit learning, but also direct the operations of memory. Is the same mechanism concerned with experimental learning and innate learning or instinct behaviour? At birth, many types of innate behaviour are potentially ready, even if not in overt activity. Support a newborn baby in an erect position, let the feet press against the floor and drag the baby forward. He will make walking and even stepping movements (Teitelbaum, 1967). Many patterns of behaviour in animals are clearly inherited and

D

need only a key factor in the environment to be triggered. Thus the conclusion cannot be escaped that they are programmed in some way in the sequence of bases in DNA. But if so, the mechanism of expression is much more complicated than the synthesis of a protein.

The following questions arise then: what do instinct and innate behaviour and experientially acquired behaviour have in common with respect to acquisition, storage and activating mechanisms? In which respect do they differ? If storage of memory is the equivalent of differentiation and growth, how far can that analogy be carried?

Interest is currently focused on molecular events as the main operative mechanism underlying learning and remembering. The macromolecules—RNA and proteins—are the most likely candidates to mediate storage of information in the brain. They have recognition sites. The neurons are rich in RNA and proteins. No other somatic cell can compete with the neuron as an RNA producer (Hydén, 1960).

TABLE I *RNA response of neurons to increased sensory stimulation.*

Stimulus	Cell type	RNA increase in μμg per cell	%	P
Intermitt.[1] horizont. rot. 25 min/day, 7 days	Deiters' neurons rabbit	1550 ⟶ 1750	10	0.01
Intermitt.[2] horizont. rot. 25 min/day, 7 days	Deiters' neurons rat	680 ⟶ 750	10	0.02
Intermitt.[2] vertical rot. 30 min.	Deiters' neurons rat	680 ⟶ 850	25	0.001
Sodium chloride[3] 1.5% 30 days	neurons of N. supraopticus rat	68 ⟶ 121	80	0.01
Thirst[4] for 7 days	neurons of N. supraopticus rat	52 ⟶ 129		0.001

[1] Hydén-Pigon, 1960
[2] Hydén-Egyhazi, 1963
[3] Edström-Eichner, 1958
[4] Edström-Eichner-Schor, 1961

Table 1 demonstrates how the amount of RNA per nerve cell varies with species and localization. The data given also show that various types of physiological stimulation lead to an increased amount of RNA per cell. The response varied from a 10 % to more than a 100 % increase. This reflects an increased synthesis of RNA which has been induced by the stimulation. So far, no base-ratio changes have been found in this newly-synthesized RNA in mammalian brain cells: the RNA formed has had the base ratio characteristics of ribosomal RNA.

At this point, I would like to stress that no data support the view that brain cells contain "memory molecules" that store information in a linear way, i.e. record and reproduce it tape-recorder fashion. This is biological nonsense. In current literature, such views are seen, but they do little more than add to the confusion. As will be seen, RNA and proteins in brain cells do respond to the establishment of new behaviour, but the mechanism and its regulation still remain to be elucidated.

Proteins are probable candidates as executive molecules in a mechanism for acquisition and retrieval of memory. They have a high specificity and could respond rapidly in millions of brain cells to a trigger mechanism.

Macromolecular synthesis in brain cells during establishment of a new behaviour pattern

A major question has been: do macromolecular changes occur in brain cells characteristic of the learning process, and which do not occur when there is only increased neural activity without learning? This has been tested by establishing a new behaviour in animals and analysing the brain cells. When a new behaviour is established, brain cells respond with a synthesis of RNA of a highly specific base composition and a synthesis of several types of acidic proteins (Hydén, 1962, 1963, 1964, 1965, 1968). Another line of research has been to interfere with the synthesis of macromolecules in the brain and to observe the effect on behaviour.

From our laboratory I would like to present two types of experiments performed on rats. In the first, neurons were sampled randomly from layers 5 and 6 in a small area of the sensory-motor cortex which has been shown to constitute a control area for the transfer of handedness (Peterson, 1934).

Rats are either right-handed or left-handed or—in a small number of cases—ambidextrous. When they execute complicated movements with their paws, they prefer the paw on one side. They can, however, be induced to switch over to the non-preferred paw. The performance curve—number of correct performances per training period and day—is characteristic for this type of learning. It comprises a linearly increasing part and, in this case, an asymptotic part after 8 days with 25 minute training periods twice a day.

TABLE 2   Changes in the RNA base composition of cortical neurons from the control (left) side and from the learning (right) side.

	Controls	Learning	Change in %	P
	Mean	Mean		
Adenine	18.4±0.48	20.1±0.11	+ 9.2	0.02
Guanine	26.5±0.64	28.7±0.90	+ 8.3	0.01
Cytosine	36.8±0.97	31.5±0.75	—14.4	0.01
Uracil	18.3±0.48	19.6±0.56	+ 7.1	0.05
$\frac{A+G}{C+U}$	0.81±0.27	0.95±0.035	+17.3	0.01
$\frac{G+C}{A+U}$	1.72±0.054	1.51±0.026	—12.2	0.02

The advantage of this experiment is that a paired t-test can be performed, since control cell-material is present contralaterally in the cortex of the same rat. Furthermore, care has been taken to allow for variations between individual animals in the random sampling of cells from a defined cortical area.

In the first experiment, right-handed rats were induced to use the left forelimb in retrieving food from far down a narrow glass tube (Hydén, 1964). Training periods of 2×25 minutes per day were given. Neurons from those areas of both sides of the cortex whose destruction prohibits transfer of handedness were dissected by free hand and analysed. These control centres are situated bilaterally in the sensory-motor cortex and comprise around 1 mm³ of the cortex. Layers 5–6 are the most important. These neurons have a large nucleus in relation to their cytoplasm. Therefore, the analytical result will mainly reflect nuclear RNA. Eighty-eight rats were used for the analysis of 14,000 cortical neurons. As already mentioned, the advantage of this learning experiment is that the controls are present in the same brain. Other control experiments were also performed.

We found a significant increase of the amount of RNA per cell from the learning side of the cortex. In an extension of the work published (Hydén, 1965) the amount of RNA was found to have increased from 220 $\mu\mu$g of RNA per ten nerve cells to 310 $\mu\mu$g.

When the base ratios of the neuronal RNA of the control side were compared with those of the learning side, it was found that the ratio $(G+C)/(A+U)$ had decreased significantly from 1.72 to 1.51 (Table 2).

TABLE 3 Characteristics of the RNA formed per neuron during transfer of handedness correlated to training periods and performance of the animals.

Animal	Training periods 2×25 min/day Days	Number of successful reaches	Relative increase of total RNA per neuron	ΔRNA composition
1	3	107	33%	A 25.5 G 36.1 C 9.7 U 28.7
2	5	163	23%	A 24.5 G 35.7 C 11.7 U 28.1
3	8	625	63%	A 26.2 G 34.9 C 16.1 U 22.8
4	9	1041	105%	A 21.0 G 35.2 C 24.0 U 19.8

In table 3, the data are divided into two groups (Hydén, 1965). The cell material from the cortex of animals 1 and 2 was taken on the rising part of the learning curve on the third to the fifth day, i.e. during an early part of the learning period. The material from the other two rats was taken on the asymptotic part of the curve on the ninth to the tenth day. In this case, the animals had already reached the maximum number of successful performances per training period on the sixth to seventh day. The increase of the RNA content per neuron of the 3 to 5 days' animals lies at 25 to 30%. Qualitatively, the RNA formed in the neurons is characterized by a DNA-like base ratio composition with adenine and uracil values around 26. (Rat DNA has the following base composition: A 28.6 G 21.4 C 21.5 T 28.4.) The cytosine values were remarkably low. The results were statistically significant. No such results were obtained from animals which had been trained for 8 to 9 days and performed with a maximal number of reaches, i.e. 70 to 80 reaches per period of 25 minutes. The RNA results deviated both quantitatively and qualitatively from those of group one. The relative RNA increase per neuron in the learning cortex was 60 to 100%. The base ratio composition of the RNA formed was similar to that of ribosomal RNA.

It should be added that when the nerve cells within the learning part of the cortex were removed during the early and acute part of the learning process, the relative RNA increase per neuron was small. The nuclear RNA formed at this stage had a DNA-like base ratio composition. Thus, a stimulation of the genome seems to occur early in such learning situations, i.e. situations which the animals have not encountered before. A differentiated formation of RNA occurs in the neurons engaged during a learning period and the beginning seems to be characterized by a de-repression of genic stimulation, judging by the character of the RNA formed.

In another type of rat learning experiment an increased production of nuclear RNA was found in neurons and their glia from areas functionally engaged in the process (Hydén, 1962, 1963). By analogy with the results of analysis of other cell material (Edström, 1962; Brawerman, 1963), this brain cell RNA was concluded to be a chromosomal RNA.

A preliminary study of how long the RNA with the high A/U value in the neurons lasts, gave the following result (Hydén, 1962). Twenty-four hours after stopping the experiment and returning the animals to their home cages, no such RNA fraction could be found. When the training was resumed for 45 minutes and the nuclei were analysed, the RNA fraction with high A/U ratio was found again. The disappearance of the RNA fraction did not mean that the fraction had ceased to be synthesized. It probably means that it is present in such small amounts as to be inaccessible with the present method of analysis.

A long series of studies has revealed that increase in sensory and motor stimulation (Hydén, 1967) and certain drugs (Hydén, 1964 and Hydén and Egyhazi, 1968) easily give rise to an increased synthesis of RNA in engaged nerve cells. But no change of the proportions of the adenine and uracil components of the new synthesized RNA occurred—as it did at the establishment of new behaviour. Neither did such changes occur in brain cells during stress experiments.

In the closely-arranged nerve cells of the hippocampus—an area of the old part of the brain necessary for the formation of long-term memory—the synthesis of two fractions of acidic proteins occurred at a 100% increased rate at learning.

These proteins are interesting because some of them are produced only in the brain and in no other organ (Moore and McGregor, 1968; Moore, 1965). They have a molecular weight of *circ.* 30,000 and a composition characterized by a high percentage of glutamic acid and also aspartic acid, but are almost devoid of tryptophane. Proteins of such a small size could constitute an electrogenic protein (Schmitt and Davison, 1966) which could respond to electrical fields in 10^{-4} sec. and undergo conformational changes, activate transmitters and be incorporated in membranes in a more stable configuration.

A question for the future is whether such acidic proteins are the specific executive molecules which operate at the biochemical differentiation of the brain during development and whether they are also used in mechanisms for storage and retrieval of information.

Another type of learning experiment which involves motor and sensory functions has been performed on goldfish by Shashoua (1968). A piece of plastic foam fastened below the jaw turns the fish upside down and lifts the head out of the water. After a while, the fish learns to compensate for the effect of the foam and to swim normally. During this learning process new RNA is synthesized in the brain of same type as we had found during learning in rats, and with similar ratios of the uracil/cytosine values. No such RNA was produced when the fish performed the same amount of swimming without the foam device; and, most important, stress did not produce these RNA changes. If protein synthesis in the goldfish brain was inhibited by puromycin, no learning occurred.

At this point, I would like to return to the question of the relationship between macromolecules and the establishment of new behaviour: we can conclude that in learning small amounts of nuclear RNA with a highly specific base composition are synthesized in brain cells. Similar RNA responses have not been found up to now at other types of physiological stimulation in brain cells in mammals. The synthesis of certain acidic proteins at learning seems also to constitute a specific response.

The RNA response can be interpreted as reflecting an activation of hitherto silent gene areas in brain cells when the animal was faced with a situation it had not encountered before, and which required learning (Hydén and Lange, 1966). The task in both of the examples quoted was obviously within the capacity of the species. As examples of learning involving problem solving, the tasks were not difficult—judged by performance and time curves.

The neuron and its cellular environment in learning

Can the glia take part jointly with the neuron in the learning process? In commenting recently on brain correlates of learning, Galambos (1967) observed that some investigators give something less than enthusiastic support to this idea. They cannot accept the notion that, even secondarily, learning is tied to macromolecular changes in brain cells.

For ten years we have analysed the neuron and its glia in mammals and have come to the conclusion that together they constitute a metabolic and functional unit. When the demand on energy in the neuron is increased the capacity of the electron transfer system in its respiratory chain increases (Hydén and Lange, 1962). In the glia surrounding the neuron, there occurs a simultaneous switch-over to the less-efficient anaerobic glycolysis to cover the glial energy demand (Hamberger and Hydén, 1963). In certain pathological processes there occurs first an alteration of the polynucleotide pattern of the glia (Gomirato and Hydén, 1963). Later on, a similar change can be seen in the neuron. Only then do the nervous symptoms become overt. Furthermore, acidic proteins are localized in the nuclei of neurons and in the cell bodies of glia (McEwen and Hydén, 1966). There is what we consider good evidence that governing molecules are transferred between the glia and its neuron. RNA seems to flow between these cells (Hydén and Lange, 1966). During learning, the glia responded in a corresponding way as did the neurons with respect to synthesis of specific nuclear RNA (Hydén and Egyhazi, 1963). Therefore, the glia may be as potent in their function as are the synapses.

The neuron and its glia seem to represent a two-cell metabolic and functional collaboration, a stable system from a cybernetic point of view. The glia may stabilize, influence and programme synthesis in the neurons by transfer of molecules, and may also modulate electrical properties of the neuron.

Interference with macromolecular synthesis in brain cells at the establishment of new behaviour

In another line of research, the effect of certain antibiotics on behaviour has been studied. They have all tended to block the production of protein or nucleic acid in the brain in a selective way (Flexner, 1967; Agranoff, 1967).

The conclusion of these experiments has been that brain protein synthesis during learning is necessary for the formation of long-term memory, as is intact nuclear DNA and RNA-mediated protein synthesis in the brain cell bodies. Short-term memory, on the other hand does not seem dependent on protein synthesis. It can persist concomitantly with the long-term memory and for as long as six hours.

The gap between electrical and biochemical phenomena

How can the gap in knowledge between the electrophysiological and biochemical data on brain cells be bridged? This is a question of fundamental importance. There is no straight way to be seen, but we move in the best circles.

Some recent data seem to open up a new lead. Adey and his colleagues have studied intra- and extracellular electrical phenomena as well as the EEG during learning experiments (Adey, 1967; Elul and Adey, 1966). They have advanced the view that there exist pathways for electrical currents outside the neurons in the extracellular spaces which may modulate the neurons. They have suggested that information could be processed in wave-front fashion by a primary system of waves generated in the neurons (50 to 100 μV). These waves are supposed not to be dependent on synaptic connections. The EEG reflects, in part, such non-linear waves. In Adey's view, the pulse-coded nerve impulses represent a secondary system for information processing. It is interesting that clear changes of the EEG pattern have been found to accompany the establishment of a new behaviour. There was a decreasing scatter of phase relations in the wave trains of the hippocampus apparently correlated with higher levels of performance.

A tentative model for learning and remembering

At birth, brain cells have undergone the main differentiation which made them nerve cells. This reflects selected properties collected in the genome of the individual and which have been incorporated during evolution.

There are three observations within genetics which are particularly interesting for a neurobiologist. The first is that only part of the genome is active at any time (2% to 5%). The rest of the gene areas seem to be silent. Secondly, such silent gene areas can be activated by external factors which have the capacity to penetrate to the genome. Examples are the cells of the reproductive tracts which are induced to synthesize their specific protein products by hormones (Kidson and Kirby, 1964). Thirdly, large populations of similar nucleotide sequences exist in the DNA complements of higher organisms. Furthermore, the RNA formed by replication on the DNA is of a different composition from ribosomal RNA. This is interesting in view of the fact that brain cells at stimulation produce smaller RNA types in great amounts (Egyhazi and Hydén, 1966). One million similar nucleotide sequences seem to constitute 10% of the repeated sequences in the mouse (Bolton, *et al.* 1965–6) and 15% consisted of 1,000 to 100,000 similar nucleotide copies. It was suggested that the sudden appearance of, say, 10,000 slightly different nucleotide sequences may lead to an evolutionary event. The products of a set of genes formed from a family of repeated sequences may provide the rich variation of surface proteins upon which the intercellular relationship relies.

I would like to suggest that the redundancy of the gene products provides a basis for the necessary richness of proteins serving as a mechanism for the storage of memories. Learning and experience lead to a further differentiation of brain cells. Additional gene areas become active under the impact of environmental factors. This leads to a synthesis of a great number of RNA species and proteins slightly differing from each other. The detailed mechanism which leads to these events may be the following (see scheme in Figure 3). The external factors give rise to firing at modulated frequencies in sensory ganglia and brain stem areas including the limbic area. The modulated frequencies can be split into a quadrature component and a vector component. I suggest that the $90°$ vector has a phase shift relative to the quadrature component. The two variables, the frequency and the phase shift impart a high information content to such an electric pattern which may be decisive for the specificity of its effect. In principle, this effect consists in the production of a change in ionic equilibrium through field changes. In the nuclear compartment these field changes give rise to conformational changes: activating enzyme proteins which in turn induce synthesis of RNA from new gene areas. Even instructional changes could occur. RNA quite easily undergoes hysteresis changes following local changes in pH; these are time-independent but dependent on the history of the system.

The RNA synthesis will result in a synthesis of specific proteins which will soon be present in the whole neuron, in the cell body as well as in the synapses. These specific proteins could be incorporated in a more permanent way in the membrane.

With time and experience, the protein pattern of the neuron will become modulated. A differentiation has occurred. RNA and specific proteins are constantly synthesized to replace and reconstitute this pattern. Each group of neurons will have its own unique protein composition. *The uniqueness of the protein pattern decides whether the neuron will respond at retrieval or not.* A neuron can therefore respond as a member of a great number of networks; or according to its topographical position. Each neuron may be a member in changing Gestalt-combinations.

In this reaction the glia may play an active role. The glia serve not only as supports and energy suppliers to their neurons. At learning, the RNA response of the glia is complementary to that of the neuron, and the RNA response is very rapid.

There occurs a transfer of acidic proteins and a flow of RNA from

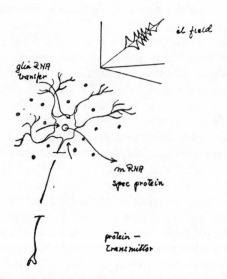

FIGURE 3   *Differentiation of a neuron during learning. A specific
information seen here in the form of an electric field and field changes
activates the neuron-glia unit. A polarization and phase angle change of
the electrical field increases the information content. The impact of the
neuron and the environmental change is to induce RNA and eventually
specific protein synthesis. The proteins are present all over the neuron and
may be incorporated in membranes of the neuron including the synaptic
areas in a more stable form.*
Concomitantly, a transfer of RNA takes place from the glia to the neuron.

glia to neuron. These acidic proteins plus RNA may block histones
in the nuclei of neurons and maintain gene activation. The induction
of RNA-synthesis in glia at learning may occur in the same way
as in neurons.

Whatever the flow of RNA from glia to neuron means, it is clear
that the uptake by neurons of specific macromolecules with recognition sites means a programming. It may be an important mechanism
because it adds to the already complicated structure of the nervous
tissue and the thousands of synapses a third type of informational
flow between brain cells and glia. This would constitute a lock-in
mechanism between two interdependent units.

According to this view, the molecular specificity is placed one step
before the transmitters. This means that only a few types of transmitters are needed.

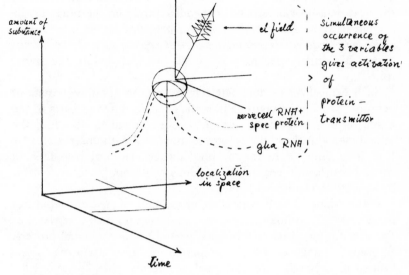

FIGURE 4 *Retrieval.*

At learning, the following sequence of events leads to the fixation of memory: information-rich modulated frequencies, field changes, transcription into messenger RNA in both neuron and glia, synthesis of new proteins in the neuron resulting in a multiple biochemical differentiation of neuron-glia units, each ready to respond to a shared, specific type or stimulus. (Figure 3.)

At retrieval, it is the simultaneous occurrence of the three variables; a specific electrical pattern, the presence of the unique proteins in the neuron, and the effects produced by the transfer of RNA from glia to neurons, which decides whether the individual neuron will respond or not (see Figure 4).

If these three conditions are fulfilled, the neuron with the specific protein pattern—or a sufficiently large part of the pattern—will respond to the stimulus within the Gestalt.

Whether the protein is in a membrane or free, it will respond with an activation of a specific transmitter since it reacts as in an antigen-antibody reaction to the incoming electrical pattern. In such a model, it is irrelevant whether the instructional electrical changes consist of pulse-coded nerve impulses, or of non-linear wave fronts generated by dendrites and proceeding in a parallel fashion. Whether the next neuron will respond, depends entirely on the range of the protein pattern. *The specific protein is activated by the same stimuli which once induced its synthesis and will rapidly activate the transmitter.* Molecular specificity is placed one step before the transmitter.

Such a mechanism does not rely merely on continuous nets of neurons. Millions of neurons plus their glia in different parts of the brain, forming a hierarchical system, can respond to stimuli of the same type. This, by the way, agrees quite well with observations by Roy John (1967). He found simultaneous electrical responses in cortical neurons on conditioning, which could not be accounted for by synaptic transmission alone.

If the whole Gestalt consisting of millions of neurons—and not necessarily in continuous nets—gives a response, then a retrieval of the stored information will become part of the perceptional process. A graded change of all of the parameters may occur when the functional expression of the phylogenetically older parts of the brain and the neocortex are considered. Some of the neuron-glia units may have only part of the protein pattern in common with the rest of the Gestalt-units. They nevertheless will constitute part of the response. With time, part of the original protein pattern will probably become changed. I would like to suggest that this explains the well-known phenomenon that the past is constantly being remade by the present.

Experiential learning by additional gene activation and by an instructional mechanism based on recognition sites of macromolecules and hysteresis would both require redundancy of molecular species in the delicate interplay between the various parts of the brain in order to integrate the functional processes.

Some aspects of future brain research

The macromolecular response in brain cells when a new behaviour is started and consolidated was interpreted as reflecting gene-activation in the cells in question. Motivation and the set of environmental factors served as triggers, and the intraneuronal genetic mechanisms responded according to their set of rules.

One may then ask what possible consequences, if any, such neurobiological data may have in the future, if we turn to the problems of development and early childhood experience. More specifically, one may ask whether this type of research could have implications for the way in which the realization of an individual's potentialities should be planned.

The concept of intelligence, as the expression of brain capacity, has changed greatly since the early 1930s. At that time, the capacity to succeed in the environment was largely looked upon as a fixed individual characteristic. It was believed to show little fluctuation from early childhood to maturity, since it was related to parental genetic traits and was relatively uninfluenced by the impact of environment. This view is no longer upheld, owing to the results of longitudinal studies of children. I would like to quote the highly important studies by Skeels (1966) which covered a period of 30 years. Skeels has convincingly shown how children around one year old, classified by conventional test methods as evidently mentally retarded, were transformed to mentally normal children. This was achieved by increasing the amount of developmental stimulation, and the intensity of relationships between the children and mother-surrogates. During a two-year period, these children made a gain of around 30 % IQ points. They were then placed in foster homes, grew up and showed emotional, educational and occupational achievements during 21 years which compared favourably with the 1960 U.S. Census figures for the United States in general. A control group of small children, initially higher in intelligence than the experimental group, were transformed into mentally retarded individuals by being left in a relatively non-stimulating environment over a prolonged period of time. The 12 children of the control group, all except one, suffered a tragic fate, ultimately being institutionalized in mental hospitals. It can no doubt be postulated that *if* the children of the control group had received the same stimulation during the critical early period of their lives, they would have achieved the normal range of development, as did the experimental children.

By inference, I would like to advance the view that a critical stimulation during the right period of early life produces gene activation in brain cells in addition to those gene areas which are already active at the time of birth. This leads to a biochemical, structural and functional differentiation of the brain.

From research on other species, it is also known how animals living in a restricted and poor milieu will develop a brain cortex which is thinner and biochemically less developed than the cortex of their litter mates living in a stimulating environment (Diamond *et al.* 1966). If young animals are reared in the dark, the retina and its ganglion cells will not differentiate functionally, or biochemically (Riesen, 1947, Brattgård, 1952).

We ought to be able, eventually, to answer questions such as: what is the critical age for training for definite achievements, abstraction and problem solving, or complicated sensory-motor tasks? What is the time-integrated pattern and type of stimulation to realize maximally the potentialities of an individual?

Neurons, orderliness and age

My last comments will deal with orderliness and age of brain cells. The nerve cells share a rare characteristic with a few other cells of the organism. They do not divide. Over the life-cycle, therefore, orderliness is not sustained by means of cell division. There will be an increasing danger of errors at synthesis in the cells, which means errors in function. Evidence has been presented that neurons may renew their DNA without preparing for division (Pelc, 1964). Future studies will have to elucidate what happens in ageing neurons, since hybridization analyses have suggested that DNA from an old animal may not be identical with DNA from a young animal (Hahn, 1966).

Old animals do not learn and consolidate memory with the same alacrity and efficiency as do young animals (Oliverio and Bovet, 1966).

At old age, some organs or systems may be in good shape, including the circulation, although the brain functions reflected in memory and the higher intellectual performances may falter. In such cases, can we increase orderliness in the brain cells in order to harmonize its functional output with that of other organs? If learning and the expression of memory ultimately reflect the actualization of latent potentialities of the genome, then, addition of extraneous gene material with high intrinsic orderliness to such ageing brain cells would be logical to attempt. We have, therefore, as a first step. prepared brain DNA of varying degrees of purity. This DNA was injected into the ventricles of the brain of other animals of the same species. In one hour, protein synthesis was found to increase significantly (Egyhazi and Hydén, 1967). Biochemical analyses showed that the DNA had been incorporated in the recipient's brain cells in a polymerized state. It is not surprising *per se* that brain cells can take up macromolecules like DNA, since it is by now a well-known fact that many other somatic cells have this capacity (Robins and Taylor, 1968). It was made highly probable by a series of control tests that the incorporation of polymerized DNA had caused the stimulation of protein synthesis in the brain cells, and RNA seemed to mediate the effect. More hard facts are needed. What is the nature of the protein formed? Is it functionally valuable or nonsense protein? How long does the effect last, for hours, days, or years?

How could this observation be utilized in future brain research, and what are its possible social uses? I would like to suggest that a way may eventually be found to increase orderliness in the brain cells of one individual by incorporation of gene material from another. The question is how this could be accomplished. There are several possibilities. The most direct would be to infect the brain with genes attached to a harmless virus entering the brain. Viruses have the capacity of penetrating into host cells, and could thus act as transporting agents. A successful counteracting of entropy increase in brain cells could change the whole structure of our society.

DISCUSSION

KOESTLER   I have three questions. According to your theory, the establishing of a protein pattern in a neuron makes it selectively responsive to information. What order of magnitude of information? I mean how many bits, or thousands of bits, of information could be carried in these frequency modulation impulses?

HYDÉN   Well, one's estimate would be something like between 10^8 and 10^{10} bits, in such a complex system.

KOESTLER   Second, when you injected the extraneous DNA from the "donor", did you look for behavioural changes in the recipient— or have you not looked so far?

HYDÉN   No, we looked for strictly biochemical effects.

KOESTLER   Rats and rabbits, or . . . ?

HYDÉN   Rabbits and rats, yes.

KOESTLER   The third question is a very stupid one. At what stage of phylogeny do the glia arise?

HYDÉN   Well, even in insects you have cells which are called satellite cells, but they look more like sort of primitive connective cells.

WEISS   I have a question. Difference between DNA and RNA. Have you or anybody else tested whether the increase in production of RNA during learning likewise drops off with age?

HYDÉN   We don't know yet for sure.

HAYEK   I am worried about the meaning of the idea of "storage" at one point of the network. It is the basic contrast between synaptic connections and localized "engrams" that has worried me ever since I was reading Semon. Would it solve anything if you found a mechanism which could be interpreted as a storage at one point? The question is: what is being stored? What you assume is being stored is not just something, but something which must possess peculiar qualitative differences from a great many other things. I do not think you can hope to find as many different proteins which differ in as many ways as the different sensory qualities. But even if you found the same number of different proteins as there are sensory qualities, etc., you would still have to show that these differences of the proteins correspond in some sense to the differences of sensory qualities; namely that they differ in their distinct effects in exactly the same manner in which we know that sensory qualities differ from each other. Now that means you would have to show that the reappearance of an excitation of a neuron producing a particular protein has just the same effect on other neural processes which go on at the same time, i.e., alters them in the same manner; or you would still have to construct a system of connections with other processes to explain the different action of this kind of memory from another kind of memory. But once you have constructed such a system of connections with others, you have already explained all you have to explain. You do not need any special residue in a single neuron or which is concentrated at a particular point. It seems to me, though I am not competent to talk about any of your factual evidence, that it is a conceptual problem which arises here. Can you explain a difference of action of something at one point in an extremely complex network of connections by an attribute possessed by that point? Must you not explain it in terms of the connection of that point with all the other points? And once you have established the connections with all the other points, you don't need any particular quality at the initial point.

HYDÉN I do not agree with you there, because if it were only a matter of the number of synapses on a surface then the only thing you would have to worry about would be whether the incoming impulse was strong enough. But whether the next neuron responds or not cannot depend only on the number of synapses. It must have differential qualities. A third dimension may be introduced by the flow of information in the form of molecule transfer between the cells. I tried to point out this type of interaction between the two types of brain cells, since the glia cells also respond with a macromolecular synthesis at the establishment of a new behaviour. Therefore, millions of brain cells share part of the same protein pattern and respond, all of them, to a certain stimulus.

HAYEK Yes, but although you assume here that the excitation of a single neuron has a distinctive mental value—let me use that word— a distinctive quality of the excitation seems to me clearly not necessary. It is entirely possible that all that is required is the excitation of a certain combination of neurons, perhaps distributed over the whole cortex. Those synaptic connections, without anything else they need have in common, may determine the further process, and by this you can develop any degree of complication of the effects. Any new establishment of a connection can alter the whole system. I do not see how any attribute or quality possessed by a particular point in the nervous system can really help. It still needs a change in the connections, and once you have the change in the connections you do not need a change in the quality of the particular point.

HYDÉN   What you call a "point" in the brain is a whole neuron plus its glia.

HAYEK   Yes.

HYDÉN   The whole neuron, not only just the synapse.

HAYEK   Yes, I meant the point in the network in that sense.

HYDÉN   But you would also have—in order to steer a function like this—you would have to assume that different parts of the brain, would be able to respond simultaneously at the retrieval of a memory, or of an activity.

HAYEK   A great many points of the brain may be required to contribute. They may have completely diffuse distribution and no destruction of a particular part of the brain would completely eliminate a particular function if a sufficient part of them remains intact.

SMYTHIES   There are two points here; the first was made by Karl Pribram—you remember his complaint about people talking about the retrieval of memory from the brain. We always seemed to think of it as a one-step operation; there is an enormous bank—just that—and the memory just pops out. His plea was that underlying this there was a very complicated communication system and retrieval of memory in fact may depend on many complex operations. This would need, presumably, complex activity in millions of neurons. It is not just a network, an unformed network; it is a highly organized one.

HYDÉN   And it is also something more than just a network. What I do not agree with is that the neuron acts as an on-and-off switch; that it just depends on the number of synapses. I cannot see how that can operate. And I cannot see how it fits with the facts.

HAYEK   What you mean is that there are conceptual difficulties in requiring a number of afferent impulses to make a neuron fire.

HYDÉN   I do not think you could get a graded selectivity in the response. If we go back to Konorski and his ideas—if somebody had shown at that time that a synapse could grow as a result of acquired learning, and the synaptic cleft would be decreased and finally there would be a firm junction—then you could have graded responses. Konorski supposed that this was so, but nobody has as yet produced the slightest evidence that it is. But if you assume biochemical differentiations in various parts of a brain with respect to both neurons and glia and their differentiated outer parts; and if you only assume an additional extra specificity with respect to the membrane, then you can have all sorts of graded responses in the most complicated networks.

HAYEK Yes, but what you just said has further confirmed my feeling regarding the possibility of a graduation by summation. When I read about these things, one thought in terms of fifty or of a hundred synaptic knobs on one cell. Now you tell us there are about ten thousand. Then you have two possibilities of summation. You have the simultaneously-arriving impulses from the different synapses and you have summation of successive impulses. This gives you any degree of graduation you can need.

KOESTLER You are using Ockham's razor with the wrong side—the blunt side, because it would be a terribly stupid brain which would use these sophisticated neuron-cum-glia assemblies simply as on-off switches.

KETY I was especially attracted by the concept that you are moving from the storage of specific information in the macromolecule to the storage of information in circuits, but that these circuits were in a sense colour-coded to differentiate them one from the other, so that a particular neuron which might be in the centre of many circuits would somehow know which circuit was being activated in terms of the specific protein which was common to that particular circuit. Am I understood here?

HYDÉN And not only one particular protein, but a family of proteins.

KETY Well, I think that's an extremely attractive way of getting around the difficulty of keeping these various circuits, these millions of circuits, which represent individual engrams, separate from each other in this mass of networks. But I find it still a little difficult to see how this is achieved in the time that we know is available for consolidations. You are putting greater emphasis now on the nucleus as a source of these proteins than on the nerve-ending.

HYDÉN No, I am putting emphasis on the whole chain of a balanced process leading to a synthesis of protein which will soon be in all parts of the neuron, whether free or incorporated in the membrane.

KETY Well, but this specific protein has somehow to move from one end of the neuron to the other.

HYDÉN Yes, let's come back, then, to the so-called short-term memory and long-term memory and the fixation period. In experiments with rats, Barondes has shown that short-term memory can persist for 5–6 hours. And let's say that somewhere during that time a consolidation occurs. This is time enough of course, for the synthesis of the macromolecules needed in all parts of a neuron.

KETY   But isn't there a certain amount of transportation involved from the centrum to the synaptic ending, and is the transport rapid enough to be effective in a few hours?

WEISS   Yes. We looked for instance at the goldfish's optic nerve. It was done in my lab. by Bernice Grafstein, and also in other laboratories.

HYDÉN   You had a 20 cm/day flow, had you?

WEISS   You can get up to 20 cm per day, so that—particularly for short distances—the *time* factor is all right. But I would like to add to this another dimension. Putting myself squarely on the basis of your facts: you have these local changes in protein, and thereby of course the whole idea collapses that the synaptic connections are there once and for all—which for other reasons, is biologically untenable anyhow. If a reciprocal conformity between surface constitutions is basic for the maintenance of a synaptic connection, then once you change the protein in one cell and not in the other, the synaptic junction will disappear, and the cell will latch on to another cell which is more conformant. If you introduce this additional dimension of conformances by self-recognition, or mutual recognition, of neuronal units being linked together according to their family resemblances between proteins which emerge to the surface, then you have really the possibility of establishing an infinite number of new circuitries. Since we know that such processes exist in ordinary cells which are not nerve cells, I don't see why we shouldn't apply the same kind of principle to neurons. Roger Sperry's ideas about chemical affinity are essentially not different from this. However, this like everything you said, raises still the question how you prevent a supervening change in the protein from extinguishing prior experiences.

HYDÉN   You may have that mechanism in the redundance of nucleotides. Britten and Roberts have a rather interesting example of the repetition of a nucleotide of DNA. They found a repetition of almost similar nucleotides in the DNA with an uneven distribution. If you have such a selective distribution and repetition in the nucleotides of a cell, you have a family of almost identical molecules which can withstand an attack that otherwise could lead to their deletion. Such a redundancy of a certain species of molecules can give rise to a great number of end products.

WEISS   Would then mental decline with age depend on the fact that fewer and fewer unprogrammed neurons remain available?

HYDÉN   No, I think the age changes have to do with the control system.

WEISS   The thing that bothers me is that if you superimposed new inscriptions or engrams, that you should have no interference with the old ones in that system—that each one retains its identity and after fifty years can still be got out.

HYDÉN   There is a flux all the time, there is reconstitution and a constant synthesis.

WEISS   Well, I know that one. But I still don't see—you put a new one on top which ought to bring in modifications in the existing pool of proteins there.

HYDÉN   The composition is so close in similarity—speaking about nucleotides for example—that it falls within the same group. It would give the hybridization reaction if you tried one, but still it differs sufficiently from the background noise, so to speak, to be specific.

WEISS   So you would say that this is essentially similar to the molecular biology of immune reactions.

HYDÉN   But more complicated.

WEISS   More complicated, but of the same nature.

HYDÉN   Yes.

WEISS   And what do you think about the making and breaking of synapses according to the affinities between them?

HYDÉN   We could say that this is the reason why we are constantly remaking our past experiences. The whole time.

SMYTHIES   J. Z. Young holds that learning is done by shutting off alternative pathways; not by opening new pathways, but by shutting off alternative ones—in the octopus at any rate.

HYDÉN   In the octopus, because there he has what he calls memory cells. But I don't assume localized memory cells at all in mammals.

SMYTHIES   You mean you think this is just special for the octopus or special for J. Z. Young? (Laughter)

HYDÉN   There may be a special case for the octopus, but I do not think that you could assume special memory cells to exist in the mammalian brain.

SMYTHIES   Do you think one reason for the great number of synapses in the cortical cells is that they are acting on the hologram principle? Then you need a tremendous amount of information in each cell, and each cell is in some sense taking part in the overall reaction.

HYDÉN Well, it's a long question (Laughter). At any rate, it seems to be perfectly clear from old experiments that you can remove large parts of the cortex.

SMYTHIES The point about holograms is this: would it be possible for every unit in a hologram to be biochemically distinct?

HYDÉN I think it would, if you increased the complexity sufficiently.

KOESTLER When you said that you do not believe in the existence of specific memory cells in man, you meant of course that all neurons are "memory" cells. But are not some concerned more and some less with memory? In the primary receptor areas some memory function could be present, of course, but would play a relatively subordinate role.

HYDÉN Of course. Take typical motor cells . . . So there must be a gradation.

WEISS I disagree on general grounds with the assumption of the fixity of synaptic connections. If you lose 10^3 of the cells in the cortex every day—what happens to your network? You need latitude in your model—a multidimensional approach. First you have the circuitry. Then there are the chemical changes which you described, changes in the neuron and the glia. You also have the electrical field in the spaces between the neurons—and according to Ross Adey this, too, is variable. So you really have three separate variables, and you must assume that not one only is involved, but that a constellation of parameters, a certain unique constellation is the sort of thing you are looking for—the engram. In ecology, of course, it is well known that you must have a certain unique constellation to give rise to a particular effect in a biological system. You may also find a completely new type of dimension. You see what I am driving at?

HYDÉN Yes.

WEISS That you have to have a harmonious constellation between the electrical properties, the requisite protein catabolism and anabolism, the production of heat and other local changes, a sort of symbiotic relation between glia cells, and neurons—they have to be in tune in their changes. And this really would give you an infinite increase in the number of possibilities.

HYDÉN When we saw this increase of RNA of a special composition in the neurons, during activity, and saw the decrease of the RNA of the same composition in the glia outside, we started to compute. Given a certain amount of DNA, and supposing that ten per cent of the DNA areas in the eight micro-micro grams of DNA were active, what would be the maximum amount of RNA you could have synthesized in one hour's time? It was not enough to cover the increase observed, so it had to be a transport, the RNA had to come from somewhere. We concluded that it came from the glia. That means that information-rich macromolecules are fed into the neuron.

BERTALANFFY   What technique was used to identify informational RNA?

HYDÉN   It was determined by first dissecting the cells and then extracting the RNA and subjecting it to acid hydrolysis and then subjecting the hydrolysate to electrophoresis on a microscopic thread.

SMYTHIES   Would you venture a guess as to what kind of information these molecules are carrying?

HYDÉN   Take spermatides and Sertoli cells, or take egg cells and the follicular cells around them, there it has been found that desoxynucleotides are passed from the follicular cells to the egg cells. And the conclusion of the colleagues who have dealt with this problem is that it is an integral part of differentiation.

SMYTHIES   But we are talking of neurons.

HYDÉN   If you accept that experiential learning means a biochemical differentiation of the nerve cell with an additional help from glia, then you have an analogy with the egg cell, and the follicular cells, where it is clearly being found that nucleotides are being passed from one to the other.

SMYTHIES   But if it is this specification of the cells that counts, then it is still the connections between the cells which convey what is being learned, the actual content of the learning, carried by nucleotides crossing the synapse.

HYDÉN   No, not the synapse. Crossing the surface of the membrane at other places, I would say, rather than at the synapse. That involves an energy question of course, which I discussed with M. Eigen, and he was certain that there was no problem with respect to the energy for the intake of polymerized nucleic acid in a very short time.

WEISS   Does any of your injected DNA go into the mitochondria?

HYDÉN   I do not know.

KOESTLER   Would you agree to the possibility of continuous gradations between neurons which function more or less on the on-off basis in the primary receptor areas, in motor neurons, and so on, and where selectivity is very underdeveloped or hardly at all developed and the whole biochemistry is secondary; as opposed to other types of neurons where selectivity is all, and connections may be almost random, because it is selectivity which matters, primarily. Would you have a scale of such gradations——

HYDÉN It would certainly be better to have a gradation.

WEISS Arthur's question is whether we are perhaps making a mistake in dealing with the neuropil and with all sorts of highly specialized cells on the same basis. It is a good question. Would you agree with that?

HYDÉN I do.

THORPE I did not quite understand what you said about pathways in the extracellular spaces or in between the glial cells. And did I understand you to say that the EEG reflects non-linear graded waves?

HYDÉN I quoted Adey and his colleagues.

THORPE Is the evidence for this really satisfactory?

HYDÉN I think so. He has recorded extracellularly and intracellularly and he has subjected the electrical results to Fourier analysis.

WEISS I was rather convinced when I saw his actual presentation, which was one or two years ago.

THORPE Of course the EEG changes in learning have for a long time been regarded as evidence of attention rather than of actual learning. Do you think this conclusion is now suspect?

KETY I think he regarded the arousal phenomena as secondary. But he never indicated that he had satisfactory evidence that the changes were specifically related to information acquisition and not to the secondary features.

WEISS The point I tried to make before, is that a particular level in that electric activity may be the element that sets the harmony of relations, let's say between glia and neuron into that unique condition which results in the particular package, that is the unit of memory . . .

HYDÉN I think you then have to assume a phase retardation or a phase change between the two variables of an electrical signal, to increase the informational content. Very little is known about the electrical phenomena in spite of all the studies that have been made. But with respect to a Fourier analysis of the EEG where you have a quadrature component and another vector at $90°$ angle, and you may have all types of phase relations between those two—that would increase both specificity as well as information. Then you could very well assume that at a certain level, this would be the tune to listen for.

WEISS If you have a system with three grade scales, three variables, such as we discussed, you can achieve combinations that show definite singularities. This relates to Arthur's question about gradients between different types of neurons. If you have three variables, you can produce singularities, constellations which represent an optimum. It is worth while to think of systems of this kind.

HYDÉN May I come back to your question about Adey? In his papers 1966–67, he discusses arousal and says that the impedance change—that was the first one he described—had to do simply with the arousal of the animal, and he found that the changes occurred first in the ventral part and later on in the dorsal part of the hippocampus. Half a year ago, he presented a manuscript where he says that the impedance changes in the hippocampal area still may reflect the arousal of the animal, but that the activity generated by the dendritic trees of the neurons would be quite another system. He maintains this very clearly.

KOESTLER May I go back to Seymour's first question quite some time ago. You asked Holger whether the same neuron, the same group or population of neurons can respond to different types of specific stimuli. In other words, can the same neurons belong to a number of different functional systems? This would mean that those types of neurons which are primarily concerned with specific memory functions, that these neurons each belong to several clubs or speak several languages, so to speak. They could all live mixed together and could be interconnected almost at random by circuitry because they respond selectively when addressed in one of the languages they speak. Would this be acceptable?

HYDÉN This participating in several networks would be a great advantage to such a system.

WEISS I can give you an example of this, from a recent experiment not yet published, which is quite dramatic. . . . The limb of the newt is innervated from three spinal cord segments (blackboard); they then branch out and get mixed up. We have shown that these segments, their whole area, is equipotential with respect to the normal functioning of the limb. That it is not broken up into districts, so to speak. That it does not operate by point-to-point projection. We checked this recently in the laboratory by very simple surgical experiments, cutting through various tracts.

KOESTLER So you have gone one better than the old fifth limb of the salamander.

WEISS This is in a way, an elaboration of that earlier experiment.

AFTERTHOUGHTS* by Holger Hydén

Ageing strikes materia, whether living cells, individuals or solid state materia. We accept philosophically the changes in our motor activities wrought by time, and our reluctance to get involved in new enterprises which draw heavily on the capacity for initiative. When we look around, there are everywhere examples of change with age in motor behaviour. What is more striking than to see an old horse, with sagging head, motionless at a fence, and to note the contrast to the lively motions, jerky jumps and glistening vitality of the colt in the nearby pasture. We take the expression of age changes for granted. It is surprising that the transformation of fertile ideas into rigid dogmas is accepted in the same spirit, and in science we hardly notice how or when the transformation occurred. The airing, in Alpbach, of current problems within a wide cultural field drove home the need to watch carefully for symptoms of sclerosis in the various lines of approach in the neurosciences.

It became increasingly clear in the course of the Symposium how the new discipline of the genetics of behaviour, to judge by some recent books, is caught in the dogmas of Mendelian genetics, without regard to developments in modern genetics during the last ten years, and to modern experimental approaches to the genetic roots of behaviour. Books on the subject usually begin with an account of the principles of Mendelian genetics. The material on behaviour deals mainly with mutated animals and their observed changes in behaviour. That is exactly what genetic principles predict. If an important mutation should not be followed by a change in behaviour—then geneticists would have to worry about the validity of the principles.

What these books fail to pay attention to is the trend in modern genetics which deals with the activation of gene areas, with the influence of external factors on the actualization of gene-potentials and their biochemical correlates in behaviour. In Parson's book on the genetics of behaviour, for example, there is no reference to the now well-known observation that hormones can act directly on the genes and alter their activity.

Yet the paper by Kidson and Kirby in 1962 and the works by Tata (1965) and Hamilton (1968)—to mention only a few—taken together give a good view of how external factors like hormones can penetrate the genes and change cellular activity and behaviour.

* Written after the Symposium.

114

Oestrogens, for example, bind to proteins in the cytoplasm. These complexes are broken down, penetrate into the nucleus, bind to the chromatine protein. The result is a stimulation of m RNA synthesis, an acceleration of the formation of precursor ribosomes and an accumulation of new poly-ribosomes in the cytoplasm, with different properties of incorporating amino acids.

One wonders why the geneticists of behaviour refrain from discussing these well-known facts, or the observations that a synthesis of RNA and proteins in brain cells is necessary for long-term memory formation. I would venture a guess that, apart from dogma, the main reason for this silence is the fear of even the slightest suspicion that one might misinterpret such facts to mean that a Lamarckian mechanism were at work.

The Alpbach Symposium succeeded well in keeping the participants on their toes, and to look out for holes in the specialist's view of his own concepts and ideas. It also served as an illustration of Bragg's statement that the important thing in science is not so much to obtain new facts as to discover new ways of thinking about them.

September 20, 1968.

REFERENCES

ADEY, W. R. (1967) in: *The Neurosciences*, G. C. Quarton, T. Melnechuk and F. O. Schmitt (eds.), New York. p. 756.

AGRANOFF, B. W. (1967) in: *The Neurosciences*, Rockefeller Univ. Press New York. p. 615.

BENNETT, E. W. (1966) *J. Comp. Neurol*, 128, p. 117.

BOLTON, E. T., BRITTEN, R. J., COWIE, D. B., KOHNE, D. E.,ROBERTS, R. B. and SZAFRANSKI, P. (1965–66) in: *Carnegie Institution of Washington Yearbook*, Vol. 65, p. 78.

BRATTGÅRD, S. C. (1952) *Acta radiol*. Suppl. p. 96.

BRAWERMAN, C. (1963) *Biochim. biophys. Acta*, 76, p. 322.

DAHLSTRÖM, A., FUXE, K. (1964) *Acta physiol. Scand.* Suppl., p. 232.

DIAMOND, M. C., LAW, E., RHODES, R., LINDNER, B., ROSENZWEIG, M. R., KRESCH, D., BENNETT, E. L. (1966) *J. Comp Neurol.*. 128, p. 117.

EDSTRÖM, J. E., BEERMANN, W. (1962) *J. Cell. Biol.*, 14, p. 371.

———— EICHNER, D., SCHOR, N., (1961) in: *Regional Neurochemistry*, S. S. Kety and J. Elkes (eds.), Oxford, p. 274.

———— EICHNER, D. (1958) in: *Nature*, 181, p. 619.

EGYHAZI, E. and HYDEN, H. (1966), *Life Sci.*, 5, p. 1215.

———— (1967), *Proc. Symp. on Nucleic Acids and Proteins in the Neuron*, Prague, 1967.

ELUL, R., ADEY, W. R. (1966) *Nature*, 212, p. 1424.

FLEXNER, L. B. (1967) *Proc. Amer. Phil. Soc.* III, p. 343.

GALAMBOS, R. (1967) in: *The Neurosciences*, G. C. Quarton, T. Melnechuck and F. O. Schmitt (eds.), Rockefeller Univ. Press, New York, p. 637.

GOMIRATO, C., HYDEN, H. (1963) *Brain*, 86, p. 773.

HAMBERGER, A., HYDEN, H. (1963) *J. Cell. Biol.* 16, p. 521.

HARREVELD, A. VAN, CROWELL, J., MALHETRA. S. K. (1965) *J. Cell. Biol.*, 25, p. 117.

HAHN, H. P. VON (1966), *Gerontologia*, 12, p. 18.

HYDÉN, H., (1962) in: *Macromolecular Specificity and Biological Memory*, F. O. Schmitt (ed.), The M.I.T. Press, Cambridge, Mass.

———— (1960) in: *The Cell*, J. Brachet and A. Mirsky (eds.) Academic Press, New York, p. 215.

————(1967), in: *The Neurosciences*, G. C. Quarton, T. Melnechuk, and F. O. Schmitt (eds.), Rockefeller Univ. Press, New York, p. 248.

————(1964) in: *Recent Advances in Biological Psychiatry*, J. Wortin, (ed.), New York. p. 32.

———— EGYHAZI, E. (1962) *Proc. Nat. Acad. Sci.* 48, p. 1366.

———— EGYHAZI, E. (1963) *Proc. Nat. Acad. Sci.* 49, p. 618.

———— EGYHAZI, E. (1964) *Proc. Nat. Acad. Sci.*, 52, p. 1030.

———— EGYHAZI, E. (1968) *Neurology* 18. p. 732

———— LANGE, P. (1965) *Proc. Nat. Acad. Sci.*, 53, p. 946.

———— LANGE, P. (1968) *Science* 59, 1370.

———— LANGE, P. W. (1966) *Naturwiss*, 3, p. 64.

———— LANGE, P. W. (1962) *J. Cell. Biol.*, 13, p. 233.

———— PIGON, A. (1960) *J. Neurochem.* 6, p. 57.

JOHN, E. R. (1967) in: *The Neurosciences*, G. C. Quarton, T. Melnechuk, and F. O. Schmitt (eds.) Rockefeller Univ. Press, New York, p. 690

KIDSON, C., KIRBY, K. S. (1964) *Nature*, Vol. 203, p. 599.

LASHLEY, K. S. (1950) in: *Symposia of the Society for Experimental Biology*, Academic Press, New York, p. 454.

LOCKE, J. (1931) *Essay concerning Understanding, Knowledge, Opinion and Assent* R. Rand (ed.), Oxford.

MCEWEN, B. S., HYDEN, H. (1966) *J. Neurochem.*, 13, p. 823.

MOORE, B. W. (1965), *Biochem. Biophys. Res. Comm.*, 19, p. 739.

———— McGREGOR, D. J., (1965) *J. Biol. Chem.*, 240, p. 1647.

MORRELL, F. (1967) in: *The Neurosciences*, G. C. Quarton, T. Melnechuck and F. O. Schmitt (eds.), Rockefeller Univ. Press, New York, p. 452.

OLIVERIO, A., BOVET, D. (1966) *Life Sci.*, 5, p. 1317.

PELC, S. R. (1964) *J. Cell. Biol.*, 22, p. 21.

PETERSON, G. M. (1934) Comp. Psychol. Monogr. 9, p. 67.

PLATO (1951) *Meno*, R. S. Buck, (ed.). Cambridge.

RIESEN, A. H. (1947) *Science*, 106, p. 107.

ROBINS, A. B., TAYLOR, D. M. (1968) *Nature*, 217, p. 1228.

SCHMITT, F. C., DAVISON, P. F (1966) *Neurosci. Res. Progr. Bull.*, 3, p. 55.

SHASHOUA, V. E. (1968) *Nature*, 217, p. 238

SKEELS, H. (1966) *Monographs of Soc. for Res. in Child Development*, Vol. 31, No. 3.

TEITELBAUM, P. (1967) in: *The Neurosciences*, G. C. Quarton, T. Melnechuck and F. O. Schmitt (eds.) Rockefeller Univ. Press, New York, p. 557.

The gaps in empiricism*
Jean Piaget and Bärbel Inhelder
University of Geneva

Empiricism has engendered many different ideas, from the naive concepts of knowledge as a copy of reality, to the more refined forms of "functional copy" (Hull's behaviourism) to logical positivism, which aims at reducing scientific knowledge exclusively to physical experience and to language. If we look for common factors in these diverse approaches we find a central idea: the function of cognitive mechanisms is to submit to reality, copying its features as closely as possible, so that they may produce a reproduction which differs as little as possible from external reality. This idea of empiricism implies that reality can be reduced to its observable features and that knowledge must limit itself to transcribing these features.

What we should like to show briefly here is that such a conception of knowledge meets with three fundamental difficulties. Biologists have shown that the relationship between an organism and its environment (at a certain level of scientific study this dichotomy is in itself an abstraction) is one of constant interaction. The view that the organism submits passively to the influence of its environment has become untenable. How then can man, as a "knower", be simply a faithful recorder of outside events? In the second place, among fields of human knowledge and endeavour, mathematics, for one, clearly escapes from the constraints of outer reality. This discipline deals essentially with unobservable features, and with cognitive constructions in the literal sense of the word. Thirdly, as man acts upon and modifies reality, he obtains, by transforming his

* Translated by Mrs. S. Wedgwood (aided by a Ford Foundation grant) with the collaboration of Mrs. H. Sinclair de Zwart. Read by Bärbel Inhelder.

world, a deeper understanding than reproductions or copies of reality could ever provide. Furthermore, cognitive activity can be shown to have structural properties: certain broad cognitive structures underlie the thought processes at different levels of development.

Behaviourist empiricism and biology

The exact counterpart of behaviourist empiricism in biological theory is a doctrine long since abandoned by biology itself, not because it was wrong in what it maintained, but because it ignored all that has since proved essential to an understanding of the relations between the organism and its environment; we are referring to the Lamarckian theory of variation and evolution. Soon after Hume had sought the explanation of the phenomena of the mind in the mechanisms of habit and association. Lamarck, too, saw the key to the morphogenetical variations of the organism and of organ-formation in the habits adopted under the influence of the environment. Admittedly, he was also speaking of a factor of organization, but he thought of it as a capability of association and not of composition, and the essential aspect of the acquisitions was for him the way in which living beings received, in modifying their habits, the imprint of the external milieu.

These ideas were certainly not entirely wrong, and, so far as environmental influences were concerned, modern "population genetics" has in fact only replaced direct causal action of external factors on the genetic unit (heredity of the acquired characteristics in the Lamarckian sense) by the concept of probabilistic action (selection) of a complex of external factors on multi-unit systems (coefficients of survival and reproduction, etc., of the genetic pool or of differentiated genotypes). But Lamarck's theory lacks the basic principles of an endogenous possibility of mutation and recombination and above all those of an active capacity for self-regulation. When Waddington or Dobzhansky today put forward the phenotype as a "response" of the genetic pool to environmental incitements, this response does not mean that the organism has simply been marked by an external action, but that there has been interaction in the full sense of the term, i.e. that as a result of a tension or imbalance provoked by environmental changes the organism has invented an original solution by means of recombinations, resulting in a new equilibrium.

E

Thus, when we compare this concept of "response" to that used so long by behaviourism in its famous stimulus-response schema (S-R), we are amazed to find that the behaviourist psychologists have retained a strictly Lamarckian outlook, as if they had ignored the contemporary biological revolution. Of course, it may well be argued that psychological and biological phenomena belong to two different levels since the modifications of behaviour studied by behaviourism are exclusively phenotypical, whilst the variations which interest the biologist are hereditary. But we know today that such a distinction cannot be clear-cut and this for numerous reasons, of which the following are the two most important. The first is that the phenotype is the result of continuous interaction between the genes' synthesizing activity during growth and the external influences. The second is that for each environmental influence which can be sufficiently circumscribed and measured, we can determine for a given genotype its "reaction-norms", which give us the range and distribution of the possible individual variations; to ascertain similar reaction-norms for behaviour we have to make an analysis at all levels of phenotypical reactions.

Consequently, the concepts of stimulus and response must undergo very wide-reaching reorganizations which completely modify their meaning. In fact, in order for the stimulus to set off a certain response, the organism must be capable of providing it. The nature of this capability must therefore first be ascertained. It corresponds to what Waddington called "competence" in the field of embryology (where "competence" is defined as sensitivity to "inductors"). At the outset, it is therefore not the stimulus itself, but the sensitivity to the stimulus which is vital and, of course, the latter depends on the capability of giving a response. The schema must therefore not be written $S=R$ but $S \rightleftarrows R$, or, more accurately, S (A) R where A is the assimilation of the stimulus to a specific reaction-pattern which is the source of the response.

This modification of the $S-R$ schema is in no way due to a simple desire for theoretical accuracy: it immediately raises what appears to us to be the central problem of mental and, particularly, cognitive development. In the exclusively Lamarckian context of behaviourist theory, the response is simply a sort of "functional copy" (Hull) of the stimuli in their particularly succession. Consequently, the fundamental process of acquisition of knowledge is considered a learning process in the empiricist sense of obtaining information through observation of the environment. If this were true, mental development as a whole would then be thought of as the result of an uninterrupted series of "bits" of learning in the above-mentioned sense. If, on the contrary, the basic point of departure is the capability of giving certain responses, i.e. the "competence", learning would not be the same at different developmental levels, and would depend essentially on the evolution of "competences". The true problem would then be to explain their development, and for this the concept of "learning", in the classical sense of the term, would be inadequate.

In our opinion, we cannot but follow the principles discovered by contemporary biology. This means a fundamental change in the psychological interpretation of mental development. In fact, these new directions in embryology have brought the modern view of biological development much nearer to our view of psychological development. As we have long argued, a parallel between biology and psychology does not imply that all is innate in individual development. As we have said, today the phenotype is considered to be the product of an indissociable interaction between hereditary or endogenous factors and environmental influences, so that it is virtually impossible to draw a clear line of demarcation, within behaviour patterns, between the innate and the acquired. This is made even more difficult by the presence between the two of the all-important zone of self-regulations.

The linguist N. Chomsky rendered a great service to psychology by providing us with a decisive criticism of Skinner's interpretation of verbal behaviour, and by showing the impossibility of representing the process of language acquisition by classical behaviourist and empiricist models. Chomsky has finally chosen the opposite approach in so far as he assumes the existence of an innate kernel[1] within his transformational grammar. But it is not necessary to take this extreme approach; what Chomsky supposes innate in the capacity for learning language can no doubt be explained by the earlier "structuralizations" due to the development of sensory-motor intelligence (or intelligence preceding language). Generally, even if it is necessary to

[1] Mehler and Bever (who consider themselves Chomskyists), regard even the operatory conservations as innate, basing their argument on experiments which have nothing to do with quantitive conservation and which bear witness to the existence at around $2\frac{1}{2}$ and 3 years of a method of numerical evaluation which may be understood as deriving from the quasi-topological criterion of "crowding" and which precedes ordinal evaluation, i.e. evaluation by length.

invoke the endogenous factors disregarded by behaviourism[2], it does not mean that everything which is endogenous is hereditary. We still have to consider the factors of self-regulation which are also endogenous but whose content is not innate. As we have just said, between the hierarchical level of hereditary characteristics and that of the acquisitions due to environmental factors, there is a level of self-regulation or equilibration, which plays a vital role in development. This does not oblige or even authorize us to think of everything which is not due to exogenous learning as innate.

Many other lessons of contemporary biology have been ignored by behaviourism, when in fact they are of great psychological importance. For instance, in addition to homeostasis (often invoked by certain schools of psychology) Waddington has distinguished and labelled "homeorhesis", the sort of kinetic equilibration through which embryonic development, when it has been disturbed, is led back to its necessary paths (which he calls "creodes").

Empiricism and mathematics

In so far as empiricism seeks to limit knowledge to that of observable features, the problem it has failed to solve is the existence of mathematics, and this problem becomes particularly acute when it comes to explaining psychologically how the subject discovers or constructs logico-mathematical structures.

Classical empiricism, as argued by Herbert Spencer for example, considered that we derive mathematical concepts by means of abstraction from physical objects: certain Soviet schools of thought share this view, though it is in fact not consistent with the theory of dialectics. In contrast to this attitude, contemporary logical empiricism has well understood the difference between physics, on the one hand, and logic and mathematics, on the other, but instead of seeking a possible common source of knowledge in these respective fields it has maintained that there are two entirely different sources. It has thus aimed at reducing physical knowledge to experience alone

[2] It is true that the neobehaviourists are more and more aware of unobservable processes and endogenous factors. However, since they do not use developmental methods of studying these factors and thus do not take into account the different role these factors play at different levels of development, they encounter new difficulties. The theory has been expanded, new hypotheses have been added, but the main theoretical basis has not been changed.

(the root of synthetic judgements) and logico-mathematical knowledge to language alone (whose general syntactic and semantic features pertain to analytical judgements).

This view poses several problems. Firstly, from the linguistic point of view, while Bloomfield's positivism (and even earlier Watson's behaviourism) aimed at reducing all thought and, in particular, logic to a mere product of language, Chomsky's transformational structuralism reverts to the rationalist tradition of grammar and logic (in doing this, as we have just seen, he exaggerates to the point where he regards basic structures as innate). In the second place, the great logician Quine was able to show the impossibility of defending a radical dualism of analytic and synthetic judgements (this "dogma" of logical empiricism, as Quine amusingly termed it). Moreover, a collective study by our Centre for Genetic Epistemology has been able to verify Quine's objections experimentally by finding numerous intermediaries between the analytic and synthetic poles. Finally, psychogenetically, it is obvious that the roots of the logico-mathematical structures must go far deeper than language and must extend to the general coordination of actions found at the elementary behaviour levels, and even to sensory-motor intelligence; sensory-motor schemes already include order of movements, embedding of a sub-scheme into a total scheme and establishing correspondences. The basic arguments of logical empiricism are thus shown today to be refuted in all the linguistic, logical and psychological areas where one might have hoped to prove them.

As far as the connections between logico-mathematical structures and physical reality are concerned, the situation seems just as clear. It became clarified through experimental analyses of the nature of experience itself. While empiricists aimed at reducing everything to experience, and were thus obliged to explain what they meant by "experience", they have simply forgotten to prove their interpretation experimentally. In other words, we have been given no systematic experimental study on what experience actually *is*.

From our prolonged and careful studies of the development of experience and of the roles which it plays in both physical and logico-mathematical knowledge, the following facts emerge.

It is perfectly true that logico-mathematical knowledge begins with a phase in which the child needs experience because it cannot reason along deductive lines. There is an epistemological parallel: Egyptian geometry was based on land-measuring, which paved the way for the empirical discovery of the relationship between the sides of a right-angled triangle with sides of 3, 4 and 5 units, which constitutes a special case of Pythagoras' theorem. Similarly, the child at the preoperatory level (before 7–8 years) needs to make sure by actually handling objects that $3+2=2+3$ or that $A=C$ if $A=B$ and $B=C$ (when he cannot see A and C together).

But logico-mathematical experience which precedes deductive elaboration is not of the same type as physical experience. The latter bears directly on, and obtains its information from, objects as such by means of abstraction—"direct" abstraction which consists of retaining the interesting properties of the object in question by separating them from others which are ignored. For example, if one side of a rubber ball is coated with flour, the child discovers fairly early on that the further the ball drops in height the more it flattens out when it hits the ground (as indicated by the mark on the floor). He also discovers at a later age (10–11) that the more this ball flattens out the higher it bounces up; a younger child thinks it is the other way round. This is therefore a physical experience because it leads to knowledge which is derived from the objects themselves.

By contrast, in the case of logico-mathematical experience, the child also acts on the objects, but the knowledge which he gains from the experience is not derived from these objects: it is derived from the action bearing on the objects, which is not the same thing at all. In order to find out that $3+2=2+3$, he needs to introduce a certain order into the objects he is handling (pebbles, marbles, etc.), putting down first three and then two or first two and then three. He needs to put these objects together in different ways—2, 3 or 5. What he discovers is that the total remains the same whatever the order; in other words, that the product of the action of bringing together is independent of the action of ordering. If there is in fact (at this level) an experimental discovery, it is not relevant to the properties of the objects. Here the discovery stems from the subject's actions and manipulations and this is why later, when these actions are interiorized into operations (interiorized reversible actions belonging to a structure), handling becomes superfluous and the subject can combine these operations by means of a purely deductive procedure and he knows that there is no risk of them being proved wrong by contradictory physical experiences. Thus the actual properties of the objects are not relevant to such logical mathematical discoveries. By contrast, it is just these properties which are relevant when—as in one of our recent experiments—the child is asked questions about how the behaviour of pebbles (which stay where you put them) differs from that of drops of water.

The method of abstraction peculiar to logico-mathematical structures is therefore different from that in elementary physical experiences. The former can be called a "reflective abstraction", because, when the child slowly progresses from material actions to interiorized operations (by "superior" we mean both "more complex" and "chronologically later") the results of the abstractions carried out on an inferior level are reflected on a superior one. This term is also appropriate because the structures of the inferior level will be reorganized on the next one since the child can now reflect on his own thought processes. At the same time this reflection enriches the structures that are already present. For example, primitive societies and children are already aware of the one-to-one correspondence, but it needed Cantor to discover the general operations of establishing relationships by means of "reflective abstraction" and he needed a second reflective abstraction in order to establish a relationship between the series 1, 2, 3 . . . and 2, 4, 6 . . . and thus to discover transfinite numbers.

In this light we understand why mathematics, which at its outset has been shown to stem from the general coordination of actions of handling (and thus from neurological coordinations and, if we go even further back, from organic self-regulations), succeeds in constantly engendering new constructions. These constructions must of necessity have a certain form. In other words, mathematical thought builds structures which are quite different from the simple verbal tautologies in which logical empiricism would have us believe.

There is a second difference between physical experience and logico-mathematical experience or deduction. Whilst the latter, proceeding by means of reflective abstractions, leads to progressive purification (whose final stages are today those of the formalization peculiar to "pure" mathematics), physical experience is always a sort of "mixture". There is in fact no "pure" experience in the sense of a simple recording of external factors, without endogenous activity on the part of the subject. All physical experience results from actions on objects, for without actions modifying objects the latter would remain inaccessible even to our perception (since perception itself supposes a series of activities such as establishing relationships, etc.). If this is so, the actions which enable us to experiment on objects will always be dependent on the general coordinations, outside of which they would lose all coherence. This means that physical experience is always indissociable from the logico-mathematical "framework"[3] which is necessary for its "structuralization". This logico-mathematical device is in no way restricted to translating the experience into formal language—as if it were possible to have, on the one side, the experience itself and, on the other, its verbal translation.

This brings us back to the central argument of empiricism: that all knowledge should be related as closely as possible to observable facts.

In reality, in every field—from physics to psychology, sociology or linguistics—the essence of scientific knowledge consists in going beyond what is observable in order to relate it to subjacent structures. Firstly, logico-mathematical structures must go outside the scope of what is observable, i.e. what is furnished by physical experience in the broad sense (including biological, psychological experience, etc.). Infinity, continuity, logical necessity, the hierarchy of constructions and of reflective abstractions are all unobservable realities according to the empiricist, and if they had to be attributed to the simple powers of a "language", this language would have the surprising property of being infinitely richer than that which it describes. Secondly, in physics we might just be justified in regarding as observable features the repeatable relations which functional analysis strives to translate into "laws", but on examination of the actual work of scientists—and not the philosophical statements to which they so often limit themselves—we have to recognize that their systematic and unceasing need to discover why things happen forces them to break through the barriers of the observable. In these last decades, measurement has become a problem and researchers have often sought to identify the structures before attempting measurement. To take just one classical example, no one would dispute that the very widespread success of the application of the group structures in physics means that physicists also subordinate what is observable to systems or models which are not. Present-day achievements of structuralism in biology also provide an example of this and almost all the social sciences are proceeding along the same lines.

[3] Establishing relationships or logical classes, functions, counting and measuring, etc.

To sum up, the innumerable problems continually being raised by the nature of mathematics and its application to experimental science have moved us further away from, rather than towards, the empiricist ideal of scientific knowledge.

The nature of intelligence

Having thus briefly recalled the difficulties inherent in the empiricist theory of scientific knowledge, we shall now examine the cognitive mechanisms of the human subject in general. First of all, does empiricism in the form of behaviourism give us an adequate interpretation of the nature of intelligence? Behaviourists started off well by studying the subject's actions before his mental mechanisms; these could be studied afterwards by means of the same methods as those used for the actions but extended in scope in order to suffice for their new purpose; G. Miller wittily called this study "subjective behaviourism".

Let us therefore, like the behaviourists, start with the action, but without prejudging the nature of either the external reality or the subject. There is only one profitable method of avoiding such prejudices which result from the fact that we are adults, and as such have certain set forms of thought and behaviour. This is the method of developmental psychology in which we follow step by step the formation and then the progressive interiorization of actions in our child subjects. This enables us to think of our adult attitudes as the result of a long development, rather than let these attitudes condition our interpretation of the facts. When we proceed in this way we make three fundamental observations. The first is that during the first year of life all actions show an interdependence between the subjects and the objects, which are bound together with no pre-established frontier separating them. There are as yet no objects independent of the subject (object permanence only starts around 9–10 months[4]) and reciprocally, the subject does not know himself as such, but only in reference to his successive actions. This initial level of complete interaction or radical inability

[4] Bower showed that if an infant is shown an object which is then hidden behind a screen, from the age of one week he can recognize it when it is shown to him again. This, however, only serves to prove that recognition is a very early phenomenon and in no way does it confirm the permanency or totalization of the object while it is hidden.

to differentiate between the subject and the objects is important. It is not a question of trying to establish how the subject is going to adapt himself to a reality looked at from outside, but on the contrary, to understand how the succession of action patterns is going to lead to an objectivization of reality (therefore to a construction of objective relations) and to an internal organization of actions.

The second observation is that an action consists in transforming reality rather than simply discovering its existence: for each new action the acts of discovering and transforming are in fact inseparable. This is not only true of the infant whose every new action enriches his universe (from the first feed to instrumental behaviour patterns, like the use of a stick as a means of pulling an object towards one), but it remains true at all levels. The construction of an electronic machine or a sputnik not only enriches our knowledge of reality, it also enriches reality itself, which until then did not include such objects. This creative nature of action is essential. Behaviourists study behaviour, therefore actions, but too often forget the "active" and transforming characteristic of an action.

Thirdly, this active nature is inherent in all and not solely the motor aspects of an action. It would thus be very misleading to contrast perception and action as if perception informed us about the world such as it "is in reality" whilst action alone succeeded in transforming it. It can be seen, for example, that if an infant is given his bottle the wrong way up, he tries to suck the wrong end (until about 7-8 months), whilst later he first of all turns the bottle round, so that he can suck from the right end. It is only then that he seems to perceive the bottle as a reversible solid, whose visible sides hide another side which he must find. As, in this particular case, the change in perception is correlative to the discovery of object permanency (whence the "constancy" of its form), it is clear that perception is a basic part of the entire action schematism and cannot be considered a realm apart.

We must therefore conclude from these three observations that an organization of actions, much deeper and richer than a simple set of "associations" between perceptions and movements, is constituted very early on in life. An action which is repeated engenders a *"scheme"* characterized by what is repeatable and generalizable in the original action. The basic property of a scheme (which must not be confused with a figurative *"schema"*—we shall return to this point) is to ensure *"assimilations"*, which, like the judgements of which they are the precursors, may be correct or wrong—which is not the case with associations. For example, when the behaviour pattern of the stick becomes established, that is when the child becomes capable of pulling towards him a desired object with the help of a stick, he may assimilate a metal rod to the stick, which is correct, or a pliable straw to the stick, which is wrong. This system of assimilation schemes presents certain general forms of coordination (order of movements, embedding of a sub-scheme in a total scheme, establishing correspondences, intersections, etc.), and we must go right back to these forms to find the roots of what will become logico-mathematical structures.

Intelligence is thus, in the initial sensori-motor period, the progressive coordination of action schemes. By contrast, with the advent of the semiotic function[5] (deferred imitation, mental imagery, language, etc.) these actions become "interiorizable" (i.e. they develop independently of externally observable behaviour). Thus an uninterrupted series of "reflective abstractions" lead to new constructions. Between 2 and 7–8 years there are no reversible operations, i.e. coordinations which can be effected simultaneously and mentally both ways (unite-dissociate, add-substract, etc.). As a result, when there are transforming actions (transferring a liquid into a bottle of different dimensions, changing the shape of a clay ball, etc.) there are no quantitative conservations. We are, however, already witnessing the formation of a "half logic"—if we may use

[5] We follow the tradition of the great Swiss linguist de Saussure as regards the different categories of signifiers.

Signifiers that are part of, or caused by, objects or events (for example, the nipple of a baby's bottle, or the traces an animal leaves in the snow, or the smoke of the fire) are called *signals* (in French "indices").

Signifiers that have some resemblance to objects or events (for example, a stick used in play as an umbrella) are called *symbols* (in French "symboles"). Images belong to this same category as do the symbolic gestures which evoke an object, a person, an event.

Signifiers that are conventional and without any direct relation to the significate (for example, words, algebraic signs, numbers, etc.) are called *signs* (in French "signes").

There are many intermediaries between symbols and signs. Symbols become collective symbols which are symbols as well as signs, depending on how much the conventional aspect of the relationship between "signifier" and "significate" is stressed. (For example, the Lorraine Cross was a symbol but has now become a conventional sign designating Gaullism and the original symbolic connection has been almost forgotten.)

this term—made up of one-way schemes. For example, the functional dependences ($y = b^{(x)}$), characterized by directed one-way mappings, can be called qualitative but do not yet include reversibility or conservations, whence the noticeable dominance at this level of the notion of order and the widespread use of the ordinal evaluations: "longer" signifies "going further"; "quicker" for a moving object means that it overtakes another, etc.

From 7–8 years, on the other hand, this first half of logic is completed by its second half, in the sense that the child realizes that the most general transformations of action (ordering, uniting, embedding, establishing correspondences, etc.) are reversible (by inversion or reciprocity). The transforming actions are then interiorized as real operations, which combine with each other into a coherent structure and which are stable precisely because they are reversible. These are the operations which enable the construction of seriations ($A > B > C \ldots$) with transitivity, of classification with understanding of inclusion, of multiplicative matrices (two-way classifications or seriations) and above all of number and measurement (by synthesis of order and inclusion, etc.).

These are the basic operational mechanisms which, in our opinion, characterize the "concrete" intelligence of the 7 to 10–12-year olds. This intelligence, still linked to the handling of objects, serves as a basis for the later more "formal" structures, which are formed from 11–12 years by the addition of combinatory systems (such as the logic of propositions) and a more general "group" structure coordinating inversions and reciprocities.

But we shall not elaborate on subjects which we have described too often elsewhere (Piaget & Inhelder, 1966). We have to ask ourselves why the empiricist view in psychology, represented mainly by behaviourism in its many variations, has so frequently neglected the activity of the subject, and why certain authors still refuse to believe in cognitive structures, and consider them to be mere theoretical and logical models made up by psychologists, instead of underlying structures inherent in the cognitive development of the human subject. We see two basic reasons for this.

The first is that psychology is thought of as a science divorced from all others—as if there could be a discipline which is sufficient in itself, when in fact the boundaries between the different sciences are always artificial. We maintain, on the contrary, that it is one of the duties of psychology to try and find the links between behaviour and organic life in general and those between man's cognitive development and his important scientific creations. It may well be argued that these are problems of the future, but we think that it is never too early to prepare for the future. The sensory-motor action schemes of the first year of life plunge deep into organic life, and, in a sense, they constitute an intermediate zone between the organic self-regulatory mechanisms and the later logical mathematical operations and their underlying structures. To seek out the possible ways of linking biology to logic and mathematics, is thus no theorist's luxury but the developmental psychologist's duty.

The second reason why so many authors do not understand the significance of such a concept of operations and underlying structures stems, we think, from the superstition of the "observable". If every action is a transformation of reality, it thus constitutes by its very nature a constant sampling of what is observable, and a constant conquest of what is possible—which is still unobservable. Why be afraid of crediting the subject with operatory structures of which he remains unconscious and which are in this sense unobservable, when at the higher developmental levels, these structures will lead logicians and mathematicians to build increasingly rich constructions, culminating in the most far-reaching achievements of the human mind?

Perception, mental imagery and memory

We can divide the cognitive functions into two broad categories, according to whether the "figurative" or "operative" aspects of knowledge predominate.

The figurative aspects bear on static reality and its observable configurations. Into this category come perception (even when it apprehends a movement from which it then retains the total form or Gestalt), imitation (first direct sensori-motor imitation then delayed imitation of something no longer present, which ensures the advent of the semiotic function) and mental imagery (interiorized imitation). The operative aspects, by contrast, bear on the transformations of one state into another and therefore include actions and operations, which both aim at the unobservable (since we perceive or "imagine" the results of transformations and some of their intermediary states, but not the transformation as such which is "understood" by the intelligence rather than perceived or pictured).

A fundamental problem concerning the conflict of empiricism and the relational interpretation of mental development is to establish whether it is the figurative aspects of the cognitive functions which lead the way to understanding the transformations or whether it is the operative aspects which direct thought and dominate the figurative aspects.

(*a*) In the realm of perception it is already clear that the operative aspects play a much greater part than is apparent at first sight. All the New Look movement in perception tends towards this direction and J. Bruner's work (Bruner, 1956) has proved the importance of operative factors by stressing the function of "perceptual categorizing" and the vital influence of the temporal schemes inherent in its mechanism. We ourselves have underlined the role of the perceptive activities and sought to explain the field or Gestalt effects by means of the "encounters" between elements of the recording organism (retina cells, ocular micromovements, etc.) and the visible parts of the display; and by the process of complete or incomplete "couplings" (or correspondences between the encounters); incomplete displays are in fact the source of certain well-known geometrical illusions.[6]

(*b*) Classical associationist empiricism considered the image a residual product of perception and a fundamental element of thought; this was the triumph of the observable features over the unobservable transformations. We, on the other hand, through developmental study of the formation of mental imagery and its role in thought processes have come to the conclusion that mental imagery does not stem from perception, but from an interiorization of imitation and that its role is that of a symbol[7] and not of a constituent element of thought; in fact, images appear probably much later than perception and only as part of the semiotic function.

The development of symbolic imagery can, of course, never be grasped directly but must be approached through actualizations such as drawings, gestures or selections from among a series of drawings, representing both correct solutions and typical errors of children of different ages and verbal comments. Along these lines we tried to make inferences about the various kinds of imagery. In particular, we studied the different ways in which children imagine the results of a transformation (of shape or position) which is not

[6] See J. Piaget: *Les mécanismes perceptifs.*
[7] See footnote 5.

performed in front of their eyes, but only indicated by gesture. Following our developmental method, we put the same problem to children of different ages; and from comparisons and the hierarchical ordering of their ways of symbolization, we tried to extract the laws governing the development and modification of symbolic representation.

One typical example is the transformation of arcs into straight lines and vice versa.

A supple piece of wire in the shape of an arc (10 cm. long) was presented to children aged 5 to 10 years, and they were asked:

> to draw a straight line showing the length of the wire as if it were straightened out (by gesture, the experimenter suggested the action of pulling the wire straight by its extremities); to cut lengths of wire equal to the result of straightening the arc; and to draw successive intermediate stages of the transformation.

The drawings of the young children

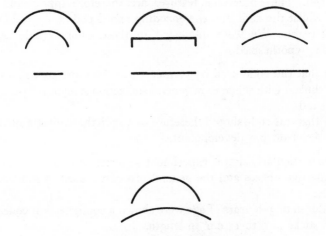

FIGURE I

as well as the cutting of the lengths of wire, were evidence of an initial difficulty of representing the result of a transformation and of a much more persistent inability to symbolize successive stages of the transformation. Centring his attention on the extremities of the figure, the child first behaved as if the result of the straightening would be equal to the chord of the arc, in other words, as if he had to "conserve" the ordinal relations of the figure's frontiers. Only very gradually, along with the development of operativity, did older children succeed in more or less correctly imagining the straight lines resulting from arcs. When this first obstacle was overcome, there remained the difficulty of representing intermediate stages of the transformation. This symbolization required an operative understanding of the ordinal change of the extremities, the seriation of intermediate stages, and the conservation of length. With progress in operative thinking, it would seem that a new form of imagery developed which captured successive moments of a continuous transformation (like the frames of a film), in order, as it were, to represent the continuity of the transformation in a schematic way.

(*c*) The next examples are chosen from our experiments on memory. Evocation memory (recall) presupposes a coding system (perceptive, etc.), a decoding system leading to a more or less adequate memory image, and a code enabling translation of what has been recorded. The observable features are therefore inputs and outputs, whilst the code and the encoding and decoding operations are unobservable. We were interested in these unobservable aspects and we hypothesized

(i) that the code used by memory must be more or less directly linked with schemes in general and action schemes in particular and

(ii) that this code should therefore vary with the subject's operatory level during development.

One particularly simple experiment will exemplify our preliminary working hypotheses and the general procedure used in most of our studies.

Children of 3–8 years of age were shown a configuration consisting of ten sticks of 9 to 15 cm. in length.

Two results appear to reveal in a striking manner the nature of the processes of memory organization. First there was a notable correspondence between the levels of memory organization and of operativity. What a child at each level of his development retained from observing the series was the manner in which the series had been assimilated to his operational schemes. We found therefore between 3 and 8 years of age a succession of memory types which closely resembled the operational stages of seriation.

Memory organization was divided according to three groups of predominant characteristics. The youngest subjects, 3 to 4 years of age, seemed to remember a great number of sticks lined up but the length of all the sticks was more or less equal. Children in the next

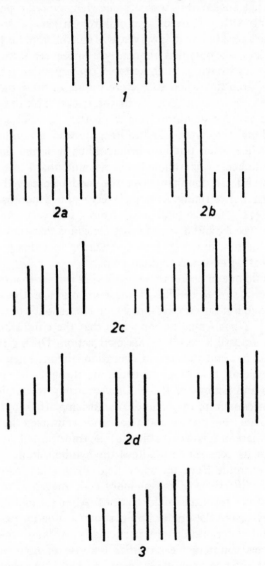

FIGURE 2

age group, 4–5 years, showed four characteristic features. (2a) The sticks were paired, a long one with a short one. (2b) All the sticks were simply divided into two classes, one long, one short. (2c) The sticks were arranged in threes or fours, using short, long and medium-sized sticks; this was a somewhat superior performance. (2d) Starting with 5 years of age, children in general succeeded in memorizing a seriation that, however, was still incomplete. (3) Starting with 6–7 years, one obtained frequently a correct serial configuration.

The memory drawings of these types corresponded rather closely to what children did when they were required, after the evocation of the series, to make a series with the sticks. The drawings were also similar to what children drew when they had not seen a series beforehand and they were asked to imagine what a number of sticks would look like when put into order. Thus it seems that memory images are linked to operational schemes and that schemes control the images and are dominant over the model perceived. A second result is even more striking. After an interval of 6–8 months, a majority of children (74% for the total group from 3 to 8 years, and 90% for children between 5 and 8 years) made drawings that showed *progress* relative to the first drawing. It is noteworthy that this progress was quite gradual without any sudden steps.

How should one interpret these results? It is clear that the memory image is not a simple residue of the perception of the model that was presented, but rather a symbol which corresponds to the schemes of the child. What seems to happen is that the child interprets the seriation as a possible result of his own action. During the interval between the first and the second evocation, the schemes themselves evolve because of their proper functioning through the spontaneous experiences and actions of the child. According to our hypothesis, the schemes constitute the code of memorizing. If true, the code is modified during the interval and a new code is utilized the next time. The code therefore is not static but is transformed and in turn according to its constraints modifies the symbolization.

Another example from the same field shows that, as operativity develops, modifications of the memory code may lead to considerable, though not fortuitous deformations. After six months children do not always, as in the first example, show a superior performance. Conflict, and deterioration are particularly striking when two or more schemes counteract each other because of non-synchronous development. In one such experiment the child was shown a configuration of matches placed as follows: four matches were placed in a horizontal broken line, and beneath it another four matches were arranged in a flattened "W". (If one presented a real "W" the child would recognize the letter of the alphabet and this would spoil the experiment.) This configuration of matches could be assimilated by the child as the result of a numerical or of a spatial relationship.

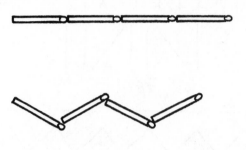

FIGURE 3

Previous research on the spontaneous geometrical notions of the child indicated that these two systems, the numerical as against the spatial relationships, produce antagonistic results before becoming coordinated in the thinking of the older child. In fact, when the child establishes numerical equivalences by a one-to-one correspondence, he is still at a level when he estimates the equality of spatial extensions by the coincidence or non-coincidence of the extremities of lines, so that "to pass" or "to go beyond" means to the child "to be longer". This characteristic of pre-operational thinking is moreover reinforced by the particularities of the mental image at this level. We have noted in our study of the image that frontiers play a predominant role.

An analysis of memory drawings as a function of age and operativity suggested five successive types. Type 1: a double line-up of a great number of matches without any numerical or figural correspondence. Type 2: (an interesting intermediary form) the coincidence of extremities dominates over numerical correspondence; the child draws one horizontal line-up of matches and another one in a zig-zag fashion, several times repeated, so that the extremities of the line-ups coincide. Type 3: shows the height of conflict with a false solution that respects the numerical correspondence but in which the length of the matches in the second line is exaggerated so that a figural coincidence is respected. Type 4: the conflict is softened; the matches in the straight line go beyond the matches in zig-zag, but the number of segments in the two lines is not the same. Finally, Type 5: the correct solution (Figure 5).

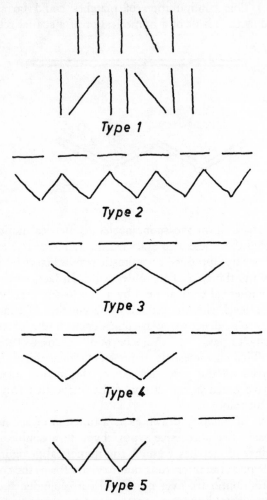

FIGURE 4

The evocation of the matches six months later was characterized by considerable deterioration. The memory, far from being faithful to the model, was organized under the dominating influence of equalizing the extremities. Apparently the developing schemes first accentuated the conflict between the numerical and the spatial systems before the schemes were sufficiently developed to surmount the conflict.

As with previous experiments, one may assume that in the memory performance pre-inferential processes intervened that are mid-way between perception and conceptual elaboration. It seems that children of a pre-operational level realize the numerical correspondence. But they do not yet possess the scheme of spatial conservation that would enable them to understand that a line broken into a zig-zag remains of the same metrical length as it was before transformation. It was as if the children made a memory pre-inference as follows: if the two lines are equal by numerical correspondence of their elements, their extremities must also coincide.

It is difficult to state exactly at which moment such pre-inferences are made. We have frequently noticed quasi-immediate deformations, and are accordingly inclined to believe that there occurs a kind of deforming conceptual assimilation right in the presence of the model, and that this deformation becomes manifest in memory. Our previous research has already shown that the image represents an object by means of its "concept" as much as, or even more than, by means of its "perception".

Thus the genetic study of memory in its relation to intelligence of which we have sketched here a few characteristic results, seems to support through a converging number of facts that the mnemonic code, far from being fixed and unchangeable, is structured and re-structured in line with general development. Such a re-structuring of the code takes place in close dependence on the schemes of intelligence. The clearest indication of this is the observation of different types of memory organization in accordance with the age level of a child so that even a longer interval of retention without any new presentations, far from causing a deterioration of memory, may actually improve it. In fact, such progress is due to and merely makes evident the general progress of intelligence during the interval concerned.

This hypothesis should not be understood in the sense that any kind of memory encoding will lead to progress with a sufficient interval of time. The progress observed was in relatively simple memory situations in which quite general schemes sufficed as a support for the child's memory. In addition, these schemes were in a state of active evolution during the ages under study. When conflicting schemes exist side by side, as is frequently the case in everyday situations, and especially when the situations are fortuitous, the schemes of the subject, as we have seen, can play a twofold role. They can bring about deformations or adaptive simplifications. Hence we are far from asserting that memory is always in progress. Our main assumption is that the figurative schemata of memory—from perceptual recognition to image evocation—are not by themselves sufficient to explain memory. They need the support of operative schemes, on which the element of comprehension—which is nearly always present in memory behaviour—depends.

Learning and development

The next example deals with the field of learning and development. The central biological question concerning learning is whether it constitutes the source or the result of development. It would certainly be the source, if development consisted merely of an accumulation of acquisitions of external origin, as empiricists suppose. On the contrary, if development consists of a continuous construction of operative structures through interaction between the subject's activities and external data, then the assimilation of these data will determine and direct the learning process which does not depend on general laws but on competence levels, i.e. on the developmental stages. Consequently, in our view, it is development that directs learning rather than vice versa.

The experiment we would like to mention concerns the acquisition of quantitative measurement in spatial configurations. One of the child's main difficulties in acquiring spatial structures resides in the fact that the elementary topological relationships (contiguity, order and enclosure) have to be transformed to fit a graded system of geometrical coordinates; such a transformation, by combining subdivision of length and displacement, leads to the concept of conservation of linear dimension, which is a prerequisite of the constitution of measurement units.

We asked ourselves the following two questions.

1. Can the acquisition of elementary spatial measurement be facilitated by exercises in which the child applies numerical operations to the evaluation of length? If so, what stages does the child go through?

2. Can the children's behaviour during learning sessions provide insight into the interval elapsing between the acquisition of number (generally at about 6 years of age) and length conservation (generally at about 9 years of age), and into the specific difficulties attached to reasoning about discrete units on the one hand and about continuous quantities and dimensions on the other?

The procedure, designed by Magali Bovet, was the following: both the experimenter and the subject have a number of matchsticks at their disposal, but the subject's matches are considerably shorter than those of the experimenter and of a different colour (7 red matches of the subject add up to the same length as 5 green matches of the experimenter). The experimenter constructs either a straight or a broken line (a "road") and asks the subject to construct a line of the same length ("just as long a road", "just as far to walk, so that two people, one on each road, would be just as tired").

Three situations are presented:

(*a*) The first situation is the most complex: the experimenter constructs a sort of zig-zag line and the subject has to construct a straight line of the same length directly underneath (Figure 5). The figural pregnancy of the situation suggests a topological solution where the beginning and end of both "roads" are congruent.

(*b*) In the second situation the subject has again to construct a straight line of the same length as the experimenter's broken line but no longer directly underneath (Figure 5); this facilitates the problem slightly, since there is no longer a perceptual pregnancy that suggests a topological solution.

(*c*) This is the easiest situation, since the experimenter's line is straight, and the subject is asked to construct a straight line directly underneath (Figure 5). Moreover, the experimenter uses the same number of matches as in the situation (*a*), that is, five, so that this situation (7 of the subject's matches are needed to make a straight line of the same length) suggests a correct solution to situations (*a*) and (*b*).

The three situations remain in front of the subject; after he has given his first three solutions, he is led to give explanations and eventually to reconsider his constructions, while the experimenter draws his attention to one situation after another.

The following reactions are typical of the behaviour of subjects during the learning sessions:

In situation (*a*) the most elementary solution is to construct a

Problems

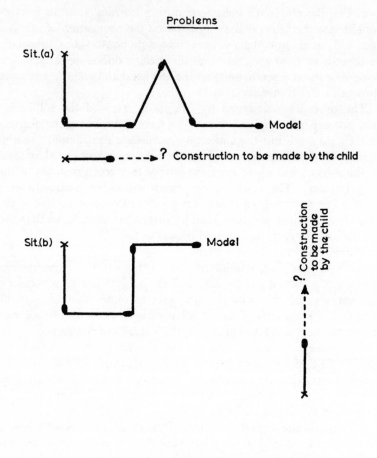

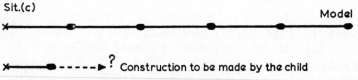

NB: The length of 5 of the model's matches
= that of 7 of the child's matches

FIGURE 5

Examples of solutions:

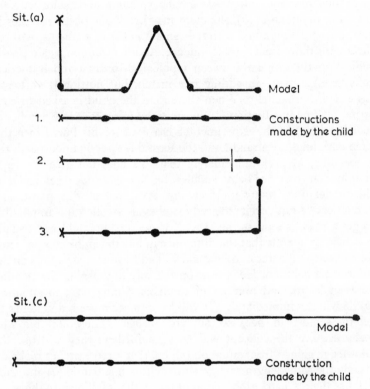

FIGURE 6

straight line with extremities congruent with those of the zig-zag line constructed by the experimenter. The child is convinced that the two lines have the same length, though his line is made up of 4 small matches, and the experimenter's line of 5 long matches. In situation (*b*) the child has no ordinal or topological point of reference, since he has to construct his line away from the experimenter's line, and he uses the numerical reference: he constructs his line with the same number of matches the experimenter has used, regardless of the fact that his matches are shorter. When the experimenter now goes back to situation (*a*), the child will notice, with some embarrassment, that there he has constructed a line he judged to be of equal length which all the same has not got the same number of elements as the model line.

At this point we often see amusing compromise solutions: for instance, in situation (*a*), the child may break one of his matches in two, thus creating a line with the same number of elements, without destroying the ordinal correspondence. Another solution, again clearly indicating the conflict between topological, ordinal and numerical references, consists in adding one match, but putting it vertically instead of horizontally. When thereupon the child is asked to construct his line in situation (*c*), he starts by using the same number of matches (five) as the experimenter has used for his line. Since this time both lines are straight and the second is directly underneath the model line, he sees immediately that this does not give the right solution; his matches being smaller, his line does not reach as far as the model line. Many children are perplexed at this point, and announce: "I can't do it; the red matches are smaller, it is impossible to get a road as long as the other." However, the child will usually realize after a while that the difference in length can be compensated for by using a greater number, and will add two matches, thus giving the correct solution. He is now on the way to grasping the relation between length and number of elements. Going back to situation (*b*) he will say immediately: "I will have to use more red matches to get the same road, because they are smaller." Those children who really acquire the concept will go even further: they will use the knowledge gained from situation (*c*), i.e. that 7 red matches make up exactly the same length as 5 green matches, in situation (*a*) and they will use their 7 matches in a straight line without falling into the error of paying attention only to the ordinal aspect.

The main interest of this experiment resides in the fact that, as in all our learning experiments, the reactions of the children during the learning sessions show us, as through a magnifying glass, both the developmental processes and the specific difficulties encountered by them. Cross-sectional experiments only rarely catch the fleeting, transitional stages, and though the general theory allows us to infer these gradual steps and the specific obstacles, the learning experiments show the development in a clearer and more detailed fashion.

As regards the specific problem of the links between numerical conservation and elementary measurement of continuous length, and the connected problem of the relationship between numerical and geometrical fields of knowledge, we seem justified in concluding that it is indeed possible to make use of already acquired numerical operations to lead the child to spatial operations of measurement. However, even in the case of total acquisition, progress is slow, and many obstacles are encountered in fitting the number-concept to conservation-of-length problems. In some situations, the ordinal and topological references are misleading, and before the two operational structures can strengthen each other, there is a period of conflict, which can be overcome only through a constructive effort by the child who discovers compensatory and coordinating actions. It is the feedback from these actions themselves which finally results in the acquisition of a structure of a higher order, and not the passive reading-off of a result which is presented to him.

These and similar experiments have thus produced some instructive results. Firstly, learning is definitely dependent upon the subject's developmental level. Generally, in all this research it has been shown that the child never manages to accomplish more than the passage from one sub-stage to the next without ever jumping a stage.

In the second place, we can now reply to a question which we are often asked (and particularly in the U.S.A.): is it possible to accelerate the passage from one stage to the next? That acceleration is possible is evident and the comparison of operatory experiments carried out in different milieux showed this long ago: for example, in young illiterate children from a village in Iran, Mohseni found our same stages but with two or three years retardation in comparison with her Teheran subjects. But whether unlimited acceleration is possible is another question and first we must find out whether high acceleration is desirable or harmful. When H. Gruber, using kittens as subjects, carried out our experiments on the permanency of the object, he found our first stages. They appeared, moreover, much earlier since the kitten achieves at four months what the human baby can only do at nine months; but the kitten stops there, whilst the child then goes much further, which shows that a certain slowness may be useful in developing the capacities for assimilation. If mechanisms in mental development can be compared to what Waddington (in embryology) calls "creodes" or necessary paths with a "time-tally", it appears obvious that development always has an *optimum* rate, neither too slow nor too fast.

In fact, it is on this problem of speed that the above-mentioned learning experiments are indirectly shedding light; the same number of learning sessions results in more marked progress for those subjects who are already at a more advanced stage. Thus hardly 10 % of subjects who are still at a frank preoperatory level show progress (this was found in almost all our learning experiments) whilst a higher percentage of subjects who start off from a slightly more advanced level improve. In other words, the ease of assimilation (which is not, we repeat, a complete jump up to the correct result, but a passage from one stage to the next) varies with the developmental level, which leads us to favour the idea of an *optimum* speed.

Thirdly, we now know more about the mechanisms of development. The study of the possible influence of one sector of knowledge on another (for example, the influence of conservation of number on that of length or vice versa, or the influence of inclusion of classes on conservation or quantity, or vice versa) shows the part played by what might be called the cross-sectional relations between the various structures, as compared with the longitudinal or internal development of each structure taken separately. These relations have in fact been proved essential, particularly for structures which are generally not acquired simultaneously. There are here two quite different problems. On the one hand, there is the case of structures which can act as tools for the construction of others—this is the case regarding the concept of inclusion, which is an instrument of quantification. On the other hand, there is the question of an overall equilibrium between structures: the conflicts between different sectors of knowledge can thus help to create a disequilibrium which in turn engenders new equilibration by means of reflective abstractions and new self-regulations.

To sum up, these new experiments on learning seem to confirm fairly accurately that developmental laws dominate learning, not only because the effects of our learning procedures depend on the level of "competence", but also because analysis of the subject's reactions to these procedures enables us to rediscover, but this time more accurately, the same formative mechanisms which we had found through analysis of the different answers and arguments given in our usual operatory tests.

Language and thought

Closely linked to the problem of learning is the question of the relation between language and thought. Many authors affirm, and many educators behave as if it were true, that verbal learning can bring about progress in cognitive structure. It can be shown experimentally[8] (Sinclair, 1967), however, that this is not the case. If a child has to learn to deal, for instance, with a logical multiplication of dimensions (length and width) we find that to teach him to use the words "long" and "short", and "wide" and "narrow" in a consistent way is easy, even if at first he has no idea of doing so. But the result of such verbal training is limited to what Soviet psychologists call "orientation of attention"; in an experiment concerning conservation of quantity, the child now correctly describes a ball of clay which has been transformed into a sausage as "long and thin". However, this does not lead him *ipso facto* to the multiplication which is involved in compensation and conservation: "longer but thinner than the ball" and thus having the same quantity.

Thus it is not language which governs thought or forms operations. If empiricism and behaviourism have both maintained the contrary to the point where Watson wanted to reduce internal thought to movements of the larynx, it is once again because language is an observable feature and because thought structures appear unobservable. But the irony of the story is that although linguistics has long adhered to the observable system of language, the transformational structuralism of Harris, and mainly that of Chomsky today, presents us with a completely different picture. These linguists are establishing algorithms and systems of transformational rules. It is therefore curious to see psychologists who, when they are considering language, think they are remaining faithful to what is observable, and refuse us the right to infer operatory structures—when linguists are doing just that.

It seems evident that if we assume the one-and-a-half-year-old child to have acquired no more than associationist psychology credits him with, that is to say, an unorganized conglomerate of bits of learning, we have no alternative but to introduce innate linguistic structures to explain the beginnings of language acquisition. But the achievements of sensori-motor intelligence may provide the infant with just those cognitive structures that will allow language development to start.

Speaking in a general and theoretical way, it is possible to show the similarity between Piaget's descriptions of sensory-motor structure and Chomsky's descriptions (Chomsky, 1965; Sinclair, in production)

[8] *Handgebrauch und Verständigung bei Affen und Frühmenschen*, Bern und Stuttgart, Hans Huber, 1968, and my contribution to this volume.

of deep structure in language. At the age of 18 months, before the infant can talk, he can order, temporally and spatially; he can classify in action, that is to say he can use a category of objects for the same action, or apply different action-patterns to one object; and he can relate objects to objects and actions to actions. The linguistic equivalents of these capacities are concatenation, categorization and function, where categorization means the major categories (noun phrase, verb phrase, etc.) and function the grammatical relation (subject-of, object-of, etc.). These are the main operations at the base of the syntactic component which characterizes a highly restricted set of elementary structures from which actual sentences are constructed by transformational rules. These basic rules have, moreover, a particular formal property, namely that they may introduce the initial symbol "S" (sentence) into a line of derivation, so that phrase-makers can be inserted into other base phrase-makers. A psychological parallel to this so-called recursive property of the base can be found in the embedding of action-patterns one into the other, which can itself be traced back to the simple circular reactions of a much earlier stage. Finally, an interesting principle has been introduced, formulated by Katz and Postal (1964), i.e., that the only contribution of transformations to semantic interpretation is that they interrelate phrase-makers, so that the semantic interpretation is independent of all aspects of the transformation marker except in so far as the latter indicates how base structures are related. This principle adds an important link with psychology, since it supposes an invariant for the system of transformational rules. One of the most important acquisitions of the sensori-motor period is precisely a first invariant, namely the permanency of the object.

Two conclusions can sum up what has been briefly outlined in this paper. The first is that in order to understand cognitive functions we must go beyond what is observable, since in all the branches of science it is only the underlying structure that is explanatory. The second is that specialists determined to remain within the boundaries of their own discipline remain irremediably short-sighted, because it is only through contact with other branches of scientific knowledge that real progress can be achieved. Without recourse to biology, logic and mathematics, developmental psychology can again only be descriptive for the same reason: the explanatory level involves underlying structures which inevitably cut across the traditional boundaries between the different disciplines.

DISCUSSION

THORPE May I start by referring to your statement that development directs learning and not vice versa. Surely both must be true.
In the animal I would say that both are quite certainly true. In birds
for instance I think one could say that development depends directly
on learning much more than learning on development. Here it often
happens that the subsequent development of the animal depends
to an enormous extent on the chance experience which it learns to
cope with, and so both factors are obviously active all the time.
Would you like to say more about this? Do you mean to imply that
learning never directs development?

INHELDER It is understandable that an ethologist today stresses
learning, which has long been neglected by their school of thought.
However, your own research shows in a particularly striking way
that the animal does not learn just anything at just any moment of
its development. Concerning the cognitive functions of the child,
the theorist's attitude has been just the opposite. The formation of
knowledge has long been thought of from an empiricist point of
view, forgetting that the knowing subject develops according to laws
which are analogous to those of biological growth and therefore
constitute an epigenesis.

The research undertaken in Geneva over the last 40 years has
brought to light the fact that development does not occur by chance
through encounters with the physical and social environment but
follows a certain direction; in the development of thought, particularly, there are sequences or stages of progressive structuration. We
took this development to obey laws of self-regulation of endogenous
origin, but to be subject to continuous modifications under the influence of the feedback resulting from exchanges with the environment. However, this research took only the learning *sensu lato* into
account, in other words the more or less fortuitous encounters of the
subject's assimilation schemes with the environment.

It was therefore important to discover the modifications of knowledge in learning situations in the strict sense where the child is
faced systematically and during a limited period with experimental
observations, verbal training, comparisons between his own judgements, etc.

The clearest result of our research on learning in this strict sense
is in fact that the modifications thus obtained are themselves sub-

jected to the laws of development. Learning can only accelerate the development within certain limits, cf. the "time tally" in embryology brought to light by Waddington. When the learning procedure results in a deviation from the spontaneous construction of thought-operations, it seems to display a tendency to return to the ways of structuring we have noted in normal, undisturbed development.

THORPE   I do not quite understand your point here. Is there really an important difference between exposing an animal to an experiment and an animal exposing itself to the environment? How do you propose to make such a distinction?

INHELDER   From your own point of view—you said so yesterday—the animal is choosing its environment.

THORPE   It makes some choice, yes! This depends on how highly developed the animal is. Some animals can choose their environment very exactly, others are much more at the mercy of the environment.

MCNEILL   I have a comment on the influence of development on language. It is certainly a fascinating and I think profound problem to derive the general and universal form of language from universal characteristics of intellectual development. But, there is a crucial distinction here, which is generally overlooked. In all your remarks, you were assuming that linguistic universals belong to a certain type, whereas it is possible to conceive of another type of universal. Universals may appear in language because they are imposed on language by virtue of the form of thought—developments in sensory-motor intelligence, and so forth. But it is also conceivable that universals of language have a separate origin, even though, as part of a communication system evolved to express thought, they have the same form as intellectual universals. Now because such communicative universals will be paralleled by intellectual ones, simply by looking at linguistic theory, where universal linguistic concepts are set forth, you cannot possibly decide which of these categories a particular linguistic universal falls into. The question cannot be answered on the grounds of formal similarity. How to go about answering or deciding this question is very problematical, but it seems that you gave one case at the very end that suggests a solution. A general strategy would be to look for a considerable asynchrony in the time of emergence of the corresponding intellectual and linguistic universals. That would suggest independent causation of the latter. For example, you mentioned the recursive coupling of sensori-motor schemas. All languages have a capacity for recursion, although in certain languages it is more marked than in others. Here is a possibility of a linguistic universal that might be derived from an intellectual universal. But the interesting fact is that linguistic recursion does not appear until around two and a half or three years of age—that is to say, considerably later than the intellectual universal. The discrepancy suggests that recursion in language is a universal of the second type—namely, one that is peculiar to the communication system, not imposed by the development of thought. If so, it would be a mistake to explain it as an imposition on language from cognitive causes. On the contrary, a different kind of theory is required, one dealing with the evolution of a communication system that has this abstract property. Moreover, even when the emergence of linguistic and intellectual universals coincide, the classification is not obvious; the linguistic universal could still be of either type. You mentioned the fact that deep structures always include noun phrases, verb phrases, and so on, and that organized action schemas include actions, actors, and things acted upon. All of this is quite true. But there is a vast gap between these two. I do not think it is bridged simply by pointing out their analogy. What is linguistically universal is something that can be written, consistently, in linguistic notation, such as S, NP, D, etc., and which has linguistic significance: for example, that it has decisive control over the form of the lexicon. There is a rather coherent linguistic theory and also a rather coherent cognitive theory. The two might or might not be directly related; whether they are is a matter of fact, not assumption.

INHELDER   It is rather difficult to comment on McNeill's views because I am not quite certain I have understood them correctly. However, one confusion should be cleared up: McNeill says that we are making an effort to "derive a general and universal *form* (my emphasis) of language from universal characteristics of intellectual development". Such is not our intention. Our hypothesis concerns the existence of parallel *mechanisms*, of parallel constructive *processes* in cognitive development and in the acquisition of syntactic structures. We would never speak about "a particular cognitive universal"; there are no isolated operations, nor isolated action-schemes; on the contrary, their characteristic is that they form groups, groupings, structured wholes, which can be formalized at different stages. Then there are the processes by which these structures are acquired; and here we see the derivation and the parallelism; not in

F

a *form* of language. The time-lag points, in our opinion, exactly in that direction: the infant has to acquire coordinations of sensory-motor schemes, which will later develop into operational structures, rebuilt on a representational level, before he can begin to understand and produce syntactic structures; and the acquisition processes involved in linguistic structuring seem to be parallel to the structuring of the action-schemes that has taken place much earlier.

THORPE   It strikes me that this is a really important point which does need working out between you three. May I therefore make the suggestion that you arrange some time in which to do this and then present a paragraph to the editor for possible inclusion in the discussion. I am sure this ought not to be left in the air, but on the other hand to work it out now is obviously impossible.

INHELDER   Yes, it is for the more physiologically minded people here.

MCNEILL   It's a burning issue for those who aren't (laughter).

BERTALANFFY   As I mentioned in my paper I am delighted with the far-reaching agreement between Piaget and myself, considering the different starting points. May I enlarge on a few items following Bärbel's presentation:

(1) Cognition as an essentially "active" process, as contrasted to passive receptivity. This, of course, is becoming the dominant conception in various recent developments in psychology (e.g., "new look" in perception, Gibson, Pribram, etc.). In system-theoretical terms, this is part of "personality as an active system", applying not only to behavioural (motor) but to cognitive aspects. Piaget has, of course, given a broad scheme empirically substantiated.

With respect to "discovery stemming from action", it is perhaps interesting to note that this notion has its history in philosophy. The idea that we can only "understand" things when we can "make" them, is, I believe, first found in Nicholas of Cusa and later, of course, in Giambattista Vico.

In a very real sense we create the world around us. For this very reason, knowledge is never "truth" but a "perspective" of reality.

(2) Regarding mathematical structures, reference may be made to Lorenz' remark that already basic psychological phenomena, such as the constancy phenomenon, imply an enormous amount of "computation", at the unconscious level. Therefore logico-mathematical structures certainly have their biological roots (ultimately in the structure and function of the nervous system) and are their development at the conscious-symbolic-linguistic level.

(3) In a similar line, linguistically formulated concepts certainly have precursors in the "pre-verbal concepts" of animals. Some discussion of the point can be found in B. Rensch, ed. (1968).

(4) I find particularly important Piaget's statement that "I" and "outside world" are not ultimately given but differentiate from what Piaget terms "primitive adualism". This has enormous implications for the criticism of the Cartesian dualism and the re-evaluation of the mind-body problem (cf. L. von Bertalanffy, 1963, 1966, 1967)[9].

KOESTLER As a professional scribe, my main headache in life is of course the relation between language and thought. So with commendable restraint I shall not comment on this hoary question. Instead I would suggest we consider the connections between what Inhelder said about the transformations of memories in the course of the development of the child and what Hydén said yesterday about the transformations of the engram—the present remaking the past. This, if you remember, was followed by a discussion in which Paul Weiss wondered what happens to the stability of memory if developmental processes change it. There is a wonderful field here for analogy.

INHELDER (to Hydén) I wanted to ask you this question yesterday—whether this is within the realm of your hypothesis about the underlying biochemical structures or whether it is out of it. Can the data which I have discussed be fitted in, or is the gap too large?

HYDÉN The gap is too large to say something precise—quite obviously it is at present.

SMYTHIES Let me just make one small point about the use of images. You say two things about images that surprised me and made me feel that you mean something different about images from what I would. You said "images are not constituent items of thought" and "images are not derived from percepts". I however consider images in the phenomenological sense—visual images of various types. Surely it is true that in one sense the visual image derives from a percept, it is a strictly phenomenological derivative. And secondly, obviously in the case of day-dreaming and non-verbal thinking, there is an enormous amount of thinking which is carried out in ordinary visual images. But I suspect that you are using images in a different sense to this ordinary phenomenological sense, is that right?

[9] "The mind-body problem—A new view", *Pyschosomatic Medicine*, 24, 29–45, 1963; "Mind and body re-examined", *Journal of Humanistic Psychology*, 1966, pp. 113–138; and for general consideration of all above points, *Robots, Men and Minds*, New York, Braziller, 1967.

INHELDER   I am being asked two quite different questions: Do images derive from perception and, secondly, are they not constituent elements of thought?

Regarding the first point, no one now upholds this theory in psychology; it was in fact the empiricist theory. If images derived from perception, we should be able to observe behaviour indicating their presence right from birth, whereas we have no proof of their intervention before the advent of the symbolic function, around 1½ years.

Secondly, from the neuropsychological point of view using the EEG or electro-myographic methods, we find in the imagined representations the same motor impulses as in the actual carrying out of actions: these two sorts of facts seem to confirm our hypothesis (put forward by Piaget since 1935) that the image is an interiorized imitation which tends naturally to copy perception but which does not derive from it. (In particular, the laws of imagery are different from those of the retinal after-image.)

As far as the relationship between image and thought is concerned, it goes without saying that in the dream and in the case of the mentally ill we find symbolic thought permeated by images, but in this case the image acts as a symbol whereas language offers to thought a system of signs.[10] You cannot therefore consider the image an element of thought any more than you can consider words constituents of thought: in both cases you must distinguish the signifier and the significate. To make the image an element of thought would bring us back to the empiricist thesis (Taine compared thought to a polypary of images), whilst psychologists from the Würzburg school and Binet in France (at the time when he was just getting over empiricist associationism) showed the independence of thought from images.

Finally, if the oneiric and pathological forms of thought need the image as a support and a symbol, for lack of sufficient abstraction, we must remember that the progress of rational and scientific thought consisted precisely in getting away from the grip of the image, as Gustave Juvet pointed out in his excellent book on the structure of new physical theories: One does not do physics with images any more than one plays chess with one's heart.

SMYTHIES   When you say images—do you mean we *think* in images? Would this be an acceptable way of putting it? I do not agree with

[10] See footnote 5. *Eds.*

the statements you made here. I think images are essentially what many people think with. Some people think only in words, and some people think in images as well, and if you make a very strong statement like that, it is misleading.

KOESTLER   To clarify it, let me put the same question in a more extreme form. Would you admit that there is a type of mentation which consists in manipulations and transformations of visual images—a type of mentation into which verbalizations hardly ever enter? As I said it is put in an extreme form.

INHELDER   It is quite possible, of course, yes.

SMYTHIES   In day-dreaming?

INHELDER   Oh, yes.

SMYTHIES   That is what I was wanting to stress. If you put it your way, one might interpret it differently . . .

THORPE   I would say it is not only possible, it is essential. This I think, brings us back to what Professor Bertalanffy was saying as a biologist—that a high proportion of the behaviour of the higher animals is absolutely inconceivable, without there being something, some internalization, which can be manipulated and changed. Such animals have no language. You can call this thinking or not as you wish, but it seems to me it is legitimate to call it thinking for the time being. In my view thinking does take place in higher animals without language. I would say there is no question whatever about this.

As a matter of fact, all the work of Otto Koehler, on the number perception of birds—number perceived on a prelinguistic level—points, I think conclusively, to the reality of preverbal thinking.

KOESTLER   And there is the classic work of Jacques Hadamard, who circulated a questionnaire among mathematicians, and discovered that they do not think in words and do not think in numbers, but in vague, hazy images . . .

WEISS   How about the work of Ploog and his group? There is a preverbal language, the monkey has a language.

MCNEILL   You cannot escape from the literal sense, if you want to use the term "language" in a well-defined sense.

BERTALANFFY   Animal communication (e.g., bee "language") does not exhibit the criterion of being "freely or arbitrarily created".

MCNEILL   If we do not define our terms we cannot discuss anything. The definition, in current theory, takes the form of a transformational grammar. Now it is interesting that no animal has a transformational grammar, as far as I know. The description of animal communication is sufficiently vague so that it is difficult to characterize it, but at least you can say what it is not.

INHELDER I should like to add two simple remarks to what has already been said:

Firstly, we need to explain why language appears at one specific developmental level and not at another: it is no accident that language acquisition in the child occurs after sensori-motor intelligence has reached its final stage. This seems to indicate firstly that the progressive and systematic coordinations of the sensori-motor schemes play an important part in the formation of language. Secondly, this seems to show that language acquisition is based on cognitive developmental mechanisms and not vice versa: consequently, there is no need to suppose innate linguistic programming, since the self-regulations are sufficient to explain the development of intelligence and its influence over language.

Secondly, we should like to repeat that language is only one of the ways in which the semiotic function reveals itself; in addition, there are symbolic play, deferred imitation, which becomes interiorized in mental imagery, gestural expression, etc. Indeed, when we come to explain the transition from sensori-motor intelligence to thought or representative intelligence, we naturally consider language to be one of the decisive factors. Firstly, however, it is far from being the only one, because it is the semiotic function as a whole which should be invoked and secondly, it is not a necessary factor since deaf-mute children manage to acquire logico-mathematical operations. The study of the development of language would be further advanced, in our opinion, if the dissociation between the semantic and the syntactic aspects, though necessary in linguistic theory, had not gained such importance in psycholinguistic studies.

AFTERTHOUGHTS* by Jean Piaget

To conclude this discussion I shall try to reply to one of the questions which has been raised. The question is really of a more general nature than is at first apparent since it concerns the significance of interdisciplinary relations: how much of what I have actually learned during my training as a biologist and in zoology, which was my first interest, have I actually been able to apply to my psychological studies? In fact I drew three lessons from biology which have never ceased to dominate my thinking.

The first is that all adaptations of the organism, particularly during the development of the phenotype, imply the closest interaction between the organism and the environment; on the one hand, the phenotype, as we have already said, is not simply a copy of the environment, since it constitutes a reaction of the genotype; in addition, one never sees a genotype itself since it is always embodied in phenotypes. When one considers the organism's behaviour on all the levels, (including the highest cognitive levels), one finds the connection between the subject (product of the organism) and the objects (indissoluble sectors of its "environment"): no subject without action on the objects and no objects without a structuration contributed by the subject. In other words, it is in the strictest sense of the word that knowledge is a special case of biological adaptation.

Whence a second analogy. Any biological adaptation implies two poles, by virtue of the interactions we have just mentioned. On the one hand, it is an "accommodation" i.e. (by definition) a temporary or lasting modification of the organism's structures under the influence of external factors. But if it were "nothing but" that, the organism would be solely dependent upon the environment, as Lamarck maintained, and one would only encounter products of accommodation ("accommodats") whose direct heredity (heredity of acquired characteristics) would have to be admitted. Any adaptation, even momentary, implies a complementary pole which, in very general terms, could be called the "assimilation" pole, and which has the task of integrating external factors into the organism's structures; this necessarily implies a continuity between earlier and later structures. This is why any reaction or response is the expression of its continuous structuralization due to the organism as much as it is due to the pressures from the stimuli, the environment. This

* Written after the Symposium.

157

polarity of assimilation and accommodation is found at all levels of cognitive as well as of organic development.

At the sensory-motor level, the baby spends his days accommodating his action schemes to the unforeseen situations which he encounters, but he knows nothing about these situations except by assimilation to his schemes, which are in the process of being built: an object for him is simply something to suck, to handle, to pull, etc. before he becomes able to find it again, locate it, etc. and each of his actions consists of assimilating the information to the schemes he has available, which become continuously richer thanks to the accommodations. At the other extreme, rational thought consists of continuously assimilating reality to concepts or operations which are the instruments of more and more advanced assimilation, whilst experience of reality necessitates a complementary accommodation. The error in empiricism is to have believed in accommodations without assimilations, whilst the fundamental concept of apriorism was based solely on the latter, forgetting that they are constructed and modified through accommodation.

A third analogy is obviously necessary: if biological or cognitive adaptation requires two poles, they both tend towards total harmony by means of successive equilibrations, a harmony which would only be possible through the assimilation of the whole universe and the attainment of a permanent accommodation. If that has been perhaps the dream of certain theoretical thought-systems (Hegel and present-day phenomenology), the reality (and not the dream) of the equilibria and continual re-equilibrations is no less worthy of most careful study because it shows us everywhere the processes of regulation and self-regulation at work. From the embryological regulations, whose fundamental stage Paul Weiss called "reintegration", or from the numerous cybernetic circuits described by Waddington at the heart of his "epigenetic landscape", up to the self-regulations which the study of mental development is continually bringing to light, we find a quite remarkable functional continuity. And if we remember that the thought operations, with their anticipations and retroactions (operational reversibility) also constitute regulations, but that they are "perfect" regulations (with precorrection of errors and not only correction after the event), we are struck by the generality of these vital fundamental processes, whose knowledge is just as indispensable to the psychologist as to the biologist.

REFERENCES

APOSTEL, MAYS, MORF and PIAGET (1957) *Etudes d'épistémologie génétique IV, Les liaisons analytiques et synthétiques dans les comportements du sujet.* P.U.F., Paris.

BLOOMFIELD, L. (1933) *Language*, Holt, New York.

BOWER, T. G. (1966) "The Visual World of Infants", in: *The Scientific American*, 215, pp. 80–92.

———— (1967) "Phenomenal Identity and Form Perception in an Infant", in: *Perception Psychophysics*, 2, pp. 74–76.

BRUNER, J., et al. (1956) *A Study of Thinking*, John Wiley, New York.

CHOMSKY, N. (1959) A review of Skinner's "Verbal Behaviour", in: *Language*, 35, p. 26.

———— (1965) *Aspects of the Theory of Syntax*, M.I.T. Press, Cambridge, Mass.

DOBZHANSKY, T. (1951), *Genetics and the Origin of Species*, John Wiley, New York; (1964, 3rd edition) Columbia University Press.

———— (1960) "Die Ursachen der Evolution", in: *Hundert Jahre Evolutionsforschung*, pp. 32–44, Herberer und Franz Schwanitz, Stuttgart.

———— (1964, 5th print) *Mankind Evolving*, Yale University Press, New Haven.

INHELDER, B. (1965) "Operational thought and symbolic imagery", in: *European Research in Cognitive Development*, Monographs of the Society for Research in Child Development, Serial No. 100, Vol. 20, No. 2, pp. 4–18.

———— (1966) "Développement, régulation et apprentissage", in: *Psychologie et épistémologie génétiques, thémes piagétiens*, Dunod, Paris, pp. 177–188.

———— (with Bovet, M. and Sinclair de Zwart, H.) (1967), "Développement et apprentissage", in: *Revue Suisse de Psychologie pure et appliquée*. Vol. 26, 1, pp. 1–23.

———— (with Sinclair de Zwart, H.) (in press) "Learning and Development" in: *New Directions in Developmental Psychology*, P. Mussen (ed.), Holt, New York.

———— (1969), "Memory and Intelligence", in: *Studies in cognitive development, Essays in honour of Jean Piaget*, Oxford University Press, New York, pp. 337–364.

KATZ, J. and POSTAL, P. (1964) *An Integrated Theory of Linguistic Descriptions*, M.I.T. Press, Cambridge, Mass.

MEHLER, J. and BEVER, T. (1967) "Cognitive Capacity of Very Young Children", in: *Science*, Vol. 158, No. 3797.

MILLER, G. (1962), *Psychology, the Science of Mental Life*, Harper and Row, New York; Hutchinson, London.

PIAGET, J. (1937) *La construction du réel chez l'enfant*. Delachaux et Niestlé, Neuchâtel et Paris.

———— (1954 translation) *The Construction of Reality in the Child*, Basic Books, New York; (1955) Routledge and Kegan Paul, London.

———— (1946), *La formation du symbole chez l'enfant*, Delachaux et Niestlé, Neuchâtel.

———— (1951 translation), *Play, Dreams and Imitation in Childhood*, Norton, New York; Heinemann, London.

———— (1961) *Les mécanismes perceptifs*, P.U.F., Paris; translation (in press): *Perceptual Mechanisms*, Routledge and Kegan Paul, London, and Basic Books, New York.

———— (1966) "Nécessité des recherches comparatives en psychologie génétique", in: *Journal international de psychologie*, 1, pp. 3–13.

———— (1967) *Biologie et connaissance*, Gallimard, Paris.

———— (1968) "Quantification, Conservation and Nativism", in: *Science*, Vol. 162, pp. 976–979.

————(1968) *On the Development of Memory and Identity*, Clark University Press with Barre Publishers, Barre (Mass.).

———— (1968) *Le structuralisme*, Collection Que sais-je?, P.U.F., Paris.

PIAGET, J. and INHELDER, B. (1959) *La genèse des structures logiques élémentaires (classifications et sériations)*, Delachaux et Niestlé, Neuchâtel et Paris.

———— (1963 translation) *The Early Growth of Logic in the Child (Classification and Seriation)*, Routledge and Kegan Paul, London; (1964) Harper, New York.

———— (1966) *L'image mentale chez l'enfant, étude sur le développement des réprésentations imagées*, P.U.F., Paris.

———— (1966) *La psychologie de l'enfant*, Collection Que sais-je?, P.U.F., Paris.

———— (translation, in press), *Child Psychology*, Basic Books, New York.

PIAGET, J., INHELDER, B. and SINCLAIR de ZWART, H. (1968) *Mémoire et intelligence*, P.U.F., Paris.

QUINE, S. (1953) *From a Logical Point of View*, Ch. 11 "Two Dogmas of Empiricism". Harvard Press, Cambridge, Mass.

———— (1956) "In Defence of a Dogma", in: *Phil. Review*, pp. 141–158.

SINCLAIR de ZWART, H. (1967) *Acquisition du language et développement de la pensée*, Dunod, Paris.

———— (in press) "Sensory-motor action-scheme as a preparation for language acquisition", in Ciba Foundation Publication.

WADDINGTON, C. H. (1958) *The Strategy of the Genes*, Allen and Unwin, London.

———— (1961) *The Nature of Life*, Allen and Unwin, London.

On voluntary action and its hierarchical structure*

Jerome S. Bruner

Center for Cognitive Studies, Harvard University

It is quite apparent that many biological systems operate from the outset as hierarchically organized wholes by their very nature. But it is also true that some systems achieve such structure slowly and haltingly. In early human growth, the initially well-organized systems seem to be predominantly of the automatic or "over-controlled" type—as with breathing, swallowing, and initial sucking. With a minimum of initial priming, all three of these are potentiated easily and "go off" in appropriate ways to appropriate stimulation. Moreover, they appear from the start to be embedded smoothly into a larger context of action. A newborn can suck, swallow, and breathe not contemporaneously, but in a fashion that is so beautifully intercalated as to leave little doubt of the role of a pre-established central control system in the production of this synergy.

Indeed, it can properly be argued without unseemly teleology that crucial biological systems requiring little plasticity generally have such pre-established organization, selected precisely to assure regulation of functions crucial for gross survival within fairly narrow limits of environmental variability. They are, in the main, systems involving relatively little information processing and in their most specialized version, are regulated by the autonomic nervous system. Such systems characteristically show little awkwardness during growth.

Obviously, certain general principles of action do not vary from one type of system to the other. Even at the lowest level, it is necessary to distinguish between ex-afference and re-afference. Von

* Read by Blanche Bruner.

161

Holst (1950) put the matter well, "If I shake the branch of a tree, various receptors of my skin and joints produce a re-afference, but if I place my hand on a branch shaken by the wind, the stimuli of the same receptors produce an ex-afference. We can see that this distinction has nothing to do with the difference between the so-called proprio- and extero-receptors. The same receptor can serve both the re- and ex-afference. The CNS must, however, possess the ability to distinguish one from the other. . . . Stimuli resulting from its own movements must not be interpreted as movements of the environment." And, of course, it is to the credit of von Holst and Mittelstädt (1950), of Sperry (1950), and others to have demonstrated that even the primitive nervous system of the fish can in fact distinguish the stimulation produced on a receptor by the self-induced action and by environmentally induced change.

But plainly, there are some systems of action in which the re-afference system, its capacity for control by corollary discharge to related systems, and therefore its "skill" is virtually zero at the outset. The visually guided use of the hands in human beings is somewhat like this and it is particularly interesting to examine its growth because it reaches such delicate virtuosity after so awkward a start. It appears to be based on the development of programmes of action that are plainly quite different from those, for example, that characterize such a system as eye-movements which, virtually from the start, are smooth and controlled by reafference principles and corollary discharge. Obviously the CNS can distinguish from the start the movement of objects across the stationary retinal field from the movement of objects across this field produced by the eye itself moving—the former producing an optokinetic pursuit, the latter not. The hand, on the contrary, cannot be so distinguished and an infant is capable of startling himself by a quick swipe of the hand into the central visual field. The corollary discharge from the effector is plainly not getting a wide distribution in the case of the hand movement, but *is* in the case of the eye-movement.

We know extraordinarily little about systems that acquire their organization in contrast to those that have much of it *built in* from the start. I believe that it is of great importance to examine the former type of system with especial care, for it is in such systems that one finds maximum plasticity, a maximum modelling of the most variable features of the environment, and a maximum amount of information processing. These are the systems of action that, I believe, become *generative* in the linguistic sense—capable of being employed in a variety of contexts by the use of a minimum set of elementary operations combined and recombined by rule-governed programmes. I believe that it is the open quality of these systems that allows for their incorporation of prosthetic devices and tools on the one side, and of language as a medium for programming action on the other.

So what I should like to do in these pages is to explore some of the features of organization of a couple of highly plastic, voluntary systems of action that we have been studying with a view to learning something about their growth and structure. My plan is simple. I should like to begin with some behaviour that initially seems to be quite automatic and reflexive and is gradually converted to voluntary control. I have chosen the phenomenon of nutritive and comfort-giving sucking as the vehicle for this discussion. The exercise will, perhaps, give us an idea of what happens in the transition from a low to a high information processing system. Then we shall sample some episodes in the growth of visually guided manipulation of objects better to appreciate how a programme of skilled action may be put together from the outset. Our progress will be from mouth to hand rather than from hand to mouth!

Consider now the growth and integration of human nutritional activity during the early months of human life. Let me begin by reminding you of the facts of sucking and the role of the mouth in early infancy.

Sucking serves several functions. It can be observed as early as the third gestational month (Peiper, 1963). Though it is instinctive, it requires some priming to get started in the neonate, as we know from the work of Gunther (1961), and if not early exercised, may become difficult to evoke. At birth the infant may have certain difficulties beginning to suck, grinding jaws back and forth, missing the pressure, etc. Once he has "connected", so to speak, and I have observed as many as four five-second periods of trying before he does, the sucking is immediately highly expert.

In addition to its role in feeding, sucking serves non-nutritively, according to the studies of Jensen (1932) and Wolff and Simmons (1967) to fulfil an antidistractant or analgesic function or both. Pin pricks and tickling of the face by a feather increase·the sucking rate or lead to initiation of sucking. Indeed, it is now standard practice in some hospitals to carry out circumcision while the child is sucking on a favoured pacifier. Sucking also inhibits the newborn infant's level of general activity which may help relieve distress (Kessen and Leutzendorff, 1963), with effective suckers showing the greatest quietening (Kessen, 1967). A variety of studies indicate that infants suck non-nutritively at about a constant individual rate of 40–80 a minute whether hungry or not (Balint, 1948; Bridger, 1962). The third function of mouthing is for exploration, and its importance in the organization of behaviour will concern us later.

The mouth, from the start, is embedded functionally in several systems. For one thing, it is the aiming point in the head-turning system. A touch to the edge of the jaw or the side of the cheek will produce a rooting reflex with mouth moved towards the touch. It is also mapped into the arm-and-trunk system, as indicated by the Babkin and the palmomental reflexes; pressing the palm will produce mouth movements in the newborn, as will pressure on the ball at the base of the thumb produce contractions in the mentalis muscle of the jaw.

There is little question either in detail or in general functional terms that sucking is part of a hierarchical structure or, indeed, of several hierarchically organized systems. At the outset, it has about it a captive quality within a digestive system that renders it quite inseparable and rigid; the sucking of the premature, for example, being part of a general digestive activity of stomach and gut that can be observed easily through the almost translucent abdominal wall not yet obstructed by the fatty omentum. Sucking can be observed to be one part of a general action of the digestive system, and a quite inseparable part. In the full-term infant as well, the role of sucking is crucial for maintaining a proper motility in the stomach.

But, over and beyond this autonomic and reflexive embedding, sucking is also destined to become a highly voluntary activity. It plainly becomes detached from the obligatory and compulsive sucking-feeding-comfort cycle and, in time, becomes part of such symbolic actions as lifting iced tea from glass to mouth through a straw or partaking in such physiologically exotic activities as the smoking of marijuana, tobacco, or opium. We have been much interested in this transition.

Consider a few observations. The first has to do with the nature of the flexibility and voluntary control that gradually permeates this originally quite reflexive system of action. While, at the outset, sucking has a very compulsive property, closer examination of it shows in what measure it is also, even in the first day of life, quite sensitive to changes in the environment that relates to it. A word about how one records sucking. Figure 1 provides a diagrammatic sketch of this system, which measures suctioning pressure on a polygraph, as well as the positive pressure of mouth and pressing the nipple with the gums and tongue. At the same time we are enabled to deliver milk directly through the nipple to the baby in response either to positive or negative sucking or to some combination, and with whatever contingency we choose. For not only does a record of the baby's sucking register on the polygraph, but also on a programming device that can be set to activate a milk-pulsing system each time the baby sucks in a specified way, every other time, etc., or at specified intervals of time after the baby has sucked. The device builds upon similar devices that have been used in recent years by Kron, Stein, and Goddard (1963) and by Sameroff (1965). Complicated though the instrument may seem in the context of infancy as lived, it is quite indistinguishable from an ordinary nursing bottle to the infant and mother.

MILK DELIVERY / SUCKING RECORDING METHOD

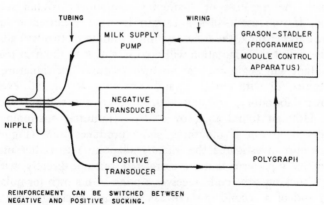

FIGURE 1 *Diagram of sucking apparatus.*

When closely examined, sucking appears to have two modes of control, one quite automatic, rhythmic, and relatively invariable; the other very much more open to the variations of the environment.

The best way of picturing the automatic component is to show you the beautiful sucking record of a sleeping infant of four weeks of post-natal age—physiologically, I should say, closer to one week, since this baby was three weeks premature (Figure 2). This splendid performance went on for minutes on end, accompanied by regular breathing and swallowing.

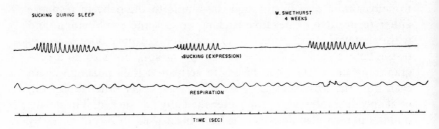

FIGURE 2   *Sucking pattern of a sleeping infant.*

Yet, for all this automatic quality of sucking, it is also capable from the start of prodigies of adaptation. Sameroff (1965) has shown that when milk is delivered exclusively for mouthing, the neonate will diminish the negative or suctioning component of his sucking. Indeed, if one establishes a certain level of required mouthing pressure to get milk, the infant within a minute or two will adapt to that level. But the adaptation will not carry over to the next feeding. The infant will begin anew at his own "natural" or signature level of pressure (or with mouthing and suctioning at original level). In our own laboratory, working with children a month of age and older, Hillman found that, over fifteen minutes, suctioning will virtually drop out if mouthing alone produces milk (Figure 3). He has also investigated the infant's adaptation to rather unusual milk delivery patterns that are correlated, but not directly, with the infant's sucking—the child receives a pulse of his own formula milk at the end of a second if there has been any sucking during that second, or for every other suck or every third suck. The learning that ensues is very interesting indeed, being much more akin to strategy-learning than to specific response acquisition. When a pulse of milk is delivered each second or every two seconds in which a suck has occurred, the effect on some babies is to shorten their sucking bursts—while the pauses between lengthen.

CHANGE IN NEGATIVE SUCKING
WHEN POSITIVE REINFORCED

JEFFREY STAFFORD
17 WEEKS

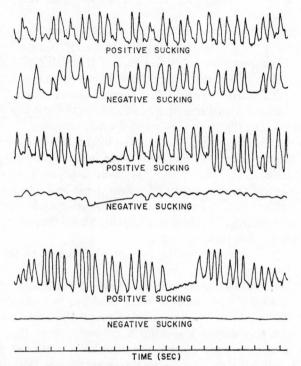

FIGURE 3 *Disruption of negative sucking when positive sucking alone produces milk.*

But what is especially interesting for our present purposes is that when the requirements of the task exceed the infant's capacity to adapt, the pattern goes back to the full-fledged, automatized sucking pattern with no account taken of the nature of the environmental situation. If, for example, the adaptation has involved the dropping out of suctioning and the modulation of the mouthing pattern, frustration will induce full rhythmic activity at a signature rate, involving both mouthing and suctioning again. It is as if the automatic system provides something akin to a shield or emergency system.

Can we say then that voluntary control is a modulation of an initially reflex system, a modulation that requires confirming feed-back to continue? I think not. I believe that reflex sucking is integral and is differently controlled and paced. We shall return to this issue in a moment.

Consider another route to the capture of sucking by a voluntary system; its intercalation with another system. "Normal" sucking, we know from observations by Kessen (1967) and Wolff (1968), can be observed in brain stem infants. What does it take to intercalate this primitive activity with such a higher-order information-pro-cessing system as visual scanning? Let me report some observations, with the warning that it is a report of work in progress.

At birth, and for some days after, the infant sucks with eyes tight shut as already noted. If the infant looks or tracks, sucking is dis-rupted—indeed, the disruption has been used as a measure of attending, as in the classic study by Bronshtein and Petrova (1967). With the three- to five-week-old baby, the eyes may be open while the infant is sucking, but there is a high likelihood that when fixation or tracking occurs, sucking stops.

There is a new pattern usually by nine weeks based on the pheno-menon of sucking in bursts, interspersed by pauses. The child now sucks in bursts, and looks during pauses. He remains generally oriented toward the stimulation while sucking, but not fixated or "locked on" while sucking goes on. Around this age, a stimulus change occurring during a sucking burst will disrupt the burst or bring it to an end sooner. But if the stimulus is presented during a pause, it will have no effect on subsequent bursts (Figure 4). It appears that the pauses are being used to process information, a matter that we shall wish to investigate much more thoroughly before letting it rest at that.

Finally, usually before four months and often as early as two months, the baby appears to be able to suck and look at once. But when one examines the sucking record, it turns out not to be the case. For now, the act of looking inhibits suctioning, while mouthing or positive sucking goes right on through, though with reduced amplitude (Figure 5). We refer to this phenomenon as *place holding*, maintenance of the structure of a more inclusive act while the various parts of it are separately executed. Once the infant has inspected and has become habituated to the object that has caught his regard, the suctioning returns.

So we may say that the integration of sucking and looking goes

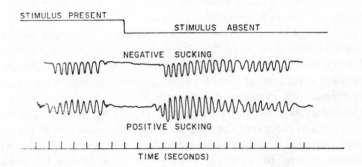

FIGURE 4 *Absence of any disruption of sucking in an eight-week-old infant when stimulus disappears during a pause.*

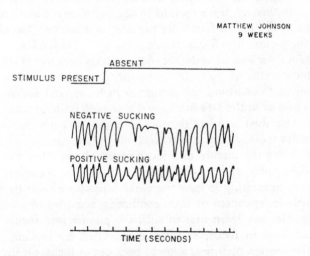

FIGURE 5 *Positive sucking being used as "place-holder" during disruption of negative sucking by a stimulus.*

through at least three stages: (1) initial *suppression* of one by the other, (2) an arrangement whereby there can be a *succession* of the two without interference, (3) and finally a *place-holding operation* where the two can hold their places in a larger act. Note that the achievement of intercalation of this kind requires that the autonomy of sucking should no longer be complete, that it should become part of a new programme of action. Note also that this new system of action is quite separate from the original systemic embedding of sucking in the feeding-comfort complex.

This brings us directly to a final experiment concerned with the voluntarization of sucking, this time sucking in the interest of a quite arbitrary goal. We can properly argue that one of the features of voluntary control of an action system is the degree to which it can be utilized as a means to a new end. We had been impressed by experiments conducted by Siqueland (1968) at Brown indicating that infants of three months were quite capable of sucking to increase the illumination of a picture on a backlighted screen in an otherwise darkened room. In an experiment in progress in our laboratory, Kalnins has altered this procedure in one crucial respect to assure that what was involved was not the comfort of the young child provided by a lighted environment. Her infants, varying from one month through three months in age, are shown a motion picture that is initially out of focus. By sucking on a pacifier, the child can bring the picture into focus. If this sucking should fall below the rate of one suck per two seconds, the picture starts back out of focus. The brightness remains virtually constant throughout. In a second experimental condition, the picture is in focus, and sucking drives it out of focus at the rate mentioned above. Refraining from sucking at the prescribed rate lets the picture come back into focus.

First let me say that a four-week-old infant can in fact learn to suck to bring the picture into focus and to desist when his sucking blurs the display. Infants plainly will work for visual clarity. What is especially interesting is how the child learns to *coordinate* the two ordinarily independent or even conflicting activities of sucking and looking. He may learn first to suck the picture into focus, but the moment it is in focus, sucking is inhibited by looking and the attractive motion picture is allowed back out of focus. Or the picture may be in focus and the irresistible pacifier leads him to suck, which drives the picture into a fog. Gradually, the amount of time during which he can suck and look or desist from sucking increases. The child seems to be learning not so much a *specific* response, but rather a hierarchically organized, adaptive strategy of responses. He shows a quite individual pattern in achieving this competence, no mere matter of gradual, incremental change. One nine-week-old keeps a picture in focus by a series of short bursts (Figure 6); another achieves his ends by a long sustained burst that holds the focus (Figure 7); and yet another manages the task of shifting his approach to a mouthing pattern (Figure 8). Or compare these two infants dealing with the inhibition of sucking that is required to keep the picture in focus. The first sucks compulsively while looking away (Figure 9), then stops sucking and looks. The second, to the accompaniment of much wriggling, inhibits sucking almost totally, inspecting the picture the while (Figure 10).

Let me draw a few simple conclusions from this series of experiments. The first and most obvious is that there are discernible steps in the process of organization for voluntary control. What Piaget (1952) years ago referred to as the *structure d'ensemble* of an operation does not emerge in one great leap forward but in several smaller ones. The second feature of the genesis of organization is inhibition as a resolution of conflict between systems, a matter that Sherrington (1906) understood particularly clearly. This is well illustrated by the pattern of succession already noted. Finally, the establishment of higher order organization involves, in the most general terms,

Figures 6, 7, 8, 9, 10 show individual strategies of four- to ten-week-olds for sucking and looking at a picture.

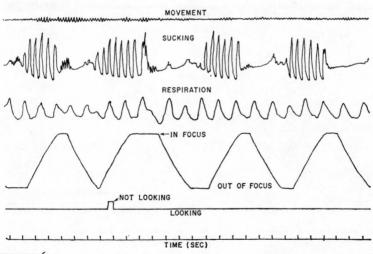

SUCKING INTO FOCUS BY MEANS OF SHORT
FREQUENT BURSTS OF SUCKING

E. DES MAISONS
9 WEEKS

MOVEMENT

SUCKING

RESPIRATION

←─ IN FOCUS

OUT OF FOCUS

NOT LOOKING

LOOKING

TIME (SEC)

FIGURE 6

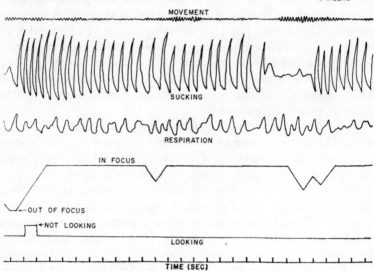

FIGURE 7

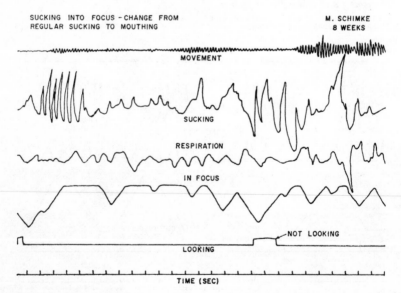

FIGURE 8

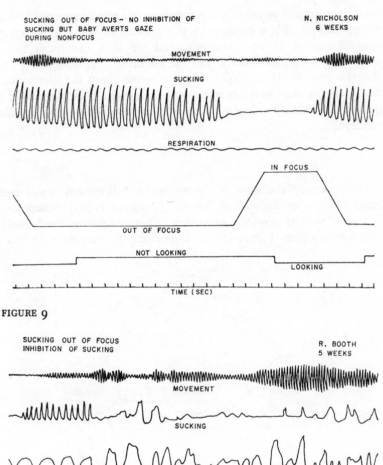

FIGURE 9

FIGURE 10

what linguists call *privilege of occurrence*—a place in a sequence where an act can fit. Place-holding in the present instance is a highly simplified form of this principle, and we shall encounter more interesting examples shortly. Let me comment, before moving on, that there is something strikingly non-specific about the "response" that emerges once new organization is achieved: it is more as if a generic way of responding is being learned rather than some specific reaction.

Consider now the issue of the organized skill system of eye and hand. The formulations of Nicolai Bernstein (1967) concerning voluntary activity are very suggestive. They are summarized in the following diagram (Figure 11). Note that *activity* contrasts with mere

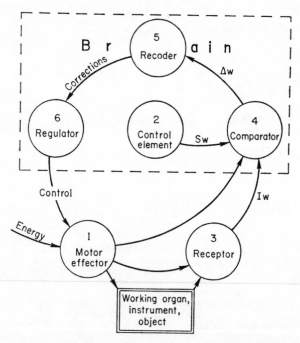

FIGURE 11 *N. Bernstein's model for a system capable of voluntary activity directed toward objects or states of the environment. (From Bernstein, 1697.)*

movement in that the former requires the coordination and regulation of the latter in the *attainment of some particular objective*. A ball is to be thrown a certain distance and has a certain weight, or a screwdriver to be turned requires the application through the hand and arm of a certain torque. To quote from Bernstein (1967):

> All systems that are self-regulating for any given parameter, constant or variable, must incorporate the following elements as minimum requirements:
>
> (1) *effector* (motor) activity, which is to be regulated along the given parameter,
>
> (2) *a control element* which conveys to the system in one way or another the *required value* of the parameter which is to be regulated,
>
> (3) *a receptor* which perceives the factual course of the *value* of the parameter and signals it by some means to
>
> (4) *a comparator device* which perceives the discrepancy between the *factual* and *required* values with its magnitude and sign;
>
> (5) an apparatus which *recodes* the data provided by the comparator device into correctional impulses which are transmitted by feedback linkages to
>
> (6) *a regulator* which controls the function of the *effector* along the given parameter (pp. 128–129).

For Bernstein, the achievement of control always involves a reduction of or "mastery" over degrees of freedom in the action-system being regulated. There are joints and tendons in fingers, wrists, elbows, shoulders, and trunk that can operate independently of each other. A hammer or a screwdriver or a thrown ball can slip this way or that.

I shall argue in what follows that the mastery of intelligent, visually-guided manipulation in infancy and childhood involves a cycle of brute restriction of forms of movement and of programmatic skill formation within the limits of that restriction, with skill moving to a next step only when restriction is altered. Any *given* programme of skilled voluntary action is gradually consolidated within its own restrictions. Its consolidation is signalled by the well-known plateau in the learning curve. Progress points in the infant's development are qualitative rather than quantitative changes of skill. These involve not consolidation but the formulation of new strategies or programmes of action which in turn must be consolidated. Each new programme of action involves an increment or actuation of degrees of freedom. The process, moreover, continues throughout life. The difference between "good skiing" and "bad skiing" is, alas, qualitative, not quantitative.

Now return to Bernstein's diagram. What it lacks is some specifications about the nature of programmes in the so-called command system. These must be generative in several ways. They must contain equivalence rules concerning substitutable types of acts. They must specify sequence: *types* of acts to be carried out in certain privileged orders. They must specify delay procedures and rules for bracketing in the event of encountering difficulty. As Piaget (1954) would also rightly insist, the programmes must be premised on some stored and/or innate model of space and objects.

But before such programmatic voluntary action can get under way, I believe that there must first be a recession of reflex control over the acts that are governed by a programme. One sees this recession from one month forward, with the emergence of athetoid, diffuse, undifferentiated action of the infant's arms, hands, and trunk when visual objects approach his "participatory" space of about 15 inches away and closer. There are also present the beautifully precise reflex patterns so carefully described by Twitchell (1965)—shoulder-traction grasping, the touch evoked grasp reflex, touch evoked avoidance-withdrawal, pronating and supinating groping without visual guidance. But in no sense are these the paving blocks from which skilled programmes of visually guided action are constructed. The beginning of skill is diffusely organized awkwardness guided by a small number of directional specifications.

Diffuse awkwardness is then supplanted by stiff awkwardness. There is, for example, a restriction in the fluid action of joints—fingers are now spread wide, elbow rigid. The midline becomes dominant. Reaching more likely occurs when head and trunk are in line at the midline and in line with a near object. Eye closing and gaze aversion are the reactions to troubles in reaching. The system gets stiff, but it gets vastly more effective. And when it does so, when it becomes "reinforced", it then changes. It is this that led R. W. White (1959) to postulate a competence motive as crucial to growth. Reaching, in keeping with this stiffness, initially serves the retrieval of objects to the mouth. In time it serves other goals—banging, dropping, etc. When this multipurpose reaching gets started there occurs also a modularization of the components of the act—reach, hand-close, lift, retrieve, etc. The time and effort per component approach uniformity. By eight or nine months, a reach takes about the same time for near or for far objects, the lift of objects is about equal in time whether the object is heavy or light, one-handed or two-handed, etc. Does this modularization permit more predictable incorporation of acts into different plans? I think so, although it is hard still to prove it.

Consider the infant of five or six months, reaching now with hands wide spread open and mouth wide open. The open mouth during this early reaching keeps the terminus of the act in evidence during the running off of the component parts. In Lashley's terms (1951), it maintains an "atemporal" organization through the sequence of the act. The rigidly opened hand is a measure against intrusive, anticipatory fist-closing. As with so much early development, processes that later become internal have initially an external and motoric representation.

We have done close analyses of a seven-month-old child coping with a cup for the first time. It is a film record shot at fifty frames per second.

The child's cup-using behaviour seems more playful than purposeful. In play, ends are altered to suit means, rather than means being altered to achieve an end held constant, as in problem solving. Thorndike (1913) long ago and others since believe that the child's early behaviour is shaped by trial and error, but that is surely an adult-centered view. Trial and error implies the capacity to hold an end constant while varying means. The segments in which this is possible are very short in duration for the child. What thwarts is distraction, not error. The placeholding techniques only slowly extend in time. And new sequences are only slowly established with the aid of what Miller, Galanter and Pribram (1960) described as EMD's—external memory devices=strategically positioned external place holders for the task at hand.

Let me state a few conclusions very briefly in order to relate these matters to the main theme set for this Symposium.

I have tried to sketch out by examples some of the ways in which man achieves adaptive, voluntary control of his interaction with the environment. In doing so, I have concentrated on the initially helpless, reflex-protected infant and his various vicissitudes en route to becoming a strategist and problem-solver. I have purposely left out of account the central role of language and of its internalization in achieving this end, since I would like to argue strongly that there are many, many species-specific features of human behaviour that precede the acquisition of language proper. Besides, David McNeill will treat these matters better than I can.

There is adaptive, generic, intelligent organization of behaviour at every age. It is attained neither by an unfolding of mysterious inner structures nor by the gradual accretion of shaping through reinforcement. It is, instead, uneven, imperfect, spasmodic, incomplete. The programmes that guide voluntary action do not achieve full control. Even spoken language, treated by linguists as if based upon a totally mastered competence, is in fact a faltering performance and in its internalized version is even more loosely controlled. Skinner attributes such order as there is to control of behaviour by orderly stimuli in the world. Piaget sees more order than Skinner, so he attributes it to the inherent logical structures of mind, accommodating only slightly to the lessons of the environment. Skinner's solution has some of the monotony of nature. But it is much more fortuitous than nature could afford to be. The generativeness in behaviour that makes it enonomical vanishes at Skinner's touch—even from language. With Piaget it is the opposite. The order and generativeness are all there from the start, like the shape of the mollusc, ready to eat so long as nutriment enters the system.

I have no complaint either against universality in human intellectual functioning, or against the shaping power of the environment. But I do have a complaint about theories that opt for internal universality in surfeit or for environmental shaping as the mould of such uniformity. Human beings are the most awkward species on earth, the most uneven in development, the most beset by obstacles that are not intrinsic to the task. I urge a new functional analysis of what it takes to grow up intelligent—a job description of growing up. It is no obvious matter. It does not suffice to say that the central issue is the question of how we achieve deeper and deeper power over the idea of invariance across transformations in surface properties, for that leaves the issue of action and control unattended. Nor does it do to say that internal processes are so much phantasmagorical nonsense and action selected by reinforcement is the real thing. That leaves order to fortuity. Rather, the questions are more mixed. How do strategies for coping with complexity develop? How are limited resources used for achieving goals the means to which vary? How do we sustain action over sufficiently long trains of striving? How do we recognize (when we do) invariant features of events so that we are not constantly ensnared in learning? I suggest we proceed not with a bang, not with a whimper, but with a good close look.

DISCUSSION

KOESTLER   Before the real discussion starts, a technical question: you mentioned the Miller-Galanter-Pribram book on *Plans and the Structure of Behavior*. How does your hierarchy scheme compare to their TOTE—Test Operate Test Exit—scheme?

BRUNER*   It seems to me very similar, but I have not analysed the difference; it certainly made me think of TOTE.

MCNEILL   I think they are just two variations on the same theme of feedback.

WEISS   I would like to contribute an experiment which has been somewhat lost sight of, but which is much to the point under discussion: how much in behavioural performance is rigidly performed and how much of it is plastic. In 1940 we constructed the first electro-myograph; we could get a really good record of twelve points on the same muscle, and we went on to study twenty-two cases with partial paralysis, some of them spastics, most of them polio cases. We selected a number of cases where the quadriceps was nearly, but not entirely, out of action, so they had pretty good flexors of the knee, but could not extend the leg. Now a classical surgical procedure is to take the biceps out of its flexor group and transplant it to the tendon of the quadriceps, and it is well known that these people learn to develop a perfectly good functional walking pattern. As a matter of fact, one girl who had it in both her legs became an expert dancer. We studied these cases with the electro-myograph before the operation, immediately afterwards, and later on through the so-called learning phase. What we found is that the transplanted biceps at first participates both in flexor and extensor operations. Then out of this generalized activity the patient—unconsciously of course—selects only the extensor phase and so the biceps becomes perfectly coordinated with what residue there is of the quadriceps; it acts as an extensor and drops its previous association with the flexors. We studied all kinds of postures, normal walking, treadmill, etc. Then the girl who had become a dancer complained to us that when she gets tired during dancing she suddenly jack-knifes right on the dance-floor. After that we studied several cases when their attention was distracted by drugs, alcohol, and so on. Then it became clear to us that the automatic patterns of walking—inward and outward rotation—that these basic performances had not at all been

* Mrs Blanche Bruner, wife and collaborater of Jerome Bruner, *Eds.*

modified. They were still there, but the reflex reactions had been overlaid or superseded by a second system; they had been temporarily superseded, not remodelled, by a cortico-spinal operation, a second apparatus which had entered into action and which presented the plastic reaction.

And the same interpretation applies to all cases of the surgical transplantation of nerves—for instance, in cases of facial paralysis, where you use the deltoid as a substitute, there are of course at first simultaneous contractions in the shoulder and the face. Then, gradually the shoulder contraction drops out because the patient had learned to use the antagonist to neutralize the deltoid's action in the shoulder. Then we had a girl with a reduplicated hand. In such cases, as in amphibians, the identical fingers always move in synchrony. But we did teach her somehow to dissociate these motions and to use the hand effectively. But this was entirely due to her having learned to counteract the symmetrical movements, and it again stopped immediately when you distracted her attention. My point is that these experiments are very close to the contents of your presentation. Here you have a lower centre operating according to a rigidly programmed scheme like an electric piano, and there is not anything you can do about that stencil. And then you can have on top of it a different system which works with much greater plasticity.

BRUNER   The point is that it is overlaid, not obliterated.

KETY   I am not sure how much you and Jerry are willing to attribute to reinforcement. There was one point there where reinforcement was not the important, compelling, overriding consideration, but certainly reinforcement does play an important role.

BRUNER   And speaking of reinforcement, success in carrying out the activity itself, without regard to pleasant or extrinsic consequences can be the reinforcement.

KETY   You will agree that reinforcement still plays an important role?

BRUNER   Oh yes, if you are specific about what you mean in each instance by reinforcement.

SMYTHIES   Can I categorize, perhaps in a naïve way by making a suggestion for the essential difference between the Piaget, Bruner, and Skinner position. You can regard the brain at birth as not having any logical programme or any programming in logic how it should work. It has to do two things: one is that its whole logical programme— how to deal with information that comes in, what kind of operations to carry out on this inflow—has to be learned. And once these basic rules have been learned, it can then use these rules to work on the actual sensory information that comes in all the time. So you have to programme first the logical operations of the brain, because it is not born with any, or very few, and secondly the programme, the data, for it to operate on. Now perhaps Piaget's theory is that this development of the logical programme is due to the unfolding of ontogenetic mechanisms at birth; such as myelination for example—myelination takes place very late. During the first year the brain exhibits continual development, in an anatomical sense, of extra mechanisms being put into operation so that more and more complicated logical things can be done. Perhaps in the Bruner theory much depends on learning—much more depends on learning and less on anatomy. And it seems to me Skinner denied that the first process takes place at all, i.e., programming is not necessary, all you have is a simple reflex which he works on. Would that be a way of looking at it?

BRUNER   Yes, although I think it overstates the case.

SMYTHIES   Well, the point at issue is: why can't the brain operate complex logical programmes when it starts? Is it just that it does not have the necessary mechanisms, e.g., myelination is not yet complete, the midbrain is working, but the cortex really is not working? Or is it that you cannot develop your complex logical operations unless you have been through these simple ones that we have seen on the board here? It seems to me probably the case is that both are operating. But I think the emphasis is slightly different.

BRUNER   (to Inhelder): I would like to ask a question about your experiment. After showing the child that figure in the upper right-hand corner of the blackboard, representing a path, you say to him: "Draw a path so that you will have the same distance to walk as I do on my path." Isn't it possible that he measures the distance between the ends of your path in terms of his own body image and the length of one of his own steps, and therefore draws a line, or "path", of his own with the ends about the same distance apart as yours? Would such a response be proof that he is not yet capable of a different "more logical" response? Is it possible that miniature trees drawn along the borders of your path might free him from his own body-image version, so that he would make a logical abstraction with equal conviction?

INHELDER   We must try it. But just now I am fascinated by your ingenious experiments on sucking, vision and prehension which show clearly that from a very early age the infant exhibits mechanisms, which are originally reflexes, outside the context of innate releasing mechanisms and without specific external reinforcement.

Indeed, "sucking in the void" as Piaget has called it (Piaget, 1936), becomes very soon a primary circular reaction and as such it acts as a particularly important cognitive scheme. This scheme becomes quickly generalized: the infant does not only suck the teat of the breast or the bottle but also all kinds of objects which happen to touch his lips as well as his own fingers. This scheme leads necessarily to discriminations and recognitions which progressively become differentiated. Such a scheme is the source of, on the one hand, cognitive adaptation with its double process of assimilation and accommodation and, on the other hand, of functional play in which assimilation dominates and which is directed towards functional pleasure (*Funktionslust,* according to Karl Bühler).

Bruner shows in a striking manner that the mechanisms of sucking, vision and prehension are at first under the influence of antagonistic inhibitory mechanisms. Parallel to this, Piaget has shown that they become progressively coordinated. Around the age of $4\frac{1}{2}$ months the infant becomes capable of grasping what he sees and looking at what he grasps, etc. These coordinations lead to a first unification of near space which at this level is still only apprehended in relation to the infant's own acts. Objects only exist as objects to be sucked, touched and looked at.

BRUNER I think so, yes. . . . I think I understand what you are saying. I think that is true, but we think that the development of strategy starts at birth, that it does not depend upon some prior learning to make it possible.

INHELDER I feel that Bruner wrongly attributes to Piaget a one-sided genetic explanation which he has never given and which points to a rather serious misunderstanding. While Piaget (1936) has always shown the inadequacies of the empiricist and Lamarckian explanations, he has just as vehemently and consistently denounced the errors arising from *a priorist* preformism. In his book *Biologie et Connaissance* (1967) he develops a position which goes beyond such a dichotomy and shows the existence of a tertium; the original aspect of this lies in the fact that Piaget seeks to explain the *necessary* character of the products of cognitive development without recourse to predetermination: he tries to show that the concept of a process of

G

construction which is in itself the source of new realities (not pre-formed) can account for this necessary character. In this constructive process it is the continuous equilibrations, i.e., the self-regulations, which play the most important part.

Thus in their final form, the logical structures have a necessary character without being predetermined. Despite his interest in molluscs, Piaget has never believed in permanent structures, even in zoology. The concept of equilibration, and Bruner has not really grasped this, explains the necessity of a form of knowledge without this form being determined in advance.

It is the functioning of the self-regulatory mechanisms and not their programming that is innate in the development of cognitive functions. Of course, there is a certain continuity (all relative) between the hereditary and the acquired programming, but it is functional and not structural. Continuity is not the same as identity, because the acquired structures are far richer than the innate structures. To say that the self-regulatory mechanisms are of endogenous origin in no way prejudges the nature of the structural and in particular logical programming. In fact, this programming is the result of a construction in which both exogenous and endogenous factors play a part.

BERTALANFFY I think you can put this in terms of classical philosophy. Locke—and also Skinner—says: *Nihil est in intellectu quod non antea fuerit in sensu.* Leibniz corrected it: *Nihil est in intellectu quod antea fuerit in sensu, nisi intellectus ipse.* And Piaget would say: *intellectus* has self-regulating capacities, and is not a fixed but a developing entity.

KOESTLER That intermediary level is essential.

INHELDER I think so, yes. It has to be worked out more and more.

KOESTLER There is a potential that is innate, but the programming is self-regulating.

INHELDER And is self-regulating all the time if you take it from the developmental point of view, which is ours, through all kinds of exchanges of course, with the environment.

MCNEILL I would like to mention an example in the development of phonology. If you make records of children's vocalizations during the first year of life, you find the gradual upswelling of different phonetic types; it becomes quite vast towards the end of the babbling period at 11 months. This continues, but at around 12 months—it can vary with the child—speech as opposed to babbling makes its first appearance. One indication of this fact is that when the child attempts to say something, that is, when he attempts to exert voluntary control over his articulatory muscles, the entire thing collapses into a simple system of two sounds. So there arises then a contrast between vocalizations intended as messages, which have an extremely tiny repertory, and vocalizations not intended to communicate anything, which have a very large repertory. This is rather interesting, since it seems to be an example of a general capacity for volition seizing upon this new domain of vocalization. There is another interesting feature: it is always the same two sounds that children select the world over—"p" and "a", not "b" and "a", for example. "Mama" is a second step, according to Jakobson (1960). The reason so many languages of the world have Mama and Papa presumably is because they exploit the facts of children's earliest phonemic development. You might interpret this as the result of general principles of perceptual development. The interesting thing, the special thing about "p"and "a" is that they are respectively the optimum consonant and the optimum vowel. When you say "p", your mouth is open like a funnel pointing backwards, and when you say "a", your mouth is like a funnel pointing forward. If you think the child is attempting to work out a system of oppositional contrasts, these sounds are a natural place to begin. That seems much more like a specific linguistic universal, which is distinct from the way it is seized on by a perhaps general voluntary programme. The second limits the repertory to a few sounds at first; the first determines what the repertory is.

WEISS How much carry-over is there from what is known about the development of animal behaviour into this field? This is a question addressed essentially to Thorpe. Is there, for instance, a really clear counterpart to what we have just heard about the development of the human baby in the innate capacity for flying in birds, with improvement with experience. Isn't that a similar case, or is that a different story?

THORPE I do not know that there is any very good evidence of what you might call supersession of reflex control. But it is clear that fine adjustments are certainly added to the innate equipment as a result of experience. Thus a bird can perform its first flying movements successfully—even if its wings have been restrained until long past the stage when, in unrestricted birds, you see what look to be the first steps in the practice of flying, namely the flapping of the wings. This confinement, this inhibition of "practice movements", does not seem to halt the development of the action at all. You can restrain birds in such a manner for a fortnight or longer, and when released one finds the birds are at the same stage in the process of development of the basic flight actions as a bird which has had till then what we hitherto regarded as the "advantage" of practice. But although in this sense learning is not necessary to achieve the ability to flap the wings in the right manner, birds when they start to fly, do quickly learn all sorts of other things which involve fine adjustment. They learn of course, to take off in the wind which is one of the elementary lessons of flying. A small bird will also learn to take off from some sort of vantage point. Again, they learn how to control their turning in the air; and other niceties of aero-dynamic adjustment.

WEISS   Gliding, taking advantage of currents?

THORPE   Yes, gliding and landing too. A young bird does a very bad job of its first attempts at landing. It may tend to fall over, it is clumsy, and so on. All that sort of thing has to be learned. But as I said you can restrict the wing movements for a very long time without what appears to be any sort of breakdown of the reflex control mechanism. Although it is possible that this itself could also be destroyed in the end by too long and too complete deprivation.

WEISS   That is very like raising Amphibia in anaesthesia. So they could not try, and the first time they come out they are perfect.

KOESTLER   (to Thorpe): An even more striking example is your description of bird-song developing in birds which were prevented from imitating it, and then the "filling in" process—but you can explain that better.

THORPE   Yes, if you rear young chaffinches, say, in a soundproof chamber, or if you have them deafened you get essentially the same basic pattern of vocalization being produced by all of them. But these birds only develop their song to a certain point, whereas the same species under natural circumstances seems to have its attention directed, perhaps innately, to the kind of sound which its species normally makes. So, if you have the bird in a soundproof room then the song that emerges is what you might call the most elementary schema of what is normally regarded as the song of the species; it has about the right pitch, about the right number of notes per minute and so forth; and that is all there is to it. For all the fine adjustments seem to be put in by imitative learning at a later stage. But here again there is a tendency to imitate the right kind of thing; and most song-birds seem to have at least some powers of vocal imitation. And such an ability is essential in establishing the normal song of the population. If, however, one rears birds such as parrots and mynahs as pets, one finds, as of course we all know, that they then start to imitate human beings. It is as if, not having other parrots to imitate, they direct their imitative powers on to the beings with which they are most closely associated. Mynahs, or to be more exact the Indian Hill Mynah (*Gracula religiosa intermedia*) are particularly interesting because they are the best talking birds in existence. They have been known as talkers since the early 17th century when the Mogul emperors of India kept them at their Courts for amusement. And yet no one has ever got any satisfactory evidence that in the wild they imitate anything! This lack of evidence was so puzzling that I arranged for a student of mine, Mr. Brian Bertram, to go to India for two years with the object of finding out what the mynah is doing with its astounding imitative powers. In brief, the answer seems to be that in the wild they are imitating, but not imitating any other species. In the wild they seem to confine their imitative powers to very subtle vocal inflections of other members of their own species; but of course, no human being would notice this since one must have a very exact recording of the natural voice of a number of different individual mynahs to detect any differences between them at all. Now this is what the imitative power is used for. It is used, as so often seems to happen in birds, to build up a sort of local dialect; they seem to learn to recognize their neighbours as a result of imitating certain features of those neighbours' vocalizations and incorporating these features in their own repertoire. You can only get a mynah to imitate human or other non-mynah noises when you get him as a young bird and hand-rear him. Then he appears to get quickly imprinted on human beings and this astounding ability of imitation of the human voice reveals itself.

KETY What happens when you cross-rear two species of birds? How much of the language—

THORPE In song birds you can get an almost complete adoption of the vocal patterns of the foster parents; but this as far as we know, only applies to the song birds. With doves by contrast, there seems to be absolutely no ability to change the vocalizations as a result of experience. This matter of dove vocalizations is of considerable interest. You may say in a sense, that all doves coo; that is, they all make cooing noises; but they do it with a perfectly specific rhythm—

so specific that you can identify a species of dove just by the rhythm of the call. Thus one can go to Africa and travel a thousand miles across the continent and hear (as in the case of the Emerald Spotted Wood Dove *Turtur Chalcospilos*, or the Red-eyed Dove *Streptopelia semitorquata*) the same species everywhere with exactly the same rhythm. And this rhythm they apparently cannot change. But in many other groups of birds, especially amongst the true song-birds or Oscines, the ability to imitate becomes evident. The prime function of this imitative ability seems to be to establish a local dialect or a communicity pattern of vocalization and vary this lightly within its limits as a specific signal so that it can become an individual signal as well. Thus, such birds learn to know individual members of the same species by slight variations in this basic pattern.

But to change the subject, I have a question—did somebody say just now that trial-and-error-learning is not possible for a child?

BRUNER   No, he means that that is not the main way of learning.

THORPE   I see.

BRUNER   You have to keep an end in view. If you are going to call that the main way of learning for young babies, you assume that they are trying to accomplish a particular thing . . . but they do not do this, they get distracted.

THORPE   I see. Is it just because they cannot keep one objective in view?

KETY   But isn't it possible for an individual to learn by trial and error without a specific goal in mind? Suppose a child just in random manipulation learned that some things were pleasing, without necessarily seeking those things, wouldn't he still establish . . .

MCNEILL   You must assume a goal. How do you recognize an error unless there is some goal?

KETY   Well, but then I thought Blanche was speaking about a symbolic goal.

MCNEILL   Yes, that is what she—

KETY   But if that is a goal, certainly the child has the ability to perceive pleasure and pain.

WEISS   Isn't it as in play, the play of animals, or children—they are just excursions so to speak, from which you learn to select and discover. The biological function of play is very much of that.

THORPE   Oh yes, it is a way of developing motor patterns—

MCNEILL   The notion of trial and error logically presupposes *some* kind of goal. The goal might be extremely primitive. I am speaking for Bruner, but this is certainly the traditional use of the term in psychology. In Thorndike's puzzle-box for cats, it was a firmly established goal—escape—that allowed the animals to select appropriate responses.

KOESTLER The definition of trial and error in the technical context of learning theory is that the learning occurs through random tries.

MCNEILL Random tries—that is exactly what is meant.

KOESTLER There, of course, Jerry Bruner's statement is totally correct. I mean that random trials are certainly not a major way of learning.

THORPE It is possible for a child to learn that way, but it is not a major way.

BRUNER I think Jerry has been quite specific about this. Others believe that the child's early behaviour is shaped by trial and error, and he is objecting to that.

SMYTHIES Isn't the final disproof against Skinnerian ideas, Richard Gregory's work on the importance of being able to touch something before you can see it. You may recall the famous experiment he did on someone who had been blind from birth, who was restored to sight by operation. Gregory spent a long time studying this man, and it was clear that he was unable to see properly. As soon as his bandages were removed he was shown a bus; but the only parts of the bus he could see were the parts he had previously touched. The first drawing he did of a bus was of the stairs, and the bit you touch on the left as you come in, and the handrails, and so on; this is what he saw. After several months he could see a much better bus. But still the front end of the bus he could not see at all—you normally don't walk in front of a bus, it's dangerous. He never saw this, even after some years. He drew a beautiful bus, except there was no bonnet and nothing in front.

THORPE The recent Monograph by Gregory and Wallace sets forth a case history of much interest which demonstrates that early touch-experience may link up with vision many years later. This patient was a man who had been effectively blind—almost since birth—and recovered appreciable sight by operation at the age of 52. He rapidly learned to link up visual experience with previous touch-experience, for instance to read capital letters which he had learned by touch from wood blocks. But he was much slower to interpret visual experience unconnected with touch; e.g. he was

190 On voluntary action

slow to learn to read lower-case letters which he had never handled, and was baffled by the expressions on faces.

SMYTHIES   I think this underlines Bruner's point that man actively works on reality and gets pitch and variety by his interaction with it.

REFERENCES

BALINT, M. (1948) "Individual differences of behavior in early infancy, and an objective method for recording them", *Journal of Genetic Psychology, 73*, pp. 57–117.

BERNSTEIN, N. (1967) *The coordination and regulation of movement*, Pergamon Press, New York.

BRIDGER, W. H. (1962) "Ethological concepts and human development", *Recent advances in biological psychiatry, 4.* pp. 95–107.

BRONSHTEIN, A. I., and PETROVA, E. P. (1967) "The auditory analyzer in young infants", in: Y. Brackbill & G. G. Thompson (eds.), *Behavior in infancy and early childhood*, Free Press, New York, pp. 163–172.

GUNTHER, M. (1961) "Infant behavior at the breast", in: B. M. Foss (ed.), *Determinants of infant behavior*, Vol. 1, Wiley, New York, pp. 37–44.

VON HOLST, E., and MITTELSTAEDT, H. (1950) Das Reafferenzprinzip. *Naturwissenschaften, 37*, pp. 464–476.

JAKOBSON, R. C. (1960) "Why 'mama' and 'papa'?" in B. Kaplan and S. Wagner (eds.) *Perspectives in psychological theory*, International University Press, New York.

JENSEN, K. (1932) "Differential reactions to taste and temperature stimuli in newborn infants", *Genetic Psychology Monographs, 12* pp. 361–479.

KESSEN, W. (1967) "Sucking and looking: Two organized congenital patterns of behavior in the human newborn", in: H. W. Stevenson, E. H. Hess, and H. L. Rheingold (eds.), *Early behavior: Comparative and developmental approaches*, Wiley, New York, pp. 147–179.

KESSEN, W., and LEUTZENDORFF, A.-M. (1963) "The effect of nonnutritive sucking on movement in the human newborn", *Journal of Comparative and Physiological Psychology, 56*, pp. 69–72.

KRON, R. E., STEIN, M. and GODDARD, K. E. (1963) "A method of measuring sucking behavior of newborn infants", *Psychosomatic Medicine, 25*, pp. 181–191.

LASHLEY, K. S. (1951) "The problem of serial order in behavior", in: L. A. Jeffress (ed.), *Cerebral mechanisms in behavior: The Hixon symposium*, Wiley, New York, pp. 112–146.

MILLER, G. A., GALANTER, E. and PRIBRAM, K. H. (1960) *Plans and the structure of behavior*, Holt, New York.

PEIPER, A. (1963) *Cerebral function in infancy and childhood*, Consultants Bureau, New York.

PIAGET, J. (1936) *La naissance de l'intelligence chez l'enfant*. Translation: see (1952) *The origins of intelligence in children*. International Universities Press, New York.

———— (1954) *The construction of reality in the child*. Basic Books, New York.

SAMEROFF, A. J. (1965) "An experimental study of the response components of sucking in the human newborn". Unpublished doctoral dissertation, Yale University.

SHERRINGTON, C. S. (1906) *The integrative action of the nervous system*, Scribner's, New York.

SIQUELAND, E. R. (1968) "Conditioned sucking and visual reinforcers with human infants". Paper presented at the eastern regional meeting, Society for Research in Child Development, Worcester, Mass.

SPERRY, R. W. (1950) "Neural basis of the spontaneous optokinetic response produced by visual inversion", *Journal of Comparative and Physiological Psychology*, *43*, pp. 482–489.

THORNDIKE, E. L. (1903) *Educational psychology*, Teachers College, Columbia University, New York.

TWITCHELL, T. E. (1965) "The automatic grasping responses of infants, *Neuropsychologia*, *3*, pp. 247–259.

WHITE, R. H. (1959) "Motivation reconsidered: The concept of competence". *Psychological Review*, *66*, pp. 297–333.

WOLFF, P. H. (1968) Personal communication.

WOLFF, P. H. and SIMMONS, M. A. (1967) "Nonnutritive sucking and response thresholds in young infants", *Child Development*, *38*, pp. 631–638.

Beyond atomism and holism—the concept of the holon

Arthur Koestler

This is going to be an exercise in General Systems Theory—which seems to be all the more appropriate as its founding father sits next to me, and as he launched that venture here in Alpbach twenty years ago. It seems equally appropriate that I should take as my text a sentence from von Bertalanffy's *Problems of Life* (1952); it reads: "Hierarchical organization on the one hand, and the characteristics of open systems on the other, are fundamental principles of living nature."

If we combine these two fundamental principles, and add a dash of cybernetics to them, we get a system-theoretical model of Self-regulating Open Hierarchic Order, or SOHO for short. I intend to discuss some of the properties of this SOHO model as an alternative to the S-R model of linear causation, derived from classical mechanics, which we seem to be unanimous in rejecting. I can only give here a sketchy outline of the idea, but I have tried to tabulate the axioms and propositions relating to it in a more systematic way in an appendix to my last book (1967), which I have also appended to this paper, as a sort of *Tractatus Logico Hierarchicus*. Some of these propositions may appear trivial, others rest on incomplete evidence, still others will need correcting or qualifying. But they may provide a basis for discussion.

Hierarchies and Old Hats

When one talks about hierarchic organization as a fundamental principle of life, one often encounters a strong emotional resistance.

For one thing, hierarchy is an ugly word, loaded with ecclesiastic and military associations, and conveys to some people a wrong impression of a rigid or authoritarian structure. (Perhaps the assonance with "hieratic", which is a quite different matter, plays a part in this confusion.) Apart from this, the term is often wrongly used to refer simply to order of rank on a linear scale or ladder (e.g. Clark Hull's "habit-family hierarchies"). But that is not at all what the term is meant to signify. Its correct symbol is not a rigid ladder but a living tree—a multi-levelled, stratified, out-branching pattern of organization, a system branching into sub-systems, which branch into sub-systems of a lower order, and so on; a structure encapsulating sub-structures and so on; a process activating sub-processes and so on. As Paul Weiss said yesterday: "The phenomenon of hierarchic structure is a real one, presented to us by the biological object, and not the fiction of a speculative mind." It is at the same time a conceptual tool, a way of thinking, an alternative to the linear chaining of events torn from their multidimensionally stratified contexts.

Paul Weiss's dislike of the tree image, which he expressed yesterday, seems to derive from a misunderstanding: he apparently thought the metaphoric tree was meant to be taken literally. His remark that the spatial structures in the organism are "telescoped" or "encapsulated" into each other rather than "arborized" is an objection frequently voiced by people inclined to think in concrete images rather than abstract schemata. The objection can be met by making a horizontal cross-section across the tree and drawing a schematized bird's eye view which shows how the twigs stem from branches, branches from limbs, and so on. Thus the tree-diagram is transformed into the Chinese-box diagram, but the latter is clumsy and contains less information (see figure on p.194).

All complex structures and processes of a relatively stable character display hierarchic organization, and this applies regardless whether we are considering inanimate systems, living organisms, social organizations, or patterns of behaviour. The linguist who thinks primarily in terms of Chomsky's (1965) hierarchic model experiences a *déjà vu* reaction—as McNeill yesterday expressed it—towards the physiologist's intra-cellular hierarchy; and this may equally apply to Bruner's presentation of the hierarchic structure of voluntary action. In this essential respect—and in others that I shall mention—these processes in widely different fields are indeed isomorphic.

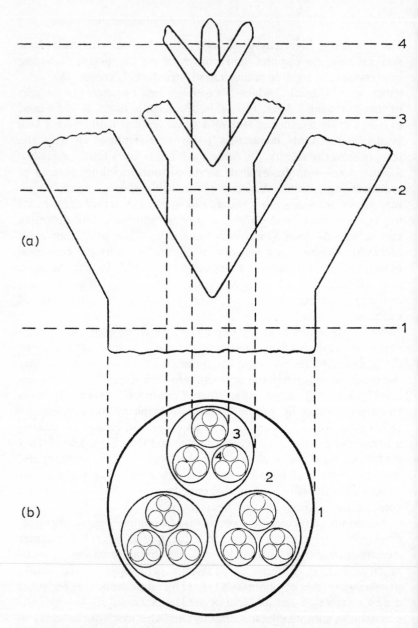

Two ways of diagramming a hierarchy of 4 levels with a "span" of 3 on each level; (a) the tree; (b) the Chinese box, derived from a cross-section through level 4 of the tree.

The hierarchic tree diagram may equally serve to represent the branching out of the evolution of species—the tree of life and its projection in taxonomy; it serves to represent the step-wise differentiation of tissues in embryonic development; it may serve as a structural diagram of the parts-within-parts architecture of organisms or galaxies, or as a functional schema for the analysis of instinctive behaviour by the ethologist (Tinbergen 1951, Thorpe 1956); or of the phrase-generating machinery by the psycholinguist. It may represent the locomotor hierarchy of limbs, joints, individual muscles, and so down to fibres, fibrils and filaments (Herrick 1961, Weiss 1950, etc.); or, in reverse direction, the filtering and processing of the sensory input in its ascent from periphery to centre. It could also be regarded as a model for the subject-index of the Library of Congress, and for the organization of knowledge in our memory-stores; lastly, as an organizational chart for government administrations, military and business organizations; and so on.

This almost universal applicability of the hierarchic model may arouse the suspicion that it is logically empty; and this may be a further factor in the resistance against it. It usually takes the form of what one may call the "so what" reaction: "all this is old hat, it is self-evident"—followed by the non-sequitur "and anyway, where is your evidence?" Well, hierarchy may be old hat, but I would suggest that if you handle it with some affection, it can produce quite a few lively rabbits—which can even be tested in the laboratory.

Evolution and hierarchic order

One of my favourite examples to illustrate the merits of hierarchic order is an amusing parable invented by Herbert Simon—whose absence we all regret. I have quoted it on other occasions, but I shall briefly quote it again. The parable concerns two watchmakers, Hora and Tempus. Both make watches consisting of a thousand parts each. Hora assembles his watches bit by bit; so when he pauses or drops a watch before it is finished, it falls to pieces and he has to start from scratch. Tempus, on the other hand, puts together sub-assemblies of ten parts each; ten of these sub-assemblies he makes into a larger sub-assembly of a hundred units; and ten of these make the whole watch. If there is a disturbance, Tempus has to repeat at worst nine assembling operations, and at best none at all. If you have a ratio of one disturbance in a hundred operations, then

Hora will take four thousand times longer to assemble a watch—instead of one day, he will take eleven years. And if, for mechanical bits, we substitute amino acids, protein molecules, organelles, and so on, the ratio between the time-scales becomes astronomical.

This is one basic advantage of employing the hierarchic method. The second is, of course, the incomparably greater stability and resilience to shock of the Tempus type of watch, and its amenability to repair and improvement. Simon concludes: "Complex systems will evolve from simple systems much more rapidly if there are stable intermediate forms than if there are not. The resulting complex forms in the former case will be hierarchic. We have only to turn the argument round to explain the observed predominance of hierarchies among the complex systems Nature presents to us. Among possible complex forms, hierarchies are the ones that had the time to evolve." If there is life on other planets, we may safely assume that, whatever its form, it must be hierarchically organized.

Motor manufacturers discovered long ago that it does not pay to design a new model from scratch by starting on the level of elementary components; they make use of already existing sub-assemblies—engines, brakes, etc.—each of which has developed from long previous experience, and then proceed by relatively small modifications of some of these. Evolution follows the same strategy. Once it has taken out a patent it sticks to it tenaciously—Thorpe remarked yesterday on its fixed, conservative ways. The patented structure, organ or device acquires a kind of autonomous existence as a sub-assembly. The same make of organelles functions in the cells of mice and men; the same make of contractile protein serves the streaming motion of amoeba and the finger muscles of the piano-player; the same homologous design is maintained in the vertebrate forelimb of man, dog, bird and whale. Geoffroy de St. Hilaire's *loi du balancement*, and d'Arcy Thompson's (1942) transformation of a baboon's skull into a human skull by harmonious deformations of a Cartesian coordinate lattice, further illustrate the hierarchic constraints imposed on evolutionary design.

Autonomous holons

The evolutionary stability of these sub-assemblies—organelles, organs, organ-systems—is reflected by their remarkable degree of *autonomy* or self-government. Each of them—a piece of tissue or a whole heart—is capable of functioning *in vitro* as a quasi-independent whole, even though isolated from the organism or transplanted into another organism. Each is a *sub-whole* which, towards its subordinated parts, behaves as a self-contained whole, and towards its superior controls as a dependent part. This relativity of the terms "part" and "whole" when applied to any of its sub-assemblies is a further general characteristic of hierarchies.

It is again the very obviousness of this feature which tends to make us overlook its implications. A part, as we generally use the word, means something fragmentary and incomplete, which by itself would have no legitimate existence. On the other hand, there is a tendency among holists to use the word "whole" or "Gestalt" as something complete in itself which needs no further explanation. But wholes and parts in this absolute sense do not exist anywhere, either in the domain of living organisms or of social organizations. What we find are intermediary structures on a series of levels in ascending order of complexity, each of which has two faces looking in opposite directions: the face turned towards the lower levels is that of an autonomous whole, the one turned upward that of a dependent part. I have elsewhere (1967) proposed the word "holon" for these Janus-faced sub-assemblies—from the Greek *holos*—whole, with the suffix *on* (cf. neutr*on*, prot*on*) suggesting a particle or part.

The concept of the holon is meant to supply the missing link between atomism and holism, and to supplant the dualistic way of thinking in terms of "parts" and "wholes", which is so deeply engrained in our mental habits, by a multi-level, stratified approach. A hierarchically-organized whole cannot be "reduced" to its elementary parts; but it can be "dissected" into its constituent branches of holons, represented by the nodes of the tree-diagram, while the lines connecting the holons stand for channels of communication, control or transportation, as the case may be.

Fixed rules and flexible strategies

The term holon may be applied to any stable sub-whole in an organismic, cognitive, or social hierarchy which displays rule-governed behaviour and/or structural Gestalt constancy. Thus biological holons are self-regulating "open systems" (von Bertalanffy 1952) governed by a set of fixed rules which account for the holon's coherence, stability and its specific pattern of structure and function. This set of rules we may call *the canon of the holon.*[1] The canon determines the fixed, invariant aspect of the open system in its steady state (*Fliessgleich-gewicht*—dynamic equilibrium); it defines its pattern and structure. In other types of hierarchies, the canon represents the codes of conduct of social holons (family, tribe, nation, etc.); it incorporates the "rules of the game" of instinctive rituals or acquired skills (behavioural holons); the rules of enunciation, grammar and syntax in the language hierarchy; Piaget's "schemes" in cognitive hierarchies, and so on. *The canon represents the constraints imposed on any rule-governed process or behaviour.* But these constraints do not exhaust the system's degrees of freedom; they leave room for more or less *flexible strategies*, guided by the contingencies in the holon's local environment.

It is essential at this point to make a sharp, categorical distinction between the fixed, invariant canon of the system and its flexible (plastic, variable) strategies. A few examples will illustrate the validity of this distinction. In *ontogeny*, the apex of the hierarchy is the zygote, and the holons at successive levels represent successive stages in the development of tissues. Each step in differentiation and specialization imposes further constraints on the genetic potential of the tissue, but at each step it retains sufficient developmental flexibility to follow this or that evolutionary pathway, within the range of its competence, guided by the contingencies of the cell's environment—Waddington's (1957) "strategy of the genes". Turning from embryonic development to the *instinctive activities* of the mature animal, we find that spiders spin webs, birds build nests according to invariant species-specific canons, but again using flexible strategies, guided by the lie of the land: the spider may suspend its web from three, four or more points of attachment, but the result will always be a regular polygon. In *acquired skills* like chess, the rules of the game define the permissible moves, but the strategic choice of the actual move depends on the environment—the distribution of the chessmen on the board. In *symbolic operations*, the holons are rule-governed cognitive structures variously called "frames of reference", "universes of discourse", "algorithms", etc., each with its specific "grammar" or canon; and the strategies increase in complexity on higher levels of each hierarchy. It seems that life in

[1] Cf. the "organizing relations" or "laws of organization" of earlier writers on hierarchic organization (e.g., Woodger (1929), Needham (1941)), and the "system-conditions" in General System Theory.

all its manifestations, from morphogenesis to symbolic thought, is governed by rules of the game which lend it order and stability but also allow for flexibility; and that these rules, whether innate or acquired, are represented in coded form on various levels of the hierarchy, from the genetic code to the structures in the nervous system responsible for symbolic thought.

Triggers and scanners

Let me discuss briefly some specific characteristics of what one might loosely call *output hierarchies*, regardless whether the "output" is a baby, or a sentence spoken in English. However much their products differ, all output hierarchies seem to have a classic mode of operation, based on the trigger-releaser principle, where an implicit coded signal which may be relatively simple, releases complex, pre-set mechanisms.

Let me again run through a few examples. In *phylogeny*, Waddington (1957) and others have convincingly shown that a single favourable gene-mutation can act as a trigger to release a kind of chain-reaction which affects a whole organ in a harmonious way. In *ontogeny*, the prick of a fine platinum needle on the unfertilized egg of a frog or sheep triggers off parthenogenesis. The genes act as chemical triggers, catalysing reactions. The implicit four-letter alphabet of the DNA chain is spelled out into the explicit, twenty-letter alphabet of amino acids; the inducers or evocators, including Spemann's "general organiser", again turn out to be relatively simple chemicals which need not even be species-specific to activate the genetic potentials of the tissue. In *instinct behaviour*, we have releasers of a very simple kind—the red belly of the stickleback, the spot under the herring-gull's beak, which trigger off the appropriate behaviour (Tinbergen 1951). In the performance of *acquired skills* you have the same process of stepwise filling in the details of implicit commands issued from the apex of the hierarchy, such as "strike a match and light this cigarette" or "sign your name", or "use your phrase-generating machine" to transform an unverbalized image into innervations of the vocal chords.

The point to emphasize is that this spelling-out process, from intent to execution, cannot be described in terms of a linear chain of S-R units, only as a series of discrete steps from one Open Sesame, activated by a combination lock, to the next. The activated holon, whether it is a government department or a living kidney, has its own canon which determines the pattern of its activity. Thus the signal from higher quarters does not have to specify what the holon is expected to do; the signal merely has to trigger the holon into action by a coded message. Once thrown into action, the holon will spell out the implicit command in explicit form by activating its sub-units in the appropriate strategic order, guided by feedbacks and feed-forwards from its environment. Generally speaking, *the holon is a system of relations which is represented on the next higher level as a unit, that is, a relatum.*

If we turn now to the *input hierarchies* of perception, the operations proceed, of course, in the reverse direction, from the peripheral twigs of the tree towards its apex; and instead of trigger-releasers we have the opposite type of mechanisms: a series of filters, scanners or classifiers through which the input traffic must pass in its ascent from periphery to cortex. First you have lateral inhibition, habituation and presumably some efferent control of receptors. On the higher levels are the mechanisms responsible for the visual and acoustic constancy phenomena, the scanning and filtering devices which account for the recognition of patterns in space and time, and enable us to abstract universals and discard particulars. The colloquial complaint: "I have a memory like a sieve" may be derived from an intuitive grasp of these filtering devices that operate first all along the input channels, then along the storage channels.

How do we pick out a single instrument in a symphony? The whole medley of sounds arriving at the ear-drum is scrambled into a linear pressure-wave with a single variable. To reconstruct the timbre of an instrument, to identify harmonies and melodies, to appreciate phrasing, style and mood, we have to abstract patterns in time as we abstract visual patterns in space. But how does the nervous system do it? It will play you the opening bars of the Archduke Trio; watch your reactions, because no textbook on psychology that I know of will give you the faintest clue (opening bars of Beethoven's Archduke Trio played). If one looks at the record with a magnifying glass, one is tempted to ask the naïve question why the nervous system does not produce engrams by this simple method of coding, instead of being so damned complicated. The answer is, of course, that a linear engram of this kind would be completely useless for the purpose of analysing, matching and recognizing input patterns. The chain is a hopeless model; we cannot do without the tree.

In motor hierarchies, an implicit intention or generalized command is particularized, spelled out, step by step, in its descent to the periphery. In perceptual hierarchies, we have the opposite process. The peripheral input is more and more de-particularized, stripped of irrelevancies during its ascent to the centre. *The output hierarchy concretizes, the input hierarchy abstracts.* The former operates by means of triggering devices, the latter by means of filtering or scanning devices. When I intend to write the letter R, a trigger activates a functional holon, an automatic pattern of muscle contractions, which produces the letter R in my own particular handwriting. When I read, a scanning device in my visual cortex identifies the letter R regardless of the particular hand that wrote it. Triggers release complex outputs by means of a simple coded signal. Scanners function the opposite way: they convert complex inputs into simple coded signals.

"Abstract" and "picture strip"

Let me briefly turn to the phenomena of *memory* and ask whether the hierarchic approach is capable of shedding some additional light on them. You watch a television play. The exact words of each actor are forgotten by the time he speaks his next line, and only their meaning remains; the next morning you can only remember the sequence of scenes which constituted the story; after a month, all you remember is that it was about a gangster on the run or about two men and a woman on a desert island. The same happens generally with the content of novels one has read, and episodes one has lived. The original experience has been stripped of detail, reduced to a schematic outline. Now this skeletonization of the input before it is put into storage, and the gradual decay of the stored material, would mean a terrible impoverishment of memory, if this were the whole story—memory would be a collection of dusty abstracts, the dehydrated sediments in the wine-glass, with all flavour gone. But there are compensating mechanisms. I can recognize a melody, regardless of the instrument on which it is played, and I can recognize the timbre of an instrument, regardless of the melody played on it. There are several interlocking hierarchies at work, each with its own criteria of relevance. One abstracts melody and treats everything else as noise, the other abstracts timbre and treats melody as noise. Thus not all the information discarded as irrelevant by one filtering system is irretrievably lost, because it may have been stored by another filtering hierarchy with different canons of relevance. Recall would then be made possible by the co-operation of several interlocking hierarchies, which may pertain to different sense modalities—sight and smell, for instance; or, what is less obvious, there may also be several distinct hierarchies with different criteria of relevance operating within the same sense modality. You may remember the words of the aria "Your Tiny Hand is Frozen" without the tune. You may remember the tune after having forgotten the words. And you may recognize the sound of Caruso's voice on a gramophone record, regardless of the words and the tune he is singing. Recall could then be compared to the process of multi-coloured printing by the superimposition of several colour-blocks. This, of course, is speculative, but some modest evidence for the hypothesis can be found in a series of experiments by J. Jenkins and myself (1965); and more tests on these lines can be designed without much difficulty.

I am aware that the hypothesis is in apparent contradiction to Penfield's (1959) experiments eliciting what looks like total recall of past experiences by electrical stimulation of points on the patient's temporal lobe. But the contradiction may be resolved if we include in the criteria of relevance also criteria of *emotional* relevance which decide whether an input is worth storing. A detail like the wart on granny's chin, or the taste of Proust's *madeleine*, or a single gesture of an actress in an otherwise long-forgotten play, might be emotionally relevant (on a conscious or unconscious level), and retained as a vivid detail of almost photographic or cinematographic clarity. One might call this the *picture-strip* type of memory which is stamped in, as distinct from *abstractive* memory which schematizes. (I have tried to coordinate this dichotomy with the Piaget-Inhelder dichotomy of "scheme" and "schema", but there are difficulties, and I am not sure whether these are merely terminological or deeper.) Picture-strip memories may be related to eidetic images; and they might even, unlike abstractive memories, originate in the limbic system (MacLean, 1958).

Arborization and reticulation

I have used the term "interlocking" or "interlacing" hierarchies. Of course hierarchies do not operate in a vacuum. This truism regarding the interdependence of processes in an organism is probably the main cause of confusion which obscured from view its hierarchic structure. It is as if the sight of the foliage of the entwined branches in a forest made us forget that the branches originate in separate trees. The trees are vertical structures. The meeting points of branches from neighbouring trees form horizontal networks at several levels. Without the trees there could be no entwining, and no network. Without the networks, each tree would be isolated, and there would be no integration of functions. Arborization and reticulation seem to be complementary principles in the architecture of organisms. In symbolic universes of discourse arborization is reflected in the "vertical" denotation (definition) of concepts, reticulation in their "horizontal" connotations in associative networks. This calls to mind Hyden's proposal that the same neuron, or population of neurons, may be a member of several functional "clubs".

Hierarchic order and feedback control

The most obvious example of interlocking hierarchies is the sensory-motor system. The sensory hierarchy processes information and transmits it in a steady upward flow, some of which reaches the conscious ego at the apex; the ego makes decisions which are spelled out by the downward stream of impulses in the motor hierarchy. But the apex is not the only point of contact between the two systems; they are connected by entwining networks on various lower levels. The network on the lowest level consists of reflexes like the patellary. They are shortcuts between the ascending and descending flow, like loops connecting opposite traffic streams on a motor highway. On the next higher level are the networks of sensory-motor skills and habits, such as touch-typing or driving a car, which do not require the attention of the highest centres—unless some disturbance throws them out of gear. But let a little dog amble across the icy road in front of the driver, and he will have to make a "top level" decision whether to slam down the brake, risking the safety of his passengers, or run over the dog. It is at this level, when the pros and cons are precariously balanced, that the subjective experience of free choice and moral responsibility arises.

But the ordinary routines of existence do not require such moral decisions, and not even much conscious attention. They operate by means of feedback loops, and loops-within-loops, which form the multi-levelled, reticulate networks between the input and output hierarchies. So long as all goes well and no dog crosses the road, the strategy of riding a bicycle or driving a car can be left to the automatic pilot in the nervous system—the cybernetic helmsman. But one must beware of using the principle of feedback control as a magic formula. The concept of feedback without the concept of hierarchic order is like the grin without the cat. All skilled routines follow a pre-set pattern according to certain rules of the game. These are fixed, but permit constant adjustments to variable environmental conditions. *Feedback can only operate within the limits set by the rules—* by the canon of the skill. The part which feedback plays is to report back on every step in the progress of the operation, whether it is over-shooting or falling short of the mark, how to keep it on an even keel, when to intensify the pace and when to stop. But it cannot alter the intrinsic pattern of the skill. To quote Paul Weiss (1951) at the Hixon Symposium: "The structure of the input does not produce the structure of the output, but merely modifies intrinsic nervous activities, which have a structural organization of their own". One of the vital differences between the S-R and SOHO concepts is that according to the former, the environment determines behaviour, whereas according to the latter, feedback from the environment merely guides or corrects or stabilizes pre-existing patterns of behaviour.

Moreover, the cross-traffic between the sensory and motor hierarchies works both ways. The input guides the output and keeps it on an even keel; but motor activity in its turn guides perception. The eye must scan; its motions, large and small—drift, flicker, tremor— are indispensable to vision; an image stabilized on the retina disintegrates into darkness (Hebb 1958). Similarly with audition: if you try to recall a tune, what do you do? You hum it. Stimuli and responses have been swallowed up by feedback loops within loops, along which impulses run in circles like kittens chasing their tails.

A hierarchy of environments

Let us carry this inquiry into the meaning of current terminology a step further, and ask just what that convenient word "environment" is meant to signify. When I am driving my car, the environment in contact with my right foot is the accelerator pedal, its elastic resistance to pressure provides a tactile feedback which helps keeping the speed of the car steady. The same applies to the "feel" of the wheel under my hands. But my eyes encompass a much larger environment than my feet and hands; they determine the overall strategy of driving. The hierarchically organized creature that I am is in fact functioning in a hierarchy of environments, guided by a hierarchy of feedbacks.

One advantage of this operational interpretation is that the hierarchy of environments can be extended indefinitely. When the chess-player stares at the board in front of him, trying to visualize various situations three moves ahead, he is guided by feedbacks from imagined environments. Most of our thinking, planning and creating operates in such imaginary environments. But—to quote Bartlett (1958)—"all our perceptions are inferential constructs," coloured by imagination, and so the difference is merely one of degrees. The hierarchy is open-ended at the top.

Regulation channels

When the centipede was asked in which order he moved his hundred legs, he became paralysed and starved to death because he had never thought of it before and had left his legs to look after themselves. When an intent is formed at the apex of the hierarchy, such as signing a letter, it does not activate individual motor units, but triggers off patterns of impulses which activate sub-patterns and so on. But this can only be done one step at a time: the higher centres do not normally have dealings with lowly ones, and vice versa. Brigadiers do not concentrate their attention on individual soldiers—if they did, the whole operation would go haywire. Commands must be transmitted through "regulation channels".

This statement looks trivial, but ignoring it carries heavy penalties of a theoretical or practical order. Attempting to short-circuit intermediary levels of the hierarchy by focusing conscious attention on physiological processes which otherwise function automatically, usually ends in the centipede's predicament, reflected in symptoms that range from impotence and frigidity to constipation and spastic colons. Frankl has coined the term hyper-reflection for these disorders and will have more to tell us about them[2].

As for theory, the S–R psychologists attempt to short-circuit

[2] Cf. Frankl in the discussion after Hayek's paper.

hierarchic levels by vague references to "intervening variables" (the O in the S.O.R. schema) is a face-saving manoeuvre to sweep all the essential problems of complex human behaviour, including language, under the laboratory carpet.

Mechanization and freedom

A skilled activity, such as writing a letter, branches into sub-skills which, on successively lower levels of the hierarchy, become increasingly mechanized, stereotyped and predictable. The choice of subjects to be discussed in a letter is vast; the next step, phrasing, still offers a great number of alternatives, but is more restricted by the rules of grammar, the limits of one's vocabulary, etc.; the rules of spelling are fixed, with no leeway for flexible strategies, and lastly, the muscle contractions which depress the typewriter keys are entirely automatized. Thus a *sub-skill or behavioural holon on the* (n) *level of the hierarchy has more degrees of freedom* (a larger variety of alternative strategic choices permitted by the canon) *than a holon on the* ($n-1$) *level*.

However, all skills tend with increasing mastery and practice to become automatized routines. While acquiring a skill we must concentrate on every detail of what we are doing; then learning begins to condense into habit as steam condenses into drops; with increasing practice we read, write, type, drive "automatically" or "mechanically". Thus we are all the time transforming "mental" into "mechanical" activities. In unexpected contingencies, however, the process can be reversed. Driving along a familiar road is an auto-matized routine; but when that little dog crosses the road, a strategic choice has to be made which is beyond the competence of automatized routine, for which the automatic pilot in my nervous system has not been programmed, and the decision must be referred to higher quarters. The *shift of control* of an on-going activity from one level to a higher level of the hierarchy—from "mechanical" to "mindful" behaviour—seems to be the essence of conscious decision-making and of the subjective experience of free will.

These considerations may have some bearing on the Mind-Body problem which, however, cannot be discussed here (cf. *The Ghost in the Machine*, pp. 197–221).

The tendency towards the progressive mechanization of skills has its positive side: it conforms to the principle of parsimony. If I could not hit the keys of the typewriter "automatically" I could not attend to meaning. On the negative side, mechanization, like rigor mortis, affects first the extremities—the lower subordinate branches of the hierarchy, but it also tends to spread upward. If a skill is practised in the same unvarying conditions, following the same unvarying course, it tends to degenerate into stereotyped routine and its degrees of freedom freeze up. Monotony accelerates enslavement to habit; and if mechanization spreads to the apex of the hierarchy, the result is the rigid pedant, Bergson's *homme automate*. As von Bertalanffy wrote, "organisms *are not* machines, but they can to a certain extent *become* machines, congeal into machines" (1952).

Vice versa, a variable environment demands flexible behaviour and reverses the trend towards mechanization. However, the challenge of the environment may exceed a critical limit where it can no longer be met by customary routines, however flexible—because the traditional "rules of the game" are no longer adequate to cope with the situation. Then a crisis arises. The outcome is either a breakdown of behaviour —or alternatively the emergence of new forms of behaviour, of original solutions. They have been observed throughout the animal kingdom, from insects onward, through rats to chimpanzees, and point to the existence of unsuspected potentials in the living organism, which are inhibited or dormant in the normal routines of existence, and only make their appearance in exceptional circumstances. They foreshadow the phenomena of human creativity which—as discussed elsewhere (1964 and 1967)—must remain incomprehensible to the S–R theorist, but appear in a new light when approached from the hierarchic point of view.

Self-assertion and integration

The holons which constitute an organismic or social hierarchy are Janus-faced entities: facing upward, toward the apex, they function as dependent parts of a larger whole; facing downward, as autonomous wholes in their own right. "Autonomy" in this context means that organelles, cells, muscles, neurons, organs, all have their intrinsic rhythm and pattern, often manifested spontaneously without external stimulation, and that they tend to persist in and assert their characteristic pattern of activity. This *self-assertive tendency* is a fundamental and universal characteristic of holons, manifested on every level of every type of hierarchy: in the regulative properties of the morphogenetic field, defying transplantation and experimental mutilation; in the stubbornness of instinct rituals, acquired habits, tribal traditions and social customs; and even in a person's handwriting, which he can modify but not sufficiently to fool the expert. Without this self-assertive tendency of their parts, organisms and societies would lose their articulation and stability.

The opposite aspect of the holon is its *integrative tendency* to function as an integral part of an existing or evolving larger whole. Its manifestations are equally ubiquitous, from the "docility" of the embryonic tissues, through the symbiosis of organelles in the cell, to the various forms of cohesive bonds, from flock to insect state and human tribe.

We thus arrive at a polarity between the self-assertive (S.A.) and the integrative (INT) tendency of holons on every level. This polarity is of fundamental importance to the SOHO concept. It is in fact implied in the model of a multi-levelled hierarchy, because the stability of the hierarchy depends on the equilibration of the two opposite tendencies of its holons. Empirically the postulated polarity can be traced in all phenomena of life; in its theoretical aspect it is not derived from any metaphysical dualism, but may rather be regarded as an application of Newton's Third Law of Motion (action and reaction) to hierarchic systems. We may even extend the polarity into inanimate nature: wherever there is a relatively stable dynamic system, from atoms to galaxies, stability is maintained by the equilibration of opposite forces, one of which may be centrifugal or separative or inertial, and the other a centripetal or attractive or cohesive force, which keep the parts in their place in the larger whole, and hold it together.

Perhaps the most fertile field of application of the SOHO schema is the study of emotions and emotional disorders on the individual and social scale. Under conditions of stress, the affected part of an organism may become overstimulated and tend to escape the restraining control of the whole (cf. Child, 1924). This can lead to pathological changes of an irreversible nature, such as malignant growths with untrammelled proliferation of tissues that have escaped from genetic restraint. On a less extreme level, practically any organ or function may get temporarily and partially out of control. In rage and panic the sympathico-adrenal apparatus takes over from the higher centres which normally co-ordinate behaviour; when sex is aroused the gonads seem to take over from the brain. The *idée fixe*, the obsession of the crank, are cognitive holons running riot. There is a whole gamut of mental disorders in which some subordinate part of the mental hierarchy exerts its tyrannical rule over the whole, from the insidious domination of "repressed" complexes to the major psychoses, in which large chunks of the personality seem to have "split off" and lead a quasi-independent existence. Aberrations of the human mind are frequently due to the obsessional pursuit of some part-truth, treated as if it were the whole truth—of a holon masquerading as a whole.

If we turn from organismic to *social hierarchies*, we again find that under normal conditions the holons (clans, tribes, nations, social classes, professional groups) live in a kind of dynamic equilibrium with their natural and social environment. However, under conditions of stress, when tensions exceed a critical limit, some social holon may get over-excited and tend to assert itself to the detriment of the whole, just like an over-excited organ. It should be noted that the canon which defines the identity and lends coherence to social holons (its laws, language, traditions, rules of conduct, systems of belief) represents not merely negative constraints imposed on its actions, but also positive precepts, maxims and moral imperatives.

The single individual constitutes the apex of the organismic hierarchy, and at the same time the lowest unit of the social hierarchy. Looking inward, he sees himself as a self-contained, unique whole, looking outward as a dependent part. No man is an island, he is a holon. His *self-assertive* tendency is the dynamic manifestation of his unique wholeness as an individual; his *integrative* tendency expresses his dependence on the larger whole to which he belongs, his partness. Under normal conditions, the two opposite tendencies are more or less evenly balanced. Under conditions of stress, the equilibrium is upset, as manifested in emotional behaviour. The emotions derived from the self-assertive tendencies are of the well-known aggressive-defensive, hunger, rage and fear type, including the possessive component of sex. The emotions derived from the integrative tendency have been largely neglected by contemporary psychology; one may call them the self-transcending or participatory type of emotions. They arise out of the human holon's need to be an integral part of some larger whole—which may be a social group, a personal bond, a belief-system, Nature or the *anima mundi*. The psychological processes through which this category of emotions operates are variously referred to as projection, identification, empathy, hypnotic rapport, devotion, love. It is one of the ironies of the human condition that both its glory and its predicament seem to derive not from the self-assertive but from the integrative potentials of the species. The glories of art and science, and the holocausts of history caused by misguided devotion, are both nurtured by the self-transcending emotions.

These scant remarks on a complex subject, which cannot be pursued here, are merely meant to indicate the wider implications of the hierarchic approach. To conclude, even this fragmentary outline ought to make it clear that in the SOHO model there is no place for such a thing as an aggressive or destructive instinct in organisms; nor does it admit the reification of the sexual instinct as the *only* integrative force in human or animal society. Freud's Eros and Thanatos are relative late-comers on the stage of evolution: a host of creatures that multiply by fission or budding are ignorant of both. In the present view, Eros is an offspring of the Integrative, destructive Thanatos of the Self-Assertive tendency, and Janus the symbol of the polarity of these two irreducible properties of living matter—that *coincidentia oppositorum* which von Bertalanffy is so fond of quoting, and which is inherent in the open-ended hierarchies of life.

APPENDIX

SOME GENERAL PROPERTIES OF SELF-REGULATING OPEN HIERARCHIC ORDER (SOHO)

1. *The holon*

1.1 The organism in its structural aspect is not an aggregation of elementary parts, and in its functional aspects not a chain of elementary units of behaviour.

1.2 The organism is to be regarded as a multi-levelled hierarchy of semi-autonomous sub-wholes, branching into sub-wholes of a lower order, and so on. Sub-wholes on any level of the hierarchy are referred to as *holons*.

1.3 Parts and wholes in an absolute sense do not exist in the domains of life. The concept of the holon is intended to reconcile the atomistic and holistic approaches.

1.4 Biological holons are self-regulating open systems which display both the autonomous properties of wholes and the dependent properties of parts. This dichotomy is present on every level of every type of hierarchic organization, and is referred to as the "Janus phenomenon".

1.5 More generally, the term "holon" may be applied to any stable biological or social sub-whole which displays rule-governed behaviour and/or structural Gestalt-constancy. Thus organelles and homologous organs are evolutionary holons; morphogenetic fields are ontogenetic holons; the ethologist's "fixed action-patterns" and the sub-routines of acquired skills are behavioural holons; phonemes, morphemes, words, phrases are linguistic holons; individuals, families, tribes, nations are social holons.

2. *Dissectibility*

2.1 Hierarchies are "dissectible" into their constituent branches, on which the holons form the nodes; the branching lines represent the channels of communication and control.

2.2 The number of levels which a hierarchy comprises is a measure of its "depth", and the number of holons on any given level is called its "span" (Simon).

3. *Rules and strategies*

3.1 Functional holons are governed by fixed sets of rules and display more or less flexible strategies.

3.2 The rules—referred to as the system's *canon*—determine its invariant properties, its structural configuration and/or functional pattern.

3.3 While the canon defines the permissible steps in the holon's activity, the strategic selection of the actual step among permissible choices is guided by the contingencies of the environment.

3.4 The canon determines the rules of the game, strategy decides the course of the game.

3.5 The evolutionary process plays variations on a limited number of canonical themes. The constraints imposed by the evolutionary canon are illustrated by the phenomena of homology, homeoplasy, parallelism, convergence and the *loi du balancement*.

3.6 In ontogeny, the holons at successive levels represent successive stages in the development of tissues. At each step in the process of differentiation, the genetic canon imposes further constraints on the holon's developmental potentials, but it retains sufficient flexibility to follow one or another alternative developmental pathway, within the range of its competence, guided by the contingencies of the environment.

3.7   Structurally, the mature organism is a hierarchy of parts within parts. Its "dissectibility" and the relative autonomy of its constituent holons are demonstrated by transplant surgery.

3.8   Functionally, the behaviour of organisms is governed by "rules of the game" which account for its coherence, stability and specific pattern.

3.9   Skills, whether inborn or acquired, are functional hierarchies, with sub-skills as holons, governed by sub-rules.

4. *Integration and self-assertion*

4.1   Every holon has the dual tendency to preserve and assert its individuality as a quasi-autonomous whole; and to function as an integrated part of an (existing or evolving) larger whole. This polarity between the Self-Assertive (S-A) and Integrative (INT) tendencies is inherent in the concept of hierarchic order; and a universal characteristic of life.

The S-A tendencies are the dynamic expression of the holon's wholeness, the INT tendencies of its partness.

4.2   An analogous polarity is found in the interplay of cohesive and separative forces in stable inorganic systems, from atoms to galaxies.

4.3   The most general manifestation of the INT tendencies is the reversal of the Second Law of Thermodynamics in open systems feeding on negative entropy (Schrödinger), and the evolutionary trend towards "spontaneously developing states of greater heterogeneity and complexity" (Herrick).

4.4   Its specific manifestations on different levels range from the symbiosis of organelles and colonial animals, through the cohesive forces in herds and flocks, to the integrative bonds in insect states and Primate societies. The complementary manifestations of the S-A tendencies are competition, individualism, and the separative forces of tribalism, nationalism, etc.

4.5   In ontogeny, the polarity is reflected in the docility and determination of growing tissues.

4.6   In adult behaviour, the self-assertive tendency of functional holons is reflected in the stubbornness of instinct rituals (fixed action-patterns), of acquired habits (handwriting, spoken accent), and in the stereotyped routines of thought; the integrative tendency is reflected in flexible adaptations, improvisations, and creative acts which initiate new forms of behaviour.

4.7   Under conditions of stress, the S-A tendency is manifested in the aggressive-defensive, adrenergic type of emotions, the INT tendency in the self-transcending (participatory, identificatory) type of emotions.

4.8 In social behaviour, the canon of a social holon represents not only constraints imposed on its actions, but also embodies maxims of conduct, moral imperatives and systems of value.

5. *Triggers and scanners*

5.1 Output hierarchies generally operate on the trigger-release principle, where a relatively simple, implicit or coded signal releases complex, pre-set mechanisms.

5.2 In phylogeny, a favourable gene-mutation may, through homeorhesis (Waddington) affect the development of a whole organ in a harmonious way.

5.3 In ontogeny, chemical triggers (enzymes, inducers, hormones) release the genetic potentials of differentiating tissues.

5.4 In instinctive behaviour, sign-releasers of a simple kind trigger off Innate Releasive Mechanisms (Lorenz).

5.5 In the performance of learnt skills, including verbal skills, a generalized implicit command is spelled out in explicit terms on successive lower echelons which, once triggered into action, activate their sub-units in the appropriate strategic order, guided by feedbacks.

5.6 A holon on the n level of an output-hierarchy is represented on the $(n+1)$ level as a unit, and triggered into action as a unit. A holon, in other words, is a system of relata which is represented on the next higher level as a relatum.

5.7 In social hierarchies (military, administrative), the same principles apply.

5.8 Input hierarchies operate on the reverse principle; instead of triggers, they are equipped with "filter"-type devices (scanners, "resonators", classifiers) which strip the input of noise, abstract and digest its relevant contents, according to that particular hierarchy's criteria of relevance. "Filters" operate on every echelon through which the flow of information must pass on its ascent from periphery to centre, in social hierarchies and in the nervous system.

5.9 Triggers convert coded signals into complex output patterns. Filters convert complex input patterns into coded signals. The former may be compared to digital-to-analogue converters, the latter to analogue-to-digital converters (Miller, *et al.* 1960).

5.10 In perceptual hierarchies, filtering devices range from habituation and the efferent control of receptors, through the constancy phenomena, to pattern-recognition in space or time, and to the decoding of linguistic and other forms of meaning.

5.11   Output hierarchies spell, concretize, particularize. Input hierarchies digest, abstract, generalize.

6.   *Arborization and reticulation*

6.1   Hierarchies can be regarded as "vertically" arborizing structures whose branches interlock with those of other hierarchies at a multiplicity of levels and form "horizontal" networks: arborization and reticulation are complementary principles in the architecture of organisms and societies.

6.2   Conscious experience is enriched by the cooperation of several perceptual hierarchies in different sense-modalities, and within the same sense-modality.

6.3   Abstractive memories are stored in skeletonized form, stripped of irrelevant detail, according to the criteria of relevance of each perceptual hierarchy.

6.4   Vivid details of quasi-eidetic clarity are stored owing to their emotive relevance.

6.5   The impoverishment of experience in memory is counteracted to some extent by the cooperation in recall of different perceptual hierarchies with different criteria of relevance.

6.6   In sensory-motor coordination, local reflexes are short-cuts on the lowest level, like loops connecting traffic streams moving in opposite directions on a highway.

6.7   Skilled sensory-motor routines operate on higher levels through networks of proprioceptive and exteroceptive feedback loops within loops, which function as servo-mechanisms and keep the rider on his bicycle in a state of self-regulating, kinetic homeostasis.

6.8   While in S-R theory the contingencies of environment determine behaviour, in O.H.S. theory they merely guide, correct and stabilize pre-existing patterns of behaviour (Weiss).

6.9   While sensory feedbacks guide motor activities, perception in its turn is dependent on these activities, such as the various scanning motions of the eye, or the humming of a tune in aid of its auditory recall. The perceptual and motor hierarchies are so intimately co-operating on every level that to draw a categorical distinction between "stimuli" and "responses" becomes meaningless; they have become "aspects of feedback loops" (Miller *et al*).

6.10   Organisms and societies operate in a hierarchy of environments, from the local environment of each holon to the "total field", which may include imaginary environments derived from extrapolation in space and time.

7. *Regulation channels*

7.1 The higher echelons in a hierarchy are not normally in direct communication with lowly ones, and vice versa; signals are transmitted through "regulation channels", one step at a time.

7.2 The pseudo-explanations of verbal behaviour and other human skills as the manipulation of words, or the chaining of operants, leaves a void between the apex of the hierarchy and its terminal branches, between thinking and spelling.

7.3 The short-circuiting of intermediary levels by directing conscious attention at processes which otherwise function automatically, tends to cause disturbances ranging from awkwardness to psychosomatic disorders.

8. *Mechanization and freedom*

8.1 Holons on successively higher levels of the hierarchy show increasingly complex, more flexible and less predictable patterns of activity, while on successive lower levels we find increasingly mechanized, stereotyped and predictable patterns.

8.2 All skills, whether innate or acquired, tend with increasing practice to become automatized routines. This process can be described as the continual transformation of "mental" into "mechanical" activities.

8.3 Other things being equal, a monotonous environment facilitates mechanization.

8.4 Conversely, new or unexpected contingencies require decisions to be referred to higher levels of the hierarchy, an upward shift of controls from "mechanical" to "mindful" activities.

8.5 Each upward shift is reflected by a more vivid and precise consciousness of the ongoing activity; and, since the variety of alternative choices increases with the increasing complexity on higher levels, each upward shift is accompanied by the subjective experience of freedom of decision.

8.6 The hierarchic approach replaces dualistic theories by a serialistic hypothesis in which "mental" and "mechanical" appear as relative attributes of a unitary process, the dominance of one or the other depending on changes in the level of control of ongoing operations.

8.7 Consciousness appears as an emergent quality in phylogeny and ontogeny, which, from primitive beginnings, evolves towards more complex and precise states. It is the highest manifestation of the Integrative Tendency (4.3) to extract order out of disorder, and information out of noise.

8.8 The self can never be completely represented in its own awareness, nor can its actions be completely predicted by any conceivable information-processing device. Both attempts lead to infinite regress.

H

9. *Equilibrium and disorder*

9.1 An organism or society is said to be in dynamic equilibrium if the S.A. and INT tendencies of its holons counter-balance each other.

9.2 The term "equilibrium" in a hierarchic system does not refer to relations between parts on the same level, but to the relation between part and whole (the whole being represented by the agency which controls the part from the next higher level).

9.3 Organisms live by transactions with their environment. Under normal conditions, the stresses set up in the holons involved in the transaction are of a transitory nature, and equilibrium will be restored on its completion.

9.4 If the challenge to the organism exceeds a critical limit, the balance may be upset, the over-excited holon may tend to get out of control, and to assert itself to the detriment of the whole, or monopolize its functions —whether the holon be an organ, a cognitive structure (*idée fixe*), an individual, or a social group. The same may happen if the coordinate powers of the whole are so weakened that it is no longer able to control its parts (Child).

9.5 The opposite type of disorder occurs when the power of the whole over its parts erodes their autonomy and individuality. This may lead to a regression of the INT tendencies from mature forms of social integration to primitive forms of identification and to the quasi-hypnotic phenomena of group psychology.

9.6 The process of identification may arouse vicarious emotions of the aggressive type.

9.7 The rules of conduct of a social holon are not reducible to the rules of conduct of its members.

9.8 The egotism of the social holon feeds on the altruism of its members.

10. *Regeneration*

10.1 Critical challenges to an organism or society can produce degenerative or regenerative effects.

10.2 The regenerative potential of organisms and societies manifests itself in fluctuations from the highest level of integration down to earlier, more primitive levels, and up again to a new, modified pattern. Processes of this type seem to play a major part in biological and mental evolution, and are symbolized in the universal death-and-rebirth motive in mythology.

DISCUSSION

BERTALANFFY As Arthur has called his presentation a study in systems theory, I would like to make a few comments. Arthur has not only presented us with a wealth of ideas but his presentation concerns many fields and levels, pointing towards the possibility of a general or interdisciplinary science, intended to cover regularities appearing in different fields. Whether or not the formal denomination, General System Theory, is adopted, or another name or no particular name, is not very important. What matters is that there appears to be a general tendency towards a superordinate conceptual system (or systems) which is felt necessary in various disciplines and, to an extent, corresponds to Leibniz' classical idea of *mathesis universalis*.

As I said in my talk and in adoption of Arthur's terms, the question of hierarchical order is that of "The Tree and the Candle"—or, mathematically speaking, questions of topology, graph and net theory on the one hand, feedback theory, open systems, etc. on the other. We do not have a synthesis or formal theory at present; but the question is more or less clearly formulated, and posing a problem may be a first step towards a solution.

This will have very pronounced effects on concrete problems. The embarrassment in the problem of biological differentiation for example, seems to be that we have an enormous amount of empirical and experimental knowledge, but the conceptual schemes applied, in the first line the Monod-Jacob scheme of repressors and de-repressors, still remain in the framework of one-way causality, and therefore by its very nature says little about the order and organization in the multitude of processes leading from the genome to the developed organism. Therefore we are left baffled just in this basic aspect of embryonic development.

WEISS I would also like to obtain general acceptance of the hierarchic concept as a realistic experience growing out of our knowledge of complex systems. I am in full agreement with practically everything Koestler said. But I do not like his tree diagram for reasons which were partly included in your own comments—because it again evokes the picture of binary decisions taken along pre-formed channels. This bothers me although I am convinced of the hierarchic concept. Yet I am almost thrown back into the linear cause-effect chain. I know that you do not mean this, but rather the opposite, yet that is what it evokes in my mind, and I can see the danger that this will happen to other people too.

Let me give you a practical example of the applications of the hierarchic concept. It is an experiment which I did more than forty-five years ago. (Blackboard.) If you cut a little piece out of the humerus of an amphibian limb, the bone will re-form an exact copy of its previous shape, and the rest of the limb does not pay any attention to what you have done. So the humerus behaves as an autonomous holon, a holon on a pretty low level of the hierarchy. But if you take the whole bone out, there remains a hole there for life because the holon as a whole has gone.

Examples of this kind can be multiplied and they should be kept in mind because they apply to all organic systems, and not just the nervous system. What bothers me about your tree diagram is that neurons happen to have linear extensions. And this linear extension, the geometric configuration of that network, has a tendency perhaps to distract our minds, and deflect it from the possibility of continua in the central nervous system which are over and above those neuron connections, as discussed by Holger Hydén yesterday. I think it is completely erroneous just to confine ourselves to a network theory of connections before we have really any answers.

KOESTLER   The lines which form the branches of the tree are purely symbolic—just as you might use an arrow in a symbolic diagram to indicate that one process leads to another. The tree can symbolize lines of communication, or the differentiation of tissues in an embryo, or Chomsky's phrase-generating mechanism, or a genealogical table, and has been used in this way from time immemorial. I really can't see your objection.

FRANKL   I should like first to read a sentence drawn from your text, and to comment on it, wondering whether or not you agree. The sentence reads that the holon is a system of relations which is represented on the next higher level as a unit, a relatum. This reminds me of something I published twenty-two years ago, and I would like you to comment on it. In a way the world as a whole may be conceived as a system of relations. More specifically, however, I added that the world could be conceived as a system of relations in ever higher dimensions. This is the point I would like to stress, on which I am going to elaborate in my paper. Because, indeed, relations —which are constitutive for the being of things—relations are, to me, intrinsically associated with this concept of dimensions. The relation between two non-dimensional points is a line. The line is one-dimensional the points are non-dimensional. Any relation between two one-dimensional lines can only be established within a plane, a two-dimensional plane. For any relation to be established between two planes we necessarily have to enter three-dimensional space. The dimension concept should be put alongside the hierarchy concept.

Now let me return to your proposition that holons are—on the next higher level—units? Do you believe that we may proceed along these lines to ever higher levels—that is to say, is this a *progressus in infinitum*? It would seem to me that the human person is characterized by both, its *oneness* and its *wholeness*, and I am defining oneness as the property of something that cannot really be split up, can no longer be divided, i.e., it is *in-dividual*. But it is also *insummabile* in so far as to wholeness nothing can be added, it cannot "*aufgehen in*", submerge, or be absorbed. And now my question is: do you really believe that the human person has no more of the property of oneness and wholeness, *individuum* and *insummabile*, than an organ? Is an organ in the same way and to the same degree a part and a whole as a person? And on the other hand, has a nation to the same extent and in the same degree the attributes of partness and wholeness as a person? I doubt that several persons can merge and enter the holon of a nation to the same degree, the same extent, in the same sense as organs are participating in the wholeness of a human organism.

KOESTLER   I agree with most of what you said, although our terminologies differ. First, I have to say that in this paper I did not go into philosophical questions, quite deliberately not because that was not on the agenda. Thus I touched on the subjective experience of free will by proposing that we have the experience of free will each time when decision-making is suddenly shifted from a lower level of the hierarchy to a higher level—from driving my car in an absent-minded way, letting the automatic pilot in my nervous system take care of the job, to conscious decision-making—whether to overtake or not. At this moment my ego at the apex of the hierarchy seems to take over from the automatic pilot situated on a lower level.

But I was deliberately begging the question about the nature of that apex. One could of course say melodiously that the apex is where everything is being integrated. But consciousness of self, and consciousness of my consciousness of myself form a receding series, and the apex vanishes in the mist. In the book on which this paper was based I have in fact come to the same conclusion as you—an infinite regress. One might think that everything is integrated here at the apex of the hierarchy—but nevertheless whenever the ego tries to attain full focal consciousness of itself, the infinite regress starts because the agency which directs the searchlight of my attention can never be caught in its focal beam. This is really the old pardaox that the subject can never be totally contained in its own awareness. Or, to put it into contemporary terms, the computer who knows everything about me, who has been fed every bit of information about me, cannot predict what I will do next because it must construct a complete model of me, including the process of constructing a model of me, and so on ad infinitum. Don Mackay has a similar formulation, and so has Karl Popper. But I have deliberately avoided all these questions.

Now to your question, what I meant by saying that a holon of the n level is represented on the $n+1$ level as a unit. I meant something quite simple. An army battalion is a complex holon consisting of companies, which consist of people and their rule-governed relations, but on the next higher level, at regimental headquarters, the whole complex battalion is simply labelled "Battalion C" and manipulated as a unit. Now you could of course again ask me where that higher level is on which the apex of the hierarchy, the complex ego, is represented as a unit, and my answer would again be: in an infinite regress.

BERTALANFFY I believe I can answer Frankl's other question in system-theoretical terms. There are holons, but there are holons which are more so (laughter). It depends on the strongness of interactions within the holon concerned and on those between this holon and the super-ordinate holon.

KOESTLER Now to Frankl's point that each individuum is also *insummabile*—it is unique, it is one and it is whole. That's fine. But I would like to add that while he is one and a whole, as you say, he is also a part. And this partness is inseparable from the individual. You cannot live as an island.

HYDEN I would like to make a brief comment on this discussion on hierarchies. Weiss's example of the holon, the biological example, is a fine one, and it is easy to grasp. It holds for many other examples. As he said, we could sit here for the whole afternoon and enumerate such examples. But the concept of a hierarchy should not be obscured just by the sight of those branches on the tree, because that is not meant as something to visualize in a concrete way, those lines are just symbolic.

But now, when we ascend in the hierarchy of holons, of sub-units or sub-packages, we come to the top, the apex, and you asked the question, what is the apex. Well, isn't the answer indicated by what von Bertalanffy said yesterday. He emphasized the flux of a system, and also that to describe an apex, the apex level of the whole hierarchy, we need a mathematical grammar just as we need it to describe the DNA's relationship with the whole system. At present we lack the means to describe it, but we can still emphasize the flux.

KOESTLER We have discussed this so often—we have not got a mathematical symbolism for this way of thinking. We have to talk in metaphors—and the tree is a useful metaphor.

BERTALANFFY We need the algorithm.

KOESTLER Whatever it is, we have not got it. And we are very conscious of this.

INHELDER In your paper you quoted Paul Weiss's statement that "the structure of the input does not produce the structure of the output, but merely modifies intrinsic nervous activities which have a structural organization of their own". We can apply this statement also to the growth of knowledge. The structure is in the transformational system from the input to the output, not necessarily in the input. We are trying to show the existence of hierarchies in the transformational structures by studying developmentally how the output changes *while the input remains the same*; i.e. assuming the input to be a certain problem or situation, we study at different age-levels the different ways children deal with these problems or situations (output). This is why I am afraid that your use of the terms "input hierarchies" and "output hierarchies" may lead to misunderstandings and may recall the terminology of certain behaviourists with whom you rightly disagree.

As far as the difference between your two types of memory, which you liken to our distinction between scheme and schema, is concerned, I think that the question of whether the distinction is one of degree or nature is still open. We are not yet in a position to say anything about eidetics which seems to be a rare phenomenon and which we have only encountered once and that was several years ago before we began our work on memory.

KOESTLER I will give you a simple example for this input-output business from adult language. When you ask me a question, say, about holons, then the structure of the input, the question you ask me, modifies the output. I would answer another question differently, but the structure of my answer is preformed in my cognitive structures. It is only modified by the form of the question you ask. Your question does not determine my answer, my answer is predetermined in its broad outlines, but it will be flexible and strategic.

INHELDER   Yes.

KOESTLER   That is what Weiss meant by the sentence you quoted.

WEISS   Arthur's paper is so beautiful that I would like to have it completely decontaminated, wherever there is a possibility of arousing misconceptions in the readers' minds. You give this example of Simon which I think is utterly misleading, because the watchmaker assembles parts from elements, he assembles them either up to the final watch or first into sub-assemblies, then he assembles sub-assemblies into super-assemblies, and I am absolutely against this kind of picture—of being able to synthesize by assembling a system, because then we are right back there where we started. This is my objection.

KOESTLER   The watchmaker's parable has of course its limitations, but it provides an exact analogy of what you said in your paper that a cell or organ does not consist of elementary bits but of autonomous, organized sub-wholes or holons.

THORPE   Arthur, you referred to hierarchy in instinctive organization. It is just about twenty years since Niko Tinbergen and I were asked to arrive at some definitions for use by ethologists when talking about instinct. We got together and started out by saying "instinct is hierarchically organized". In fact, this statement has not, over the years, been found terribly useful; although it is more useful at the level of, say, the instinctive behaviour of insects and fish, where it works pretty well, than it is at higher evolutionary levels. And it is still quite a useful guide in designing experiments. But when we come to the higher animals, it is found to be either useless or, at most, of very limited use. At this level what is required is a complicated network; not a hierarchy or at least not only or primarily a hierarchy. We found that there had to be so many connections to different centres at the same level, so many feedback loops to higher centres and so on and so forth, that the idea of hierarchy, while it was still there, became, so to speak, submerged amongst the many other means of influencing the lower levels of behaviour. So this idea of instinct being a hierarchy tends now to be dropped out when we are speaking of higher animals. Another reason for this is that we now know so much more about the action of hormones. You find in many cases that the action of hormones is not even on an intermediate centre, but may be direct, influencing perhaps the very tissues being activated in the final movement. And where hormones play a very important part in the organization of the animal, I think the concept of hierarchy has a rather limited role. So it seems to me that while one would never suggest that anything like Tinbergen's hierarchy network was found anywhere in the nervous system as such, behaviourally it does seem to be rather more characteristic of those animals where the structure of the nervous system is the prime means of organizing highly stereotyped behaviour. And one certainly gets evidence of something like a hierarchical organization working, in the case of highly elaborate sense organs such as the eye, between the periphery and the centre. When you get down to a structure such as the nerve net found in the *coelenterates*, I doubt whether the hierarchy concept is of any use at all. I would like Paul Weiss's view about this. It seems to me that you would not need the hierarchy concept in describing any behaviour I can think of amongst the *coelenterates*.

SMYTHIES Let me just make one point about models. I think there is a danger of mixing up two different types of hierarchy. Talking about the nervous system, clearly the anatomy of the nervous system is arranged in terms of hierarchies of neurons extending up to very complicated lattice works. I do not think there is any fundamental difference depending on how many end points you have, between a lattice work and a hierarchy. A hierarchy does not have to have only one end point. So the two blur into each other. During much of the discussion, talking about trees and things, one has to think of the straightforward hardware involved in the brain. There is also another hierarchy—of the logical flow-pattern of information about what the processes are doing. Starting with very simple identification, going on to classification, to categorization, and ending with decision-making—this again is a hierarchy. It is not necessary that the highest parts of the first hierarchy should be consonant with the highest parts of the second hierarchy. You can have very, very high degrees of functional hierarchy in the retina, for example, which is very lowly on the anatomical hierarchy. So one tends to get confused if you do not keep anatomy and the logical flow pattern separate.

BERTALANFFY Let us not fall into reification of concepts. Don't let us say that the nervous system *is* a hierarchy, but a hierarchy is a partial model of certain aspects and there may be other things. . . .

SMYTHIES   Well, if you say the nervous system is a hierarchy, you are not contradicting yourself by saying it also has other attributes.

KETY   I found Arthur's particular model a very intriguing and satisfying description of the machinery of the nervous system, which brought together some of the modern concepts of the programming at lower levels. But I still say it is not only compatible with mechanism, it is a model of a mechanism, and where I felt that you made a leap which you did not take the time to explain was the leap into free will and consciousness, which I do not frankly believe you can fit into your system, and certainly have not fitted in in your very brief reply to Frankl.

KOESTLER   I have just explained that I deliberately did not want to go into this today. But in the book on which this paper was based I did attempt to fit free will into the system. There are some fifty pages of discussion on free will and the mind-body problem and all the earlier chapters in the book—and all that I said today—leads up to it.

But may I briefly answer Thorpe. You see, when Niko Tinbergen published his stickleback hierarchy diagram that was really so oversimplified and coarse—and it was not really a model because the higher levels represented just categories of behaviour—"mating", "fighting" in the abstract—and not actual behaviour. Real action only came in at the lower twigs of the tree.

THORPE   It branched into appetitive behaviour when the next lower level was reached.

KOESTLER   Precisely. But before that level was reached the diagram did not represent actual behaviour, merely a classification. Incidentally, this same criticism has already been made by Miller and Pribram.

THORPE   Yes.

KOESTLER   But now if we turn to your own remarkable diagram of the nest-building actions of the tit—the lining, the trampling, and collecting, and so on—then your own diagram provides the answer to your objection. I suggested in my paper that hierarchic arborization—sort of vertical tree structures—and horizontal reticulation, networks of horizontal interactions between the branches of the trees—that these two factors are complementary. But in some systems the hierarchic aspect dominates, in other systems the reticular aspect dominates. When you walk through a very dense forest where the trees are interlocking, entwined, you may tend to forget that the trees are vertical structures because they are obscured by this horizontal reticulation of networks of entwined branches on level after level. But you could not have done your model by doing away with the tree structure.

THORPE   Yes, I agree. I said that the model was less useful not so much because hierarchy no longer applies, but because there are so many other more direct ways of control. The simple Tinbergen scheme was, in fact, too simple. It was useful to some extent for certain groups of lower organisms but was so over-simplified that it was of very little value for the higher ones.

KOESTLER   Exactly. And when you say that when we come to higher animals, then the obscuring of the tree-structures by reticulation is even more pronounced, I agree of course. But now when we get to an even higher animal, namely man, then in skills such as languages or mathematics or piano-playing, you get a very clear hierarchical tree-structure.

THORPE   Yes. To give one example of the elaborate behaviour of an insect where the Tinbergen scheme did at least seem to fit very well, I would mention the work of Baerends on the Hunting Wasps (*genus Ammophila*) and their nest provisioning. One did not need very much else to have at least a good theoretical skeleton on to which to hang one's facts. But when one comes to a higher level it seldom works out that way.

KOESTLER   Our linguists have not vocalized yet.

MCNEILL   Linguists, even psycholinguists, do not vocalize (laughter) . . . I wonder if you have considered and rejected or just not considered the implications of the fact that the structure of grammar is transformational?

KOESTLER   That is one of the headaches.

MCNEILL   The implications are very interesting. You mentioned your experiments, with number sequences on the tachistoscope, in which inversions were a common mistake. Inversion is important also in syntax. Transformations are the main source of syntactic idiosyncrasy among languages, but there are a few elementary transformations that apparently are universal. Among them is inversion. One way of regarding the fact that inversion is a universal transformation is that there is a particular sensitivity to that kind of relation. Your experiment showed an ability to recall number symbols, and also what you described as an inability to recall their correct order. Perhaps you could just invert the interpretation and say that the experiment shows an ability to recall symbols and an ability to invert order.

KOESTLER   It goes one step further. You can also have groups of three digits permutated.

MCNEILL   Okay. I was going to mention an experiment I did, which is somewhat similar to yours. It is interesting how people solve anagrams. What I did was to start with a collection of anagrams—not sentences but words, scrambled words—and have people try to figure out what they were. They were to write down everything that came to mind as they attempted to solve the anagrams. Some anagrams are seen and understood instantly; those cases are of no value. Other cases are never figured out, and they too are of no value. The interesting ones are those that take some time to be perceived but the correct solution is eventually reached. Now you can look at the history of hypotheses a person goes through in order to reach a solution. The surprising thing is that this history bears absolutely no relation to the solution. Presumably all along, in parallel to the conscious rearrangement of letters, a completely separate and unconscious process of rearrangement was also going on. Inversion of letters was taking place automatically, spontaneously.

KOESTLER   The prayer-mills were spinning round . . . I can give you the results briefly. Half the errors were transpositions—correct identification of all the digits, but combined with transpositions of digits. The more difficult the task becomes—usually when you show eight to nine digits at one-hundredth of a second, there is a steep increase of errors, and fifty per cent of these are inversions in the order of correctly identified figures.

MCNEILL   One suggestion is that the elementary transformations—like inversion—interrelate hierarchies. That's their function in language, to map hierarchical structures on to one another. It is quite astonishing that there are so few universal transformations: of all conceivable relations between hierarchies only a handful actually appear in language. The degree of specialization is extreme. The same elementary relations might be important outside language, for all we know, and the connecting of one holon to another might be understood in the light of this same specialization. Thus, one of the possible interconnections among hierarchies is by transformations, and that would be a place where they would fit into this . . .

BRUNER   I am concerned about being caught in three-dimensional thinking about this question, which I think we necessarily are when we try to concoct a picture. I think we are limited by three dimensions.

KOESTLER   The tree can also be in semantic space or wherever you want it.

WEISS   I could give you any amount of experimental examples of hierarchic order, starting with the individual muscle innervation of a single reflex and then going up to the operation of a whole limb, and the coordination of limbs in ambulation, and up to the intersensory coordination. The input is merely producing a distortion, a very definite distortion or modification of the existing pattern, which in fortunate, favourable cases has a one-to-one correlation with the output, just as in an elastic network. If you distort an elastic network of the greatest complication—I am speaking now as a person with an engineering training—every thread in that network knows exactly the dimension, the direction, the orientation, the amount of distortion to which each element has been subjected. It knows it, it reverses its orientation by the same amount and the network snaps back into the old organization.

AFTERTHOUGHTS* by Arthur Koestler

As one of the recurrent leitmotifs of the Symposium was the isomorphism of some basic laws and principles in widely different branches of science, it may be in order to inquire whether isomorph principles also operate across different domains of art, or even across the boundaries of the two cultures. I believe this to be the case, and would like to point briefly to a few examples which I have discussed at length in an earlier book (1964)[1].

Among the many criteria by which we assess the impact of a work of art there are three which seem to me of fundamental importance: *originality*, *emphasis* and *economy*.

The value of *originality* is self-evident, whether it creates a revolutionary discovery in science or opens new vistas in art. In both cultures revolutions have a constructive and a destructive aspect: old theories and conventions are ruthlessly discarded, new ones are erected in their place. What Thomas Kuhn (1962) calls "paradigm changes" in science has its counterpart in the periodic upheavals shown in the history of every type of art.

But revolutions do not happen every day. In between there are long periods of consolidation, elaboration and exploitation of the new frontiers—the equivalent of Kuhn's periods of "normal" or "routine" science. In the evolution of both cultures the experience gained during these periods is cumulative, while the paradigm changes are transformative.

Let us turn to the second criterion, *emphasis*. The artist, as the scientist, is engaged in making models of reality, whether his medium is paint, marble or ink. But the model can never be an exact copy of reality. The innate and acquired structuring of his perceptions on the one hand, and the limitations of his medium on the other, compel him at each step to make decisions: to select and emphasize those aspects of reality which he considers significant and ignore those which he considers irrelevant, to decide which features in the input should be treated as signal and which as noise. Some aspects of experience cannot be reproduced at all, some only by oversimplification or distortion; and some only at the price of sacrificing others. What I called "emphasis" thus always involves three interrelated factors: selection, exaggeration, and simplification. It is a technique found in

* Written after the Symposium.
[1] I have used a few passages from it in this text.

228

all fields of art, in the caricature as well as in the portrait (they differ only in the motivation which provides the criteria of relevance); in the narration of events, whether historic or fictitious. But the same technique is also employed by the scientist: every model, theory or schema is a deliberate and painstaking caricature of reality based on selection, exaggeration and simplification according to the model-maker's criteria of relevance, which determine where to put the emphasis. And the history of science makes us realize that the criteria of relevance are liable to as radical changes as the changes of style in art. Signals become noise, and the noise gives birth to new signals.

The third factor I mentioned is *economy*. Yesterday's discoveries are today's commonplaces. A daringly fresh metaphor soon becomes stale by repetition and loses its emotive impact. The recurrent cycles of stagnation, crisis, and new departure in the arts are to a large extent caused by the gradual saturation which any particular invention, technique or style produces in both artist and audience. When that happens, the audience is exempted from the necessity to exert its imagination, and deprived of its reward; while the artist, in growing frustration, senses that the conventional techniques have become stale. The history of art could be written in terms of the artist's struggle against the cumulative, deadening effect of saturation and habituation. If he is a genius, he will invent a new style which inaugurates a revolution, a change of paradigms. But in between these turning points one can observe a more gradual evolution of styles, which seem to proceed in two opposite directions, both intended to counteract saturation. One is a trend towards more pointed emphasis; the other towards more economy or implicitness. The first strives to recapture the artist's waning mastery over the audience by mannerisms, an overly explicit appeal to the emotions, by "rubbing it in"—symptoms of decadence which also have their parallels in the mannerisms, rigid pedantry and jargon-formation of tired branches of science—whether it be Aristotelian physics, Ptolemaic astronomy or behaviourist psychology.

The opposite trend in the evolution of art forms, the trend towards economy and implicitness, is more interesting and significant from our point of view. It is customary to credit the French Symbolist poets with having inaugurated the shift from explicit statement to implicit hint, and the French Impressionists with having initiated it in painting. In reality, this progression from the obvious to the elusive and oblique is much older, perhaps as old as art itself. All mythology is studded with symbols, veiled in allegory; the parables of Christ pose riddles which the audiences must solve; the dialogues of Plato are still being re-interpreted. Implicitness does not aim at obscuring the message, but at making it more luminous by compelling the recipient to work it out by himself—to re-create it. To this end the message must be handed to him in implied form—and implied means "folded in". To make it unfold, he must fill in the gaps, complete the hint, see through the symbolic disguise. However, the audience has a tendency to become more sophisticated with time; once it has mastered all the tricks, the excitement goes out of the game; so the message must be made even more implicit, more tightly folded in.

I believe that this development towards greater economy (meaning not brevity, but implicitness) can be traced in virtually all periods and forms of art: I have called it the "law of infolding". It is the antidote to the law of diminishing returns in emotive terms. In the book mentioned earlier I have given examples of it going back to the origins of Greek tragedy. But even art forms as recent as Victorian melodrama strike us as grotesque; and even films not more than twenty years old, which we highly enjoyed at the time, appear now surprisingly dated—overdone, obvious, over-explicit.

Emphasis sans economy, as in pre-Raphaelite painting, is a safe criterion of bad art. Economy demands that the stepping-stones of the narrative should be spaced wide enough apart to require a significant effort from the reader. To this end the writer must leave out, first, everything that he considers irrelevant; and then everything that is obvious and that the reader can supply out of his own imagination. Modern prose had to accelerate its pace, not because trains run faster than mail-coaches, but because the trains of thought run faster than a century ago, on tracks beaten smooth by popular psychology, the mass-media, and torrents of print. The novelist no longer needs to crank up the reader's imagination as if it were a model-T car; he pushes the button of the self-starter and leaves the rest to the battery. The surest symptom of decadent art is that it leaves nothing to the imagination; the Muse has bared her flabby bosom like a too-obliging harlot—there is no veiled promise, no mystery, nothing to divine.

The law of infolding can also be traced, *mutatis mutandis*, in the history of science. Aristotle believed that all possible inventions and discoveries had already been made in his time. Two thousand years later, Francis Bacon and Descartes both believed that the edifice of science would be completed within a generation or two. Ernst Haeckel, the leading German propagandist of Darwin, proclaimed that six of the seven mysteries of the universe had been solved. Only since the collapse of the Laplacian model in our century did it begin to dawn on us that the unfolding of the secrets of nature was accompanied by a parallel process of infolding. The more precise knowledge the physicist acquired, the more ambiguous and elusive symbols he had to use to express it; he could no longer make an intelligible model of sub-atomic reality, he could only allude to it by symbolic equations. Any attempt to get a direct grasp at naked reality turns out to be self-defeating: Urania, too, like the other Muses, always has a last veil left to fold herself in.

REFERENCES

BARTLETT, F. C. (1958) *Thinking.* Allen & Unwin, London.

VON BERTALANFFY, L. (1952) *Problems of Life.* Harper Torch Books, New York.

CHILD, C. M. (1925) *Physiological Foundations of Behaviour.* Henry Holt, New York.

CHOMSKY, N. (1965) *Aspects of the Theory of Syntax.* M.I.T. Press, Cambridge, Mass.

GALANTER, E. (1960) See Miller, G. A.

HEBB, D. O. (1958) *A Textbook of Psychology.* W. B. Saunders, Philadelphia and London.

HERRICK, C. J. (1961) *The Evolution of Human Nature.* Harper Torch Books, New York.

JENKINS, J. (1965) See Koestler, A.

KOESTLER, A. (1964) *The Act of Creation.* Hutchinson, London and Macmillan, New York.

———— and JENKINS, J. (1965) "Inversion Effects in the Tachistoscopic Perception of Number Sequences" in *Psychon. Sci.,* Vol. 3.

———— (1967) *The Ghost in the Machine.* Hutchinson, London and Macmillan, New York.

KUHN, T. (1962) *The Structure of Scientific Revolutions.* Univ. of Chicago Press.

MacLean, P. (1949) "Psychosomatic Disease and the 'Visceral Brain' " in *Psychosom, Med.*, Vol. II, pp. 338–53.

MacLean, P. (1958) "Contrasting Functions of Limbic and Neocortical Systems of the Brain and their Relevance to Psychophysiological Aspects of Medicine" in *Am. J. of Med.*, Vol. 25, pp. 611–26.

Miller, G. A., Galanter, E. and Pribram, K. H., (1960) *Plans and the Structure of Behaviour*. Henry Holt, New York.

Needham, J. (1941) *Time, the Refreshing River*. Allen & Unwin, London.

Penfield, W. and Roberts, L. (1959) *Speech and Brain Mechanisms*. Princeton Univ. Press.

Pribram, K. H. (1960) See Miller, F. A.

Roberts, L. (1959) See Penfield, W.

Simon, H. J. (1962) "The Architecture of Complexity" in *Proc. Am. Philos. Soc.* Vol. 106, No. 6.

Thompson, D. W. (1942) *On Growth and Form*. Cambridge Univ. Press.

Thorpe, W. H. (1956) *Learning and Instinct in Animals*. Methuen, London.

Tinbergen, N. (1951) *The Study of Instinct*. Clarendon Press, Oxford.

Waddington, C. H. (1957) *The Strategy of the Genes*. Allen & Unwin, London.

Weiss, P. (1950) (ed.) *Genetic Neurology*. Univ. of Chicago Press.

———— (1951) in *Hixon Symposium*, ed. Jeffress, L. A. Wiley, New York.

Woodger, J. H. (1929) *Biological Principles*. K. Paul, London.

Aspects of consciousness

J. R. Smythies

University of Edinburgh

The topic I would like to discuss this morning is the time-honoured problem of consciousness and its nature. This is in fact not one problem but an interwoven complex of problems—which includes the mind-brain relation, the nature of perception, as well as the more limited problem of what does consciousness itself really consist. Looked at in this way, we can see that consciousness is a problem that comes within the range of professional interest of a number of different disciplines. The philosopher has customarily dealt with mind-brain relations; many neurologists and brain scientists working with the brain itself have felt free to speculate about how this complex machine is related to the human mind; the names of Russell Brain, Eccles and Sherrington come immediately to mind here. Then psychiatrists and psychologists must clearly hold to some theory of the psyche as a basis for their own formulations of the normal and abnormal in human experience and behaviour; as formulated, for example, by Freud and Jung.

My method will be to present the various theories currently held in each of these fields and attempt very briefly to evaluate each, both within itself and how it relates to the theories in the other fields.

If we start with the problem of the mind-brain relation: the traditional theory that goes back to Egyptian civilization and beyond, was most clearly formulated in Western culture by Descartes. Man consisted of two quite distinct parts—his body and his mind. The body was an extended object in the physical world no different in essence from all the other objects in the physical world. The mind consisted of the Ego together with its thoughts and feelings as this is immediately given to introspection. Some psychologists and philosophers today state that they do not understand what is meant by the terms "Ego" and "introspection" and they deny that any mental events of this type—given to direct experience—exist. Consciousness, it is said, is an abstraction. Since it is important to clarify what is meant by these terms let me ask you to imagine that you are sitting in an easy chair with your eyes shut and I ask you to pay attention to what is going on around you—the events in your field of experience. Each sense-field will contain particular events—distant sounds, the various subtle pressures of the chair on your body, sensations from muscles and joints, a stream of thoughts and shadowy mental images. Your visual field will contain a single reddish expanse in which various after images float about, changing colours and fading away. Now when David Hume was asked his opinion about the existence of the Ego, he complained that he was certain that it did not exist, for whenever he searched around in his consciousness all he could ever find was some particular thought or image and never any "Ego". However, this very failure on Hume's part enables us to define the Ego—for Hume's Ego was that which was actually doing this searching and failing to find anything except thoughts and images. Every man is entitled to give his own account of what goes on in his own consciousness, but for me at any rate, and for many other people, consciousness is not just a collection of thoughts and images and sensations but a collection of these belonging to, related to, presented to, observed by, a central Ego, that is also aware of its own existence in a way hard to describe and which can only be experienced—all I can say is that I am aware of my own existence separate from my thoughts, my images, and my sensations and that every perception, every thought and every image is a *relation* between my Ego and my sensations, thoughts and images. This is exemplified most clearly in dreams. Dreams are experienced by the Ego and sometimes I know that I am dreaming and can make observations e.g. "Are my dreams in colour?—Why yes, they are—not very bright but colours all the same". In Eccles' (1967) words: "First of all, let us consider what I mean by 'conscious experience'. This is strangely difficult to define, but fortunately I do not think a definition is necessary. My approach to conscious experience is, in the first instance, based on my direct experience of my own self-consciousness. I believe this to be the only valid way in which I can present to you this problem that lies central to our being."

Thus the consciousness of other people may be for me an abstraction, but my own consciousness is for me reality. All my scientific observation has been obtained by a conscious process and if I deny my own consciousness I saw off the branch on which all science rests. The device used by most scientists to avoid this problem, as we shall see, is to adopt simultaneously two mutually incompatible theories of perception, one for use in their scientific work and one for use in their private lives.

Thus for Descartes a man consisted of his body—a physical extended object—and his mind—made up of the Ego and thoughts, and his criterion for deciding which was which, was *extension in space*. Physical objects including brains were extended in space, mental objects were not. The puzzle then arose of how an unextended mind could interact with an extended brain, and this has never received a satisfactory answer. I will not, however, discuss this problem for I will argue later that the basic criterion chosen by Descartes for distinguishing between mental events and brain events, was in error.

During the last hundred years or so, the rival theory of psychophysical monism has gradually become the dominant theory in Western science and philosophy. In this theory man consists solely of a physical organism and it holds that every aspect of his experience, life and behaviour can be explained and accounted for fully on the supposition that the brain operates solely as a physical mechanism. All thoughts, all feeling, all perception and the control of behaviour are mediated by the complex electro-chemical events in the brain and the Cartesian "mind" simply does not exist.

This solution leaves us, however, with a number of problems, the main one being how one can fit the complex phenomena of consciousness, as I have briefly defined them above, into this scheme? One solution given by radical behaviourists is simply to deny that experience of the kind I described takes place at all. It seems perfectly plain to me for example when I have a red, round after-image, that I am observing out there a red, round entity, that changes to green, then back to red, fades away and returns etc. and that in this I am observing a series of events in my visual field. But the radical behaviourist would deny this and would say that I am merely emitting the words "I am seeing a red, round after-image etc." in response to complex events in my brain and that there is really no such red, round entity at all. A slightly less radical behaviourist theory would run as follows: it would say that when I observe a red, round after-image, the after-image is not really any kind of entity and that all that is involved is a certain state of a brain-mechanism making an analysis of the sensory inflow and it is merely my conviction, or my belief, that it is identical with the brain state. Likewise in veridical perception my usual field of coloured geometrical shapes has no real existence—all that really exists is a complex of visual judgements carried out by the relevant mechanisms in the brain. However, I would again feel that this theory does not do justice to the phenomena. Visual sensations are more than belief. A visual sensation of a tomato would seem to be an instance of the topological concept of a closed curve and the same applies to after-images and does not apply to sounds or toothache for example. However, the theory is certainly a possible one in competition with the other theories in the field. To my mind both varieties of behaviourist theories are not very plausible and invite us to look around for alternative theories. For a psychiatrist even an hallucination is more than a delusion. I am referring here to radical behaviourism as a philosophical theory, which I reject, and not to behaviourism as a working plan in certain aspects of psychology and brain sciences, which is perfectly valid.

The second theory is the theory of psycho-neural identity. This grants that the events in our consciousness really do take place just as we experience them, and suggests that these events are identical with certain brain events. The red, round after-image merely *is* a collection of neurones in my brain, as are my thoughts and sensations. In response to the complaints that the contents of my experience do not appear to be anything like, in shape, contour, colour, etc. the neural events they are supposed to *be*, Hirst has developed his double-aspect theory. This argues that things may have very different aspects according to the conditions of observation and that we should not therefore be surprised that the contents of our consciousness which are our brains experienced from the inside are so unlike events in our brains as experienced from the outside by a brain surgeon for example.

Now I shall shift the development of my argument to consider perception. For I shall argue that this whole problem has been confused and confounded by the almost unconscious use by most theorists in this field of a theory of perception totally incompatible with the consequences of the theory of psycho-neural identity or Hirst's "double-aspect" theory. Many scientists from innocence, and many philosophers by deliberate intent, hold to the theory of direct realism. Theories of perception are concerned with the problem of just how do we perceive the real world? Although there are a number of different theories I will develop only two—direct realism and the representative theory, ignoring such doubtful starters as phenomenalism.

Most people in their ordinary lives unwittingly accept the theory of direct realism. This theory states that, when you open your eyes the coloured shapes that confront you so immediately and compellingly in your visual field *are* literally the objects of the physical world; the sounds you hear *are* outside yourself in the physical world; the collection of pressures, aches, tingles and itchings that make up the body image in consciousness, to use Paul Schilder's technical neurological term, *is* the physical body.

However, doubt has been thrown on this account by physiology, neurology and psychology. It has become clear that perception is dependent on a causal chain consisting, in the case of vision, of light waves emitted or reflected by the object and then a complex of events in the retina and brain. The only *necessary* feature of this chain is the brain event. That is to say, the events in the object, eye, optic nerve etc. do not affect vision at all except in so far as they alter events in the visual cortex. Secondly experimental psychology showed by its discovery of size constancy, and a whole host of similar phenomena, that the visual field as we experience it, is really not very like the physical world it represents, as "seen", for example, by a simple camera. Thirdly, considerations of the famous time-gap in perception make it extremely difficult to accept naïve realism as a theory of perception. Suppose I look around my room and then go outside and look at the night sky. The brightly lit coloured objects in my room give way to a velvety blue dome sparkling with tiny points of light. The experiences are similar in nature. Yet science tells me that some of the stars I am seeing now no longer really exist, since some are located millions of light years away and I am seeing them in their past—some millions of years in their past. I am certainly *not* seeing the present star—which may not exist, for it may have blown up in the interval. How, then, can we be said to be immediately aware of the object itself in our perception of it, when it is plain that we are only in contact with past states of objects —far in the past, in the case of stars, a few micromicroseconds in the case of earthly objects?

Since no naïve realist has ever given a satisfactory defence against these arguments, let me look at the only alternative theory—the representative theory. This states that all our sensations, our percepts, the sense-data present in our sensory fields in consciousness, are all constructions of the nervous system. My world of vision is thus *not* a direct view of the ordinary world as common sense bids us to believe, it is rather a representation of the external world constructed by our brain mechanisms and is located inside the brain. In other words, no mechanism can act in its entirety both as a telescope and as a television set. The principles involved are mutually contradictory. Likewise our familiar "body" in consciousness, the somatic sensory field or the "body image" in neurology, is also located inside the brain. Thus, when I sit in my chair examining my world of experience and ask myself "where is my brain?" I may be mistaken if I answer "inside this head I experience here in my consciousness". For the entire "body" that I experience in consciousness is, according to this theory, really inside my brain.

This idea is one which we may hold when theorizing about other people's brains but it feels most uncomfortable when we apply it to ourselves. To protect ourselves against this sense of unease and disquiet, the concept of "projection of sensations" was invented. The account of perception led from object to light wave to retina to brain and then the perception aroused in the brain was held to be "projected" back into the physical world. However, no such process of projection exists, as Russell Brain and I have both argued, and the only reason to invent it was to escape from the dilemma "How can the brain be in the head if the head is in the brain?" If we identify the body image in consciousness with the physical body—as most of us do—then since visual percepts are clearly outside this body, then clearly they must have got out of the brain somehow—hence "projection". However, if we ask how this "projection" is brought about, on what physiological mechanism it depends, no answers are given. Clearly the muddle has arisen of trying to combine two incompatible theories—the physiological representative theory in the case of vision and naïve realism in the case of bodily sensation.

To avoid these muddles, we must choose either a thorough-going naïve realist theory or a thorough-going representative theory. For the three reasons I gave earlier the naïve realist theory is not very satisfactory and so we are left with the representative—double-aspect theory —no matter at what cost of violence to "common sense" this entails.

Having dealt with perception, we are now in better shape to look again at Descartes' problem of trying to distinguish between mental events and brain events. You will recall that the criterion he chose was spatiality—physical events had spatial qualities, mental events did not. Now it is important to distinguish here between Cartesian ideas of mind and those put forward by Gilbert Ryle and his followers. Descartes was talking about events he could experience himself —his own thoughts, ideas etc. Ryle's account is based on quite a different use of the word "mind"—that is in its dispositional sense: for example, if we say "Jones has a brilliant mind", we are not referring to any sparkling ghost in his machine, but we are saying something of how Jones is likely to do in various circumstances. In spite of Ryle's energetic attempts, I cannot really agree with him that the dispositional use of "mind" can be extended to cover *all* its uses. It certainly covers many of its uses but common usage of English also describes its other sense. I used one such at the beginning of this paper—"the names of Russell Brain, Eccles and Sherrington come immediately to mind here"; to say "a thought or an image came into my mind" is to describe a change in my conscious state—a new fact of my experience—and it does not necessarily have any relevance to my dispositional state.

With that proviso, let us look at consciousness again. I would like to define this as consisting of our sensory fields, thoughts, emotions and images—as immediately experienced by the Ego. Now if we define it in this way, we can at once spot Descartes' error. For, as Professor Price has pointed out, some of the contents of consciousness may lack spatial properties—for example, it does not make sense to say that a thought or a smell has a left side—but other contents certainly have spatial properties. An after-image, a visual image and, of course, visual sensations (or to give them their technical name—sense data) do have spatial properties. If you observe our old friend the red, round after-image again—it clearly has spatial qualities. Its edge describes a simple curve in a topological sense, dividing all visual space into one inside and one outside. A number of such visual images can bear spatial relations such as "above", "below", "between" to each other. Note here that the relations are mainly topological rather than metric, although metric qualities are introduced by the observation that one visual image can be larger or smaller than another. Likewise bodily sensations have intrinsic and extrinsic spatial qualities. Auditory sensations, however, have only location and olfactory sensations seem to lack all spatial attributes.

Thus it is clear that we cannot, if we wish to be consistent, distinguish between mental and physical events by the Cartesian criterion of spatiality. How then are we to distinguish between them? Is the "double-aspect" version of the psycho neural theory our only *possible* alternative to radical behaviourism? The answer is "no" for in the last few years alternative theories worthy of consideration have been put forward by the British philosophers C. D. Broad of Cambridge (1923) and H. H. Price of Oxford (1965). In 1952 Professor Price suggested that the difference between mental events, as I have defined them, and physical events, lay not in the fact that the latter were spatial whereas the former were not, but in the fact that they were both spatial but had different spatial location. That is to say, he suggested that there is one physical space in the world containing stars, planets and human bodies and brains, and in addition a number of other spaces, one for each human individual, that contained his thoughts, images and sense-data. Now we have been so conditioned for centuries by the Cartesian theory that we find it difficult to grasp the possibility that there could be more than one space in the world. Price argues, however, that there is no *a priori* reason why there should not be more than one space in the world. It is fairly easy to grasp his idea that physical objects are in one space and images etc. in another space.

But how, then, do we relate these different spaces? At this point Price's and Broad's theories part company. Price suggests that there are temporal and causal relations between the events in the two systems but no spatial relations. This, of course, introduces a new twist into our idea of causal relations: since science has up to now recognized only one spatial system in the world, causal relations have always been described between objects in this system—but Price argues that this is not logically necessary and that it is logically possible to have events linked by causal but not by spatial relations.

Broad's theory differs in the following respect. He suggests that physical space-time and the various mental space-times may all form cross-sections of an n-dimensional space where n can equal any number larger than 4 since there are many different ways of assembling a number of 3 spaces into one common space. Any point in physical space-time can be specified completely by 4 co-ordinates and any point in someone's visual field can also be specified by 4 co-ordinates. Broad's theory suggests that the axes from which these co-ordinates are obtained are not the same in the two cases (except the time axis).

The easiest model for this theory is obtained if we divide a cube up into planes. There are a great number of different ways of dividing a cube into planes, likewise we can put a number of different planes together in a wide variety of ways in a 3-space, a 4-space, a 4, 5-space etc. The same applies to 3 spaces ordered in an n space.

The advantage of Broad's theory over Price's is that the causal relations postulated between brain events and mind events do not incur the concept of action across a "non-space" but merely that the geometry of space is more complex than we had thought. Let us represent the entire 3-dimensional physical world by one plan A and the space of one person's mind by a second plan B, that intersects the first (Figure 1). The man's physical body and brain will be in A and

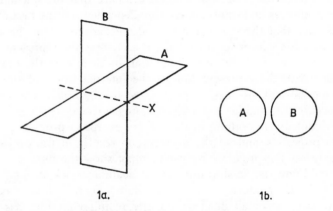

1a.   1b.

FIGURE 1 (*a*) *Broad's theory*
A represents the 3-dimensional physical world and B represents the
mental world. X represents the volume of their intersection.
          (*b*) *Price's theory*
A represents the entire physical world and B one person's entire mental
world. The "space" on the paper between them should be regarded as
non-existent.

his consciousness—his images, thoughts and sense-data—will be in B. The visual field, for example, can be regarded as being something like the screen of a television set, and similarly for the other sensory fields. The Ego's view of the physical world will thus be mediated by a mechanism working just like a television. There are some implications of interest in this theory.

(1) Events in mental space would seem to satisfy many, if not all, of the criteria for being "material" as specified for events in the physical world. The events would be, clearly, spatially distributed and organized. Thus the theory is thoroughly materialistic: it is just that there is more matter in the n-dimensional universe than there would be in a 4-dimensional one.

(2) Consciousness, as here defined, would be located not in the brain at all, as in the double aspect theory—but *outside* it altogether in a space of its own. The reason the visual field looks nothing like any part of the brain would be that it *is* not any part of the brain but a separate piece of the human mechanism in its own right. The brain thus becomes not the total mechanism of mind but merely a part of it. Since it is clear from the medical evidence that the phenomena of consciousness and mind are entirely dependent on the brain, one can deduce that the postulated mental mechanisms could have no other function except to mirror in consciousness the representation of the external world as organized in all its details by the brain— together with the representation of the internal world of thought and emotion and memory also organized by the brain. The model that occurred to me here is a cinema projector and its screen. The screen contributes nothing to the picture except to provide a locus for its projection and all the organization resides in the projector. Nevertheless the projector by itself cannot show the film.

I would now like to shift our outlook again and look at this problem from the point of view of the brain scientist. The essential unattractiveness or all dualistic theories of mind for him has lain in their untidiness. He finds that he can give a coherent account of brain function and its role in behaviour in terms of physics and chemistry. It is irritating to be asked to graft on to this elegant system the activity of "mind influences" as Eccles (1967) dubbed them, coming from some mysterious non-spatial entity and about whose nature and properties nothing could be said. Price's theory does not help us here either. If the Ego has any causal control over brain affairs, it is impossible to indicate to the brain scientist how these might appear to him. Action *across* a "non-space" is hardly more intelligible to him than action *from* a "non-space". Broad's theory, however, is more helpful. All physical events in the brain can be located by four coordinates obtained from the four space-time axes of current physics OX, OY, OZ and OT and brain interactions involve the use of four vectors based on this same system.

Broad's theory involves the addition to this system of new axes (each at right angles to *all* the first set of axes) which we can call OM, ON and OP. This defines mental space (note the way our legacy from Descartes causes the juxtaposition of "mind" and "space" to grate on our instincts). Thus all interactions of "mind" and "brain" can be represented in brain science by new vectors at right angles to all vectors in the old system OX, OY, OZ. We can depict one such vector by selecting a point in OM, ON, OP and another in OX, OY, OZ by drawing a line between them.

Thus in essence Broad's theory requires a radical change in physics, and cosmology. But unlike previous changes this is not an internal rearrangement in the present system, but the addition of a whole new system. If one surveys the present state of physics one should not be surprised at this request. Almost every preconception of classical Newtonian mechanics has been overthrown by the new quantum mechanics and relativity—e.g. distinction between space and time, the wave and particle dichotomy, the concept of parity, matter and anti-matter, etc. About the only preconception not seriously challenged by modern physicists is that space is somehow necessarily three-dimensional.

If, therefore, we examine Broad's theory, it is evidently a possible cosmological theory. There is no *a priori* reason why space must be three-dimensional—that is, we are not compelled by logic to locate every event in the universe by using only four space-time axes. Therefore the theory must be judged on its results—does it explain the phenomena currently known and does it enable us to make predictions that can be tested by experiment? The theory has both advantages and disadvantages. On the credit side it enables us to construct a simple theory of perception free of the internal contra-dictions that beset its two main rivals—naïve realism and the form of the representative theory based on the double-aspect theory. It also presents a simple theory of mind-brain relationship based on a recognized geometry and that allows us to adhere to all the tenets of scientific materialism. But the real test is its predicative power. Can we make any predictions from the theory that can be tested by experiment? I would suggest that we cannot do so at the moment, as before this can be done, the theory needs development by those qualified to do so—i.e. physicists and mathematicians. This development might take the following lines. Modern cosmology regards the 4-dimensional Universe it describes as the sum total of all there is. This may, however, if Broad is right, be merely a cross-section of the real n-dimensional universe. In other words, Broad suggests that we are like the inhabitants of Flatland—to recall E. A. Abbott's tale of 2-dimensional people, who live in a plane—who suddenly realize that their little flat universe may be surrounded by a larger cube. The physics of interaction between events in the cube and events in the plane is different from the physics of interaction of events in the plane. To a Flatlander the plane represents his total universe. We Cubelanders can see, however, that this plane is surrounded on every side by the enveloping cube and we can also see that there are ways into and out of his world, of which he is naïvely unaware—in other words, that there is an interface between the plane and the cube in a very real physical sense. Perhaps, then, we may be able to construct a physics of the possible interface phenomena between brain space and mind space that may lead eventually to predictions that can be tested experimentally, since we ourselves may be in a position just like the Flatlander.

The disadvantages of Broad's theory is that the rival theory of psycho-neural identity is more economical. The principle of Occam's razor suggests that a four-dimensional system is to be preferred if possible to an n-dimensional one. On the other hand, since Broad's is a possible theory, its very postulation clarifies the issues at stake more precisely. We can certainly abandon Cartesian dualism and we can see more clearly just what the theory of psycho-neural identity amounts to by comparing it with naïve realism on the one hand, and Broad's theory on the other. Figure 2 shows the essential difference between the three theories.

In summary, I have reviewed our current notions of consciousness, brain-mind relation and perception and I have outlined three possible theories in these fields. These are Cartesian Dualism, the theory of psycho-neural identity, and the new theories put forward by Broad and Price. These are closely linked with two theories of perception—naïve realism and the representative theory. I would suggest that these interesting new theories by Price and Broad are worthy of serious consideration by those grown weary of the eternal debate between "dualism" and "monism". For these theories are neither "dualistic" nor "monistic" in the old sense, nor do they ask us to believe in any ghosts in the machine. They merely suggest that many of the old metaphysical arguments can be avoided if we pay more attention to the geometry of the space-time in which we live.

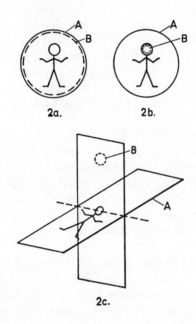

FIGURE 2
(*a*) *Naïve Realism* (*b*) *Psycho-neural Identity* (*c*) *Broad's theory*
————*Physical world A* — — — *Perceptual world* (*of consciousness*) B

DISCUSSION

MCNEILL   I have two related questions. One is to put in a good word for poor Descartes and the form of his argument which is very interesting; it is an argument against reductionism. Essentially what he said is that in order to define the domain of mental phenomena you must examine everything that human beings can do, that can be explained in physical terms, and then whatever is left over is the mind; this requires for its explanation a totally different set of laws. So it is an explicit counterargument against the reductionist explanation of mental phenomena. I think this is a very useful strategy.

This leads me on to Prof. Broad's view that you were outlining; I am not sure at all if I grasped it. It sounds to me as if mind-brain interaction has been elevated from a hyphen to a coordinate, and my question is: is this really an advance or substantial change of any sort? And do we not simply replace "mind" with "body" by essentially arguing that physics needs an additional set of field coordinates in order to achieve a complete description of the physical universe?

SMYTHIES   It isn't physics at all; it's cosmology.

MCNEILL   Well, let us never describe mind as a physical system; O.K.? That seems to me a new form of reductionism. What's the importance of the distinction between cosmology and physics?

SMYTHIES   This system has no relevance to *ordinary* physics. The additional coordinates are not used for saying anything about physical events in the world as we *now* understand it.

MCNEILL   That's an *a priori*; we have no idea.

SMYTHIES   The additional coordinates are really, as I understand the theories, a matter of geography, or cosmography. There may be extra dimensions of space which we have assumed up to now just don't exist. When we set up our four-dimensional space-time, we assume that outside of that there is nothing; well, how do we know?

MCNEILL   Also, how do you know if you establish such additional coordinates you wouldn't be led to make quite radically different statements about physical events? It seems to me that this is really an appeal for a novel type of physical description on the assumption that current descriptions are incomplete; they can't include among other things, the mind. This is almost a reductionist's dream. I shouldn't imagine you would be able to do that.

SMYTHIES I think that is a different problem. But to return to Descartes—I think the other confusion he led to concerned the primary and secondary qualities of objects; he located secondary qualities in the mind, and primary qualities in the outside world. We can clearly distinguish between primary and secondary qualities there; the shape of the bottle is distinct from its colour. But Leibniz, very shortly after Descartes, said that if anything, the primary qualities must be in the mind as well as the secondary ones, as it doesn't make sense to separate them in this way.

MCNEILL I'm not sure if that's a comment on the wisdom of Descartes' strategy as opposed to the wisdom of his particular theory; the two are quite distinct. As a matter of fact, his strategy was followed by Newton, and he arrived at the concept of gravity. What everybody did was to exhaust all the available physical descriptions and then to invent something that seemed totally fictional.

SMYTHIES All I am saying is that in order to make good sense, you must define the limits of the possible, and in doing this you ought to know what are your basic, previously unconscious assumptions: that is, you have to discover these.

HYDÉN I would like to take up one aspect of consciousness which is our appreciation of time. One could start with the old quip that in order for a child to be a child, he has to be a child for some time.

Our appreciation of time varies. I am not talking about the emotional colouring of time, or of a special event, but of physical constants which can change our perception of matter and time. Let me take three examples where the judgement of time has been changed. If we are cold, time will seem shorter; during fever, it will seem much longer. The third example has been discussed by McCulloch. He has studied people driving cars and who have escaped what looked in retrospect a serious accident. As Arthur put it yesterday, when a person is driving uneventfully his automation takes over. Let's say then, he's approaching a top of a hill and suddenly another car will approach head on. Here McCulloch claims that he has found that the time perception may change. The people who are able to react rapidly between a tenth and a hundredth of a second, and to avoid a serious accident, suddenly perceived time as in slow motion. Everything seemed to go slowly and they found they had plenty of time to make the necessary avoiding manoeuvres. They have perceived them in another way. It was so striking that they said that time took on another quality.

I

How time is perceived and judged, and how this affects our perception of structure and matter, is quite a problem.

SMYTHIES More particularly the hallucinogenic drugs enormously distort our sense of time; they usually elongate it so that one second seems like 100 years. One of the most interesting things you can do is to reverse the order in which you perceive things. One of our subjects, had a very interesting experience, which he describes thus: "I had two dimensions of time because I experienced 'having tea' before 'having lunch'." He had time reversal in a sense, which must be due to some switching around in the circuits in the brain of time-recording codes laid down in memory.

KETY I don't want to engage in a debate on the true nature of the world, because such a debate has been going on for several thousand years. It may be of value to indicate how one person resolves these problems for himself, and especially since I may not entirely agree with what John described as the assumptions which biologists make.

In the first place, I am not out to get at "truth", but at simply a personally satisfactory parsimonious and heuristic model of the universe, and applying those criteria I am forced to reject absolute idealism as being not parsimonious. Solipsism can be made a consistent philosophy, but only at the cost of a tremendously complex cosmogony and physics. To try to develop the laws of a universe which ceases to exist when I cease to exist, and which comes into being every morning, or which wobbles when I drink too much alcohol, is just not a parsimonious thing. Therefore, I am forced to feel that parsimoniously there is an outside world which, remarkably or not, gives us a considerable consensus. All these different consciousnesses somehow agree that there is a table, that there is a star up there, there is a moon and a sun. The most parsimonious explanation for that is that there *is* an outside world, but this would not necessarily be a naïve realism. It would be more like the Kantian "Ding an sich". I don't know what the outside world really is; the task of modern physics is to come closer and closer to what it really is. But in some way or other it affects my brain, and it affects my consciousness, so I am forced to a dualism. On the other hand absolute materialism is not satisfactory to me, because it rejects one of the most obvious phenomena, which is consciousness. Therefore, I am a perfectly happy dualist. With Sherrington, as a matter of fact, I think if one can reduce this whole universe to two entities, that's pretty good. I don't feel any compulsion to reduce them to one.

But now the problem arises as to how these two entities, consciousness and the material world, interact. Well, again, if we try to adopt the Berkelian concept, we have to develop tremendously complex laws. If my consciousness determines the universe, again this is not parsimonious; it does not permit us to make generalizations that seem to be borne out by experiment. If, with Leibnitz, we try to insist that there is no interaction between the two spheres, then we have to develop the unsatisfactory concept that these are two worlds which have a remarkable synchrony for all of our lifetimes without interacting with each other, which I find difficult. And so again, the most parsimonious idea is that matter affects mind; even though one is not reducible to the other, there is still a primacy in matter in affecting what the events in my brain do, in fact determine the events in my consciousness. A simple example: if I take a drug my consciousness changes tremendously, my sense of time changes tremendously; but when I awake from this drug other people around me, and clocks, and so on, insist that I was wrong. I think it is more parsimonious to agree that I was wrong and that I had an illusion of time having changed, whereas real time did not change. Therefore this is a heuristic kind of idea because one can then study the brain and one can study the interaction on the assumption that it is a mechanism and that consciousness is somehow, in the words of Huxley and Clifford, an epiphenomenon of the mechanism.

Free-will, determinism, of course, are problems which are implicit in this kind of model. I could really leave them open for the time being, saying that I don't understand how a thought can affect behaviour, because the only thing that I can see which is affecting behaviour is a chain of neurons leading to a muscle, and how a thought in this other sphere can affect a neuron is something that is quite incomprehensible to me. But as a matter of fact, equally incomprehensible to me is how one body can affect another body through gravitation outside of any material interaction between them.

So, if we can on the one hand maintain a sort of neutrality in physics, I suppose I can maintain a neutrality in this question of free will and determinism, except to say that as a scientist somehow I think it is more heuristic to believe that there *is* a mechanism and try to find it. As human beings, we all operate as if there were free will; and here again is a dualism which I find tentatively more satisfactory than trying to come up with the "truth".

SMYTHIES What I've been trying to do this morning is to ask whether Descartes was right in his particular criteria for distinguishing between the mental and the physical and to weigh up the other runners in the field. What I was talking about was, what are the *criteria* for relating these two? What are the possible ways?

KETY Well, you were trying to develop a consistent model for two domains, and this is something I just don't feel compelled to do. I am happy enough to stay in an unresolved domain.

THORPE It depends on just how far you want to carry your consistency; you want to be consistent only up to a point.

INHELDER In psychology there is only one way of speaking about consciousness without falling into the trap of introspective trivialities, and that is to place "consciousness" within the context of "behaviour". Consciousness can in fact be viewed as a special case of behaviour which may be called "the becoming conscious of". In particular—and this is a crucial point—it is possible to become conscious of part of one's self through the observation of the results of one's actions; especially, as Claparède has shown, in those cases where one's actions are as yet unadapted to reality. It is by means of a series of re-adaptations and corrections that one becomes (partially) conscious of one's self rather than by means of a direct apprehension of interior mechanisms. The development of consciousness and its deformations raise vital problems whose solution conditions all studies of normal adult consciousness.

This approach seems to hold a promise for the solution of the problem of the way connections within consciousness can be brought together with physical or physiological connections. The former concern meanings and links between meanings, whereas the latter concern causal factors as opposed to implications, which simplifies the classical mind-body problem.

KOESTLER Playing around with this problem is like playing chess, which I find aesthetically very satisfying. However, if a kind of mad Caligula were to decide that there should be a game of chess and the loser should be put to death, then suddenly it would become a very serious game. Now, in one respect the mind-body problem is that deadly game of chess—I mean in its applications to the criminal code. The code is based on the axiom of free will, of criminal responsibility, on you "ought" to have acted this way or that. Now what that "ought" means here is that the subject A in the situation X *should* have acted differently and not in the way he did—which in fact means that either the situation should have been Y not X, or the subject should have been B, not A. In contrast to our criminal code, however, our whole social philosophy is based on either a crude determinism of the behaviourist kind or a crude sort of Darwinian determinism. So, the practical interpretations of the mind-body puzzle can be really quite deadly. I have tried to propose a solution for the problem of free will as applied to concrete every-day ethics, and I have come up with a kind of schizophrenic answer. But it is a very healthy sort of schizophrenia. That is to say, that I have the subjective experience of freedom and I accept it at face value, but I deny that anybody else has a free will (laughter). For me this is the only acceptable and practically applicable ethical attitude. Because it prevents me from getting angry with John Smythies, because the poor chap can't help it. I am not allowed to hang him, I am not allowed to sentence him to death because the poor chap could not help it, because he could not act otherwise than he did. But I cannot permit myself the same indulgence. I have an excuse for you but not for me. So the maxim I propose boils down to a variant of the French maxim *tout comprendre c'est tout pardonner*. And the variant goes: *tout comprendre ne rien se pardonner*. That's all, but I think it is a useful maxim, a useful kind of schizophrenic ethics.

THORPE   I think one can hold the dual view of this same world without calling oneself a dualist, just as a physicist can hold both the wave theory, and the corpuscular theory of light as two different aspects. You can call that dualism or not, as you like, so I don't think one need be alarmed by dualism; one can employ it as a practical device and whether it holds up philosophically doesn't matter.

WEISS   You said that the fish has a different picture of the universe from mine, and the worm still another picture. And a superman would have a still wider picture. However, I can't see how I can form a concept of the universe, except within the limitations of my human structure, including my brain. This operates within a canon of three dimensions in space and a fourth dimension in time. I am then compelled to force any five-dimensional or six-dimensional scheme into the four dimensions of my perception, reducing thus the concept of the cosmos to my human powers of visualizing nature, for that's all I can do by way of spatial representation on a blackboard, even a moving one.

So a lot of these attempts that we have heard about are quite legitimate in a way, but to me they seem to be tautological. They don't help me at all because they try to squeeze and to reduce this world view until it can be shrunk down to the size, to the kind of dimensionality, and to the kind of concepts my constitution permits me to admit. Now, physics is not the same as the universe; physics can merely represent that fraction which leaks through our senses, strengthened by our instrumentation, expanded by our concepts and our logical constructs but still far short of a total replica of the universe. Similarly to what Seymour said, I'm perfectly satisfied with this, but I have to put in some counterpart which restores the unity which is my greatest experience: that even though I know I am constantly changing—all molecules are changing, everything in me is being turned over substantially—there is nevertheless my identity, my consciousness of being essentially the same that I was 20 years ago. However much I may have changed, the continuity of my identity has remained undisrupted. And this is why I am questioning the purpose of trying to contrive fictional physics—which is not real physics, dealing with what we can really measure—in order to give us the tautological satisfaction that, after all, the universe is there.

SMYTHIES Yes, this is an enormously important point—the limitations which the structure of the nervous system imposes on our ideas. We can only think in the ways we are constructed. But very few people have investigated possible ways in which our thinking is constricted by the structure of the nervous system, and I think it's a very promising field for research.

WEISS Maybe our concept of our nervous system is equally inadequate and insufficient, because so long as you use only electrical instruments, you get only electrical answers; if you use chemical detectors, you get chemical answers, and if you determine numerical and geometrical values, you get numerical and geometrical answers. So perhaps we have not yet found the particular kind of instrument that tells us the next unknown.

THORPE This may be outside the nervous system, but the trouble is we don't yet know what the potentialities of the nervous system are.

BERTALANFFY I believe a radically new approach to the mind-body problem is necessary which I tried to develop in its various ramifications elsewhere (1963; 1966). This is based on the consideration that "philosophical" problems of old must nowadays be approached by the ways of science, just as the problems of space, time, causality, substance, etc., used to belong to the realm of philosophy and metaphysics, but have become subjects of scientific research. I am not claiming that the studies mentioned present the "solution" of the mind-body problem, but I believe that the approach indicated is valid, and that it leads to a basic re-evaluation of the problem. All previous theories—such as interaction, parallelism, identity theories, epiphenomenalism, etc., including modern versions such as Ryle's and Broad's—were based on the Cartesian dualism of *res extensa* and *res cogitans*; that is, an obsolete conceptualization which was a product of 17th-century physics.

Similarly, theories of perception and of cognition in general were based upon the preconception of a "given" reality outside, which in some way is represented, mirrored, or projected in the "mind". All these theories are anthropomorphic, more strictly speaking, "adulto-morphic" (with apologies for a bastard word); that is, based upon the experience of the human adult of Western civilization. Psychology, however, shows (*a*) that other forms of experience also exist, e.g., the child's, the non-Westerner's, the mystic's, etc., which cannot be simply disregarded in favour of the supposedly "objective" cognition of post-Renaissance man. Similarly (*b*) perception and cognition—so testified by modern psychology—are not passive intakes of stimuli coming from physical things; but active processes of organizing primary experience into a universe. These processes are dependent on (*a*) biological factors (e.g. the fact that man's space is essentially visualtactile; compare the "world view" of man with that of a dog or worm); and (*b*) on cultural factors such as the structure of language used (cf. B. Whorf), the general world outlook (e.g. the technological orientation in Western civilization), etc. Hence the "ghost-in-the-machine" conception, characteristic of a certain stage in our civilization.

The distinction of "object" and "subject", "body" and "mind", "matter" and soul", is not an ultimate given. Rather, it is a product of conceptual differentiation, occurring in individual and socio-cultural development. The ultimate given is the flow of direct experience (*Erleben* in German terminology, *le monde vécu* of French phenomenologists)—conscious, but with indefinite borders to the unconscious. This flow is organized and conceptualized with the help of categories which, as has been said, are biologically and culturally determined. In other words, the Cartesian dualism is not a given disjunction; rather, a primary adualism (Piaget) is progressively differentiated, both in the development of the child and in cultural history.

So far as the question of "free will" is concerned, we *experience* ourselves as free (because immediate experience is not categorized); we *explain* behaviour (of others and our own) by application of the category of causality.

I believe that within this framework the problems mentioned by Smythies can be taken care of. A "unitary theory", in the sense of L. L. Whyte, of physico-physiological and psychological events will presumably be possible, based on isomorphies of a general and abstract nature, in the general direction of system theory.

HAYEK There is really nothing useful I could say briefly at this stage; nor anything new which I haven't explained at length in *The Sensory Order*. I will confine myself to three words: order is not an object, and that really expresses it all. What we are studying is an order of events, and an order of events cannot be represented by anything which is amorphic.

WEISS Any static concept of an event in nature, and that includes the notion of a molecule, is an abstraction; the actual thing is an ongoing dynamic process of which the static image is merely a temporarily arrested sample. Thus, static molecules can never be regarded as the elements of life; for life is a process and logically the element of a process can only be an elementary process.

SMYTHIES Of course objects are abstractions from processes.

HAYEK Except that I wouldn't use abstractions in this derogatory sense.

BRUNER It seems to me that Broad's theory opens up possibilities that we don't have at all in a three-dimensional concept, and includes things like time, motion, and space which could also allow for process. Viewed as a skeleton of a possible framework it might be very useful.

FRANKL Smythies' paper was very inspiring. I agree with him that only if mental phenomena could be explained in causal terms, would we be justified in speaking of mental phenomena in terms of epi-phenomena; otherwise, I believe, we have to take them at their face value, as phenomena with their own mode of being. Now, my contention is that the physiological basis does not *cause* anything mental, but it does *condition* it and there is a great difference between causing and conditioning. If through the hypofunction of the thyroid gland, someone becomes a moron and if you feed this person with thyroid gland, or the like, his IQ will increase; but are you therefore justified in contending that intelligence is nothing but thyroid gland substance? Certainly not. We have just removed the cause that has impaired the intelligence of this person. Similarly, an impaired sense of identity, due to impaired functioning of the adrenal cortex, can be remedied by the administration of desoxy-corti-costerone-acetate—but I do not conclude that the ego is "nothing but" desoxy-corticosterone-acetate.

As for Broad's theory, it reminds me of a similar concept—the dimensional analogy which I developed many years ago. Broad's concept of various *spaces* which as it were, interpenetrate one another, and my own concept of *dimensions* are brought together in this simple diagram. (Blackboard: see Figure below). You have here two planes—or spaces or dimensions—(*a*) and (*b*). The line L represents a series of events or phenomena pertaining to the space, or dimension, of (*b*). On the plane (*a*) this line is merely represented by four disjointed points. It is represented in a partial, fragmentary way, without indication of the interconnectedness of these points. In

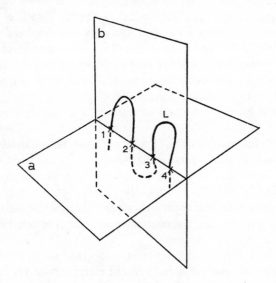

other words, on one plane the meaningful connection is not discernible, but in the other dimension it may well exist. And this meaningful connection might be a causal one, but it might also be a teleological one—as in the case of events that evolutionary theory tries to explain by mutations, and the mutations by mere chance. What I want to convey is that although not everything can *be explained in meaning-ful terms*, at least we may explain *why this is necessarily the case*; why despite the apparent meaninglessness that confronts us, we are entitled to believe in some meaning beyond it, behind it, or above it.

THORPE One point I would like to raise concerns an argument I had with Eccles about his views of "free will" in relation to his ideas of an "experiencing self". I found him very much against any tendency to concede that animals may possess an "experiencing self". It seems to me that, since we ourselves have an "experiencing self", it is extremely difficult to believe that the higher animals, with the very elaborate learning and problem-solving performances that they exhibit are not in some sense conscious; in other words that they are not also "experiencing selves". I wonder whether anyone here has considered that this kind of formulation should be restricted only to human beings, and if so, why?

SMYTHIES You can't argue with an animal; it hasn't got language.

THORPE But animals do have languages! The word "argue" can mean many things. One can certainly "argue" with a "man-im-printed" bird in gesture language! And doesn't a shepherd argue with the sheep dog in the agreed language of whistles and signs? Moreover I expect it may turn out that it is possible to "argue" with chim-panzees in deaf and dumb language. It seems to me you are in fact, really proposing another form of solipsism.

WEISS You have no method of testing this, therefore it's useless to argue.

KETY But for that matter, Paul, we have no method of testing consciousness in another individual, so to be logical I think one would have to assume that there are degrees of richness of consciousness; and of course, man is remarkably different from even the next highest primate in the ability to symbolize and conceptualize. So one would expect that his consciousness was a much richer thing than that of the highest apes.

THORPE Yes, but the difference is in degree, not in kind. I would like to quote as relevant here, a recent statement of H. L. Teuber: "It has become clear . . . that linguists are ethologists, working with man as their species for study, and ethologists linguists, working with non-verbalizing species."

KOESTLER That would lead up to a sentence that seems to me to sum up the discussion: "Consciousness is a primary datum of existence and as such cannot be fully defined. The evidence suggests that at the lower levels of the evolutionary scale consciousness, if it exists, must be of a very generalized kind, so to say unstructured. And that with the development of purposive behaviour and a powerful faculty of attention, consciousness associated with expectation will become more and more vivid and precise." I am quoting Thorpe speaking at the Eccles-Vatican Symposium (1966). I am quoting this because here we get away from consciousness as an all-or-nothing process, to consciousness as a developing process.

FRANKL Yes, but what about the possible qualitative difference between consciousness and self-consciousness?

THORPE I doubt it, but we can argue about this later.

SMYTHIES Our own consciousness varies, too.

THORPE I'd like to close with a story of Bertrand Russell who was giving a lecture on solipsism to a lay audience, and a woman got up and said she was so delighted to hear Bertrand Russell say he was a solipsist; she was one too, and she wished there were more of us (*laughter*).

REFERENCES

VON BERTALANFFY, L. (1963) "The mind-body problem: a new view", *Psychosomatic Medicine* Vol. 24, pp. 29–45: "Mind and body re-examined", *Journal of Humanistic Psychology*, Fall 1966, pp. 113–138; and elswhere.

BROAD, C. D. (1923) *Scientific Thought*. Routledge and Kegan Paul, London.

ECCLES, J. C. (ed. 1966) *Brain and Conscious Experience*, Springer Verlag, Berlin-New York.

———— (1967) in J. D. Roslansky (ed.) *The Human Mind*. North Holland Publishing Co., Amsterdam.

PRICE, H. H. (1965) "Survival and the idea of another world" in J. R. Smythies (ed.) *Brain and Mind*, Routledge and Kegan Paul, London.

The paranoid streak in man
Paul D. MacLean*

National Institute of Mental Health, Bethesda, Maryland

Behold! human beings living in an underground den; here they have been from their childhood, and have their legs and necks chained so that they cannot move, and can only see before them, being prevented by the chains from turning round their heads . . . and they see only their own shadows, or the shadows of one another, which the fire throws on the opposite wall of the cave . . . And suppose . . . that the prison had an echo which came from the other side, would they not be sure to fancy when one of the passers-by spoke that the voice which they heard came from the passing shadow?

Plato's Republic, BOOK VII

In the denouement of his recent book *The Ghost in the Machine* (1967) Koestler refers repeatedly to the inherent "paranoid streak" in man. He concludes that "the damages wrought by individual violence for selfish motives are insignificant compared to the holocausts resulting from self-transcending devotion to collectively shared belief-systems". "We have seen," he says, "that the cause underlying these pathological manifestations is the split between reason and belief—or more generally, insufficient coordination between the emotive and discriminative faculties of the mind." "I believe that if we fail to find . . . (the) cure, the old paranoid streak in man, combined with his new powers of destruction, must sooner or later lead to genosuicide." "Before the thermonuclear bomb, man had to live with the idea of his death as an individual; from now onward, mankind has to live with the idea of its death as a species" (Koestler, 1967).

* Read by J. R. Smythies.

If, as Koestler claims, the "paranoid streak" is so indelibly a part of man's nature that it now threatens his very existence, it deserves in a symposium of this kind to be singled out for special consideration. As pinpointing the site of a disorder in a mechanism is the first step in fixing it, I wish to devote this presentation to a search for the neural substrate of the "paranoid streak" in man. For conducting this search, we must first describe the demon and how he disguises himself. Then we shall try to track him down in the evolution of the mammalian brain in the hope that this will lead us to the part of the cerebrum in which he may be lurking. But we should be forewarned that there is little hope that we shall lay our hands on him. With his reputation as one of the most elusive demons of all time, let us be satisfied if we can only find his hideaway.

That paranoid reactions show up so regularly in the psychoses from generation to generation supports Koestler's contention that the "paranoid streak" is inherent in the nature of man. According to Adolph Meyer (1928) the term paranoia (disordered mind) was used as long ago as 1764 by Vogel in the broad sense of vesania, a mental disorder marked by the four stages of mania, melancholia, delusional states, and dementia. In 1879 the meaning was narrowed down by Krafft-Ebing (1904) to apply to all forms of systematized delusional insanity. The adjectival form of the word as it is now used, however, usually refers to a mental condition characterized by false beliefs which predispose to delusions of persecution or grandeur.

The emotive element. How does the emotive element, as Koestler correctly assumes, enter into the system of false beliefs? The answer requires a brief analysis of what is meant by emotion.

We commonly speak of the subjective and expressive aspects of emotion. As the subjective aspect is purely private it needs to be distinguished by some such word as "affect". Only we as individuals can experience affects. The public communication of affects requires expression through some form of verbal or other behaviour. The expression of affect is appropriately denoted by Descartes' word "emotion". Scientifically, however, it is important to recognize that regardless of whether we are dealing with affect or emotion, each is manifest to us as observers only as information. To paraphrase Berkeley and Hume in Wiener's words, "information is information, not matter or energy" (Wiener, 1948). And it also deserves emphasis that whether we are working in the behavioural or physical sciences the brain stands always between us and what we observe. The cold hard facts of science or the firm pavement underfoot are all derivatives of a soft brain.

In considering the "paranoid streak" it is important to make the distinction that it refers to a subjective state rather than to a form of behaviour. Further qualification requires the help of introspection. Introspection leads to the conclusion that it is the element of subjectivity that most clearly distinguishes psychological from other functions of the brain (MacLean, 1961). How many forms of subjective experience are there? Granted that man may still be an amateur in the penumbral science of introspection, he has through the centuries arrived at the recognition of six main kinds of psychic information that may be classified under such headings as (1) awareness, (2) sensations, (3) perceptions, (4) compulsions, (5) affects, and (6) thoughts.

Affects differ from other forms of psychic information in so far as they are subjectively recognized as being imbued with a physical quality that is either agreeable or disagreeable. There are no neutral affects, because emotionally speaking, it is impossible to feel unemotionally.

The affects can be subdivided into three main types which for purposes of discussion I refer to as basic, specific, and general (MacLean, 1966b). The basic affects are informative of basic bodily needs that are subjectively recognized as hunger, thirst, and the various urges to breathe, defecate, urinate, have sexual outlet, etc. The specific affects are those aroused by activation of specific sensory systems, as illustrated by the feeling of startle to loud sounds or the feeling of pain after a noxious stimulus.

The traditionally regarded "emotions" such as love, anger, etc., are included among the general affects. I call them general because they are feelings that may pertain to situations, individuals, or groups. All the general affects may be considered in the light of self-preservation or the preservation of the species (MacLean, 1961). (This, parenthetically, implies no debasement of the so-called cultural or aesthetic affects, as they maintain their respected position in promoting the well-being of man and his society.) Affects giving information of external threats to the self or species fall into the category of disagreeable affects. In the opposite category are affects informative of the removal of threats and the gratification of needs.

If we exclude verbal behaviour, we can identify in animals and man six types of behaviour that are inferred to be guided by the general affects (MacLean, 1966b). These six behaviours are recognized as searching, aggressive, protective, dejected, gratulant, and caressive. Verbally, they may be respectively characterized by such words as desire, anger, fear, sorrow, joy and affection. Symbolic language makes it possible to identify many variations of these affects, but in working with animals inferences about emotional states must be largely based on the six mentioned types of behaviour.

One more qualification must be made before returning to the emotive element of the "paranoid streak". The general affects, compulsions, and thoughts differ in a fundamental respect from sensations, perceptions, and the basic and specific affects. Under normal conditions the latter forms of psychic information occur only in conjunction with contemporaneous signals conveyed by the sensory systems, whereas the general affects, compulsions, and thoughts have the capacity to persist or recur long after the inciting circumstances. The unexplained process making this possible may be called mentation (see below).

The emotive element in the "paranoid streak" answers the description of a general affect. Essentially, it amounts to an unpleasant feeling of fear attached to something that cannot be clearly identified. The feeling has the capacity to persist or recur long after the inciting circumstance and may apply to a situation or thing, an individual or group of individuals. It has survival value because it puts the organism on guard against the unexpected—hence, the words apprehension and suspicion often used in describing the paranoid state. A few examples will illustrate how inadequate information received by way of one of the sensory systems may arouse fear and suspicion and how suspicion may often be quickly allayed by comparison with information from one or two other sensory systems. The horse which has a keen olfactory sense, but relatively poor vision, is uncommonly fearful of ill-defined objects. One might say it is particularly "paranoid" about paper bags! The sight of one, as riders well know, will often cause a horse to shy or come to a sudden halt. If, however, it can be coaxed to approach and smell the bag, its fear and suspicion seem suddenly to melt away, and it will proceed as though nothing had happened. Information from its most trusted sense reveals the object to be lifeless and harmless.

Man, on the contrary, relies largely on vision to relieve his uncertainty about the nature of things. Poorly visualized objects are under suspicion until he can bring them clearly into focus. But perhaps his greatest suspicions are aroused by things he cannot see at all. Consequently darkness has always held its particular terrors for him. Unexplained muted house sounds at night become amplified in the mind into the footsteps of a burglar. The faint smell of smoke explodes into a blaze of fire. Man's inherent fear of the dark is brought into clear relief under pathological conditions. Patients suffering from high fever or toxic poisoning not uncommonly become delirious with the approach of darkness. I remember a man with lead poisoning who became so terrified at the onset of darkness that he leapt from his bed, ran headlong down the hospital corridor, and was apprehended only in time to save him from jumping from a balcony.

Repeated exposure to something that cannot be seen predisposes to a state of chronic suspicion. I recall a child who was thrown into a persisting state of fear by seeing an old man pass down his street each day with a sack over his shoulder. Unable to see for himself that the sack contained nothing but empty bottles, his imagination led him to believe that the old man was a kidnapper and that any day he might be the next victim.

Some individuals become hypochondriacal in a somewhat similar manner. Unable to visualize the source of a recurring internal ache or pain, they are led, according to their social background, to suspect all manner of deleterious processes. The more primitive the individual the more primitive may be his explanation. I recall a young man of primitive background who had a belly ache and came to the hospital complaining of fighting cats and dogs in his stomach. While I was examining him and listening to his heart with a stethoscope, he asked me, "Can you hear them?" Given the seed of suspicion, the human mind is capable of developing any kind of paranoid hybrid.

There is yet another form of the unseen that besets man and distinguishes him from all other animals. This is the poorly outlined and uncertain picture of future events which he is forever striving to see. When nature gave him the prefrontal neocortex for anticipation and connected it with his visual cortical areas, she failed to provide a radar antenna and viewing screen. Consequently, all his probings into the future must be done with obscured, remembered images of the past combined with his picture of the present. As the future is always generating more "futures" *ad infinitum* it is apparent why its uncertainties are responsible for most of man's chronic forms of suspicion.

The emphasis given to the capacity of vision both to arouse and allay suspicion should not imply an insignificant role of the other senses in this respect. There are, for example, situations where one can clearly see what is taking place, but where hearing only in part, or not at all, what is happening or being said results in uncertainty and suspicion. For such reasons deaf people are especially prone to develop paranoid tendencies. Receiving a chance smile or look from someone in a group conversation, they may gain the impression that they are being talked about. In the paranoid psychoses auditory illusions and hallucinations are more frequent than those in the visual sphere. I recently saw an epileptic patient with schizophrenic symptoms who claimed to hear clicking sounds and imagined that people were snapping pictures of her.

Stripped to nakedness, the paranoid demon steps forth from the foregoing analysis as *a general affect characterized as an unpleasant feeling of fear attached to something that cannot be clearly identified.* Seen in this pristine state, he is hardly an impressive figure. The thing, obviously, that gives him mystique is his capacity to assume as many disguises as there are individuals. He could thus be compared to a plot that lends itself to as many stories as there are potential authors.

Evolutionary considerations. With this brief description and character-ization of our protagonist we are now in a position to look for signs of him in the evolution of the mammalian brain. Our starting point is 200 million years ago in the age of reptiles when animals which never learned to talk, began to work their way into the brain of man (MacLean, 1964). Perhaps the most revealing thing about the study of man's brain is that he has inherited the structure and organization of three basic cerebral types which for purposes of discussion I label reptilian, old mammalian, and new mammalian (MacLean, 1962, 1964, 1966b, 1967). Despite their great differences in structure and chemistry all three must interconnect and function together. In terms of Koestler's hierarchical concept (Koestler, 1967) each cerebral type would be comparable to a "holon" with vertical and horizontal arborization and reticulation. The "holonarchy" (Koest-ler, 1967) is schematized in Figure 1. Man's brain of oldest heritage is basically reptilian. It forms the matrix of the brain stem and com-prises much of the reticular system, midbrain and basal ganglia. The reptilian brain is characterized by greatly enlarged striatal structures which resemble the caudate-putamen-globus pallidus complex of mammals.

But in contrast to mammals there is only an incipient cortex. The evolving old mammalian brain is distinctive because of the marked expansion and differentiation of a primitive cortex which, as will be explained, is synonymous with the limbic cortex. Finally, there appears late in evolution a more highly differentiated form of cortex called neocortex which is the hallmark of the brains of higher mammals and which culminates in man to become the brain of reading, writing and arithmetic.

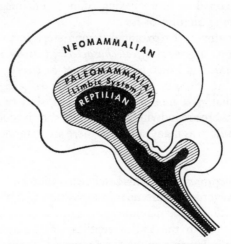

FIGURE I   *Schema of 'holonarchic' organization of the three basic brain types which, in the evolution of the mammalian brain, become part of man's inheritance. Man's counterpart of the old mammalian brain comprises the so-called limbic system which has been found to play an important role in emotional behaviour.* (*From MacLean, 1966.*)

In the popular language of today, the reptilian and the old and new mammalian brains might be regarded as biological computers, each with its own subjective, gnostic, time-measuring, memory, motor and other functions (MacLean, 1966b). On the basis of behavioural observations of ethologists, it may be inferred that the reptilian brain programmes stereotyped behaviours according to instructions based on ancestral learning and ancestral memories. In other words, it seems to play a primary role in instinctually determined functions such as establishing territory, finding shelter, hunting, homing, mating, breeding, imprinting, forming social hierarchies, selecting leaders and the like. In the experimental situation the presentation of a dummy or a fragment of a dummy may release the sequential acting out of an instinctual form of behaviour. Indeed, a mere phantom is sometimes sufficient to trigger the entire copulatory act.

The reptilian brain seems to be a slave to precedent. If, for example, having found a safe way home, it is thereafter inclined to follow that route even though it means going around Robin Hood's barn. It would be satisfying to know to what extent the reptilian counterpart of man's brain determines his obeisance to precedent in ceremonial rituals, legal actions, political persuasions, and religious convictions. Obeisance to precedent is the first step to obsessive compulsive behaviour, and this is well illustrated by the turtle's always returning to the same place year after year to lay its eggs. In stereotyped and repetitious forms of behaviour it would seem as though the reptilian brain were neurosis-bound by an ancestral superego (MacLean, 1964). At all events it appears to have inadequate machinery for learning to cope with new situations.

The evolutionary development in lower mammals of a respectable cortex might be regarded as Nature's attempt to provide the reptilian brain with a "thinking cap" and emancipate it from inappropriate stereotypes of behaviour. In all mammals most of the primitive cortex is found in a large convolution which Broca called the limbic lobe because it surrounds the brain stem (Broca, 1878). From the standpoint of behavioural implications it is significant that this lobe, as illustrated in Figure 2, is found as a common denominator in the brains of all mammals.

Because of its close relationship with the olfactory apparatus, the old mammalian brain was commonly believed to have purely olfactory functions and was accordingly identified in many texts as the rhinencephalon. Papez's classical paper of 1937 (Papez, 1937), however, struck a mortal blow to this line of thinking and since then it has become evident from clinical and experimental findings that the functions of this brain are far more encompassing, playing an important role in emotional, viscero-somatic, and endocrine functions (MacLean, 1949, 1958). In 1952 I suggested the term "limbic system" as a designation for the limbic cortex and brain stem structures with which it has primary connections (MacLean, 1952). Broca's simple descriptive term "limbic" has the advantage, as he stated, that it implies no theory as regards function. It should be emphasized that the limbic cortex has similar features in all mammals and is structurally primitive compared with the neocortex. From this it can be inferred that it continues to function at an animalistic level in man as in animals. Also in marked contrast to the neocortex, it has strong connections with the hypothalamus which plays a basic role in integrating emotional expression and viscero-somatic behaviour.

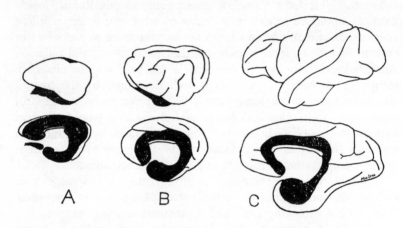

FIGURE 2   *In all mammals most of the cortex of the old mammalian brain (limbic system) is found in the great limbic lobe which surrounds the brain stem. These drawings showing the relative sizes of the rabbit (A), cat (B), and monkey (C) brains, illustrate that throughout the mammalian series this lobe, represented in black, is a common denominator of the cerebrum. The neocortex which undergoes a dramatic expansion late in phylogeny is shown in white. (From MacLean, P.D. In: Wittkower, E. Cleghorn, R., eds.,* Recent Developments in Psychosomatic Medicine. *London, Pitman,* 1954)

That the limbic brain is a functionally, as well as an anatomically integrated system, is dramatically demonstrated by mapping the propagation of hippocampal seizure discharges which show the tendency to spread throughout and be largely confined to the limbic system. Nothing brings home so dramatically the dichotomy of function, or "schizophysiology" as it has been called, of the old and new brains (MacLean, 1954, 1958).

Functions of the limbic system. The search for brain mechanisms underlying the emotive element of the "paranoid streak" requires a brief review of functions of different parts of the limbic system that have been revealed by clinical and experimental observations. This is most easily done by referring to an anatomical diagram of pathways linking three main subdivisions of the limbic system. Figure 3 focuses on three ramifications of the medial forebrain bundle that connect the ring of limbic cortex with the hypothalamus and other parts of the brain stem. The two upper branches of the bundle meet with descending fibres from the olfactory apparatus and feed into the lower and upper halves of the ring through the amygdala and septum at the points marked 1 and 2. Clinical and experimental findings suggest that the lower part of the ring fed by the amygdala (No. 1) is primarily concerned with affects and behaviour that insure self-preservation. Its circuits are kept busy with the selfish demands of feeding, fighting and self-protection (MacLean, 1952, 1958). On the other hand, structures in the upper part of the ring connected by the septal pathway (No. 2) appear to be involved in expressive and feeling states that are conducive to sociability, procreation, and the preservation of the species (MacLean, 1958, 1962).

The third pathway (No. 3), which bypasses the olfactory apparatus, connects the hypothalamus with the anterior thalamic nuclei and the cingulate gyrus in the upper half of the ring. It also articulates with the medial dorsal nucleus which projects to the prefrontal neocortex, a part of the brain, as mentioned earlier, involved in anticipation. Phylogenetically it is notable that the septal region in primates remains rather static, whereas structures comprising the third subdivision of the limbic system becomes progressively larger in primates and reach a maximum size in man. There is some evidence that this condition reflects a shifting of emphasis from olfactory to visual influences in socio-sexual behaviour (MacLean, 1962, 1966b). Significantly, in this respect, stimulation within parts of the third subdivision, as in the septal circuit, elicits sexual responses in monkeys (MacLean, 1962).

Case histories of psychomotor epilepsy provide crucial evidence that the limbic cortex is implicated in the generation of affective states. Irritative lesions in or near the limbic cortex of the lower part of the ring give rise to epileptic discharges accompanied by basic and general affects that under ordinary conditions are important for survival. The basic affects include feelings of hunger, thirst, nausea, suffocation, choking, retching, cold, warmth, and the need to defecate or urinate. Among the general affects are feelings of terror, fear, foreboding, familiarity, strangeness, unreality, wanting to be alone, sadness, and feelings of a paranoid nature. The automatisms that follow these aural affects often seem to be an acting out of the subjective state: for example, eating, drinking, vomiting, urinating, running and screaming as if afraid.

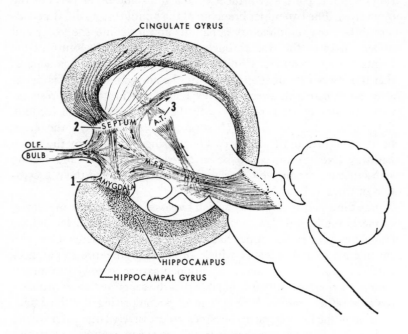

FIGURE 3   *Diagram of pathways linking three main subdivisions of the limbic system. The ring of limbic cortex is shown in light and dark stipple. See text for further anatomical details and functional significance. Abbreviations: A.T.—anterior thalamic nuclei; HYP.—hypothalamus; MFB.—medial forebrain bundle; OLF.—olfactory. (From MacLean, 1958, 1967.)*

Subsequently, with the post-ictal depression of the structures involved in the discharge, some patients are able to carry out highly complicated forms of motor and intellectual behaviour which presumably depend on a functioning neocortex. Hughlings Jackson's famous Case Z, for example, was that of a young doctor with a small cyst in the anterior hippocampal gyrus, who after the initial aura was able to examine a patient and make a correct diagnosis, but had no recollection of it afterwards (Jackson, 1898). Such individuals are temporarily like disembodied spirits, and cases of this kind recall what was said earlier about the "schizophysiology" of limbic and neocortical systems. It is entirely possible that a personal identification with happenings in the external world, as well as a memory for them, depends on a dovetailing—an integration—of internally and externally derived experience. For illustrative purposes it is sufficient to recall the "breath of recognition" or the change in heart-rate that attend the solution of a problem or the surprise meeting with an old friend. The neocortex receives its information largely from the auditory, visual, and somatosensory systems. The findings in psychomotor epilepsy suggest that without a contemporaneous integration of somato-visceral functions by the limbic brain, there is an inability to personally identify with and remember what is transpiring in the external environment. Significantly in this respect, during the initial epileptic discharge of psychomotor epilepsy a patient may have a feeling of depersonalization, or what Hughlings Jackson called "a mental diplopia", in which he feels as if he is viewing himself and what is going on from a distance. This, parenthetically, is a symptom commonly experienced by individuals under the influence of psychedelic drugs, and, as well known, may be a feature of the endogenous and toxic psychoses.

The possibility likewise suggests itself that the general affects represent an amalgamated form of internally and externally derived experience. The study of psychomotor epilepsy teaches, however, that the experience of affects does not depend on contemporaneous activation of sensory systems. The epileptic affect of fear, for example, arises centrally in conjunction with the cortical discharge, and an accompanying feeling of a racing heart may be present without actual cardiac acceleration. It is also significant under these conditions that the affects commonly occur without an attachment to any specific event or person. In other words, as I have pointed out in previous writings (MacLean, 1952), ictal affects represent "feelings out of context". The sadness felt in the epigastrium pertains to no particular situation. The feeling of the need of company is not related to particular persons. It would thus seem that the raw stuff of the affects is built into the circuitry of the limbic brain. Instead of deriving experience in terms of compulsions, as was implied in considering the reptilian brain, or in terms of abstract thoughts, as presumed in the case of the neomammalian brain, the mentation of the limbic system would appear to involve a process whereby information is encoded in terms of affective feelings that influence its decisions and course of action.

As already noted, paranoid feelings may occur in conjunction with the initial discharge of psychomotor epilepsy. I once treated a young man, for example, who at the beginning of his seizure had a feeling of fear that someone was standing behind him. If he turned to see who it was, the feeling of fear became intensified. During inter-seizure periods some patients are tormented by persistent paranoid feelings. One such patient was continually obsessed by the feeling that God was punishing her for over-eating. While she was expressing such thoughts, the basal electroencephalogram showed random spiking at the site of the electrode just underneath the temporal lobe. I referred earlier to the patient with epilepsy and schizophrenic manifestations who claimed to hear clicking sounds and imagined that people were taking pictures of her.

Questions of limbic inputs. If, as suggested, the limbic system contributes to a sense of personal identity by integrating internally and externally derived experience, how is this accomplished? Classical anatomy provides no evidence of inputs to the limbic cortex from the auditory, somatic and visual systems. Consequently it has always been puzzling that with epileptic discharges arising in or near the limbic cortex of the insula, hippocampal gyrus and hippocampal formation, patients may experience illusions, alterations of perceptions, or hallucinations involving any one of the sensory systems. Recently Malamud reported an unusual case of a twenty-six-year-old man with a small ganglioma in the uncus-amygdaloid-hippocampal region who variously experienced ictal gustatory, olfactory, auditory, visual and somatic illusions or hallucinations without loss of consciousness (Malamud, 1966). As not uncommon, this patient subsequently developed mental symptoms diagnostic of a schizophrenic reaction.

In a paper of 1949 I presented a diagram suggesting that all the sensory systems feed into the hippocampal formation (MacLean, 1949). At that time it was known that the olfactory apparatus was indirectly connected with the hippocampal formation, but there was no experimental evidence of a representation of the other senses.

During the past several years we have demonstrated by recording from single nerve cells in awake, sitting monkeys and by neuro-anatomical techniques that the posterior parahippocampal cortex receives visual connections from the geniculo-pulvinar complex. It has been found that after a lesion in the ventrolateral part of the lateral geniculate body (the main nucleus for transmitting visual impulses) a continuous band of degeneration extends into the core of the hippocampal gyrus and that some fibres enter the cortex here and neighbouring areas (MacLean, 1966a). The lower part of this degenerating band corresponds to the temporal loop in man. It has always been a mystery why the optic radiations made this long temporal detour, but it now appears on the basis of the anatomic and microelectrode findings that it travels this roundabout way in order to distribute fibres to the posterior limbic cortex. The inferior pulvinar, considered to be a visual association nucleus, also contributes fibres by way of the band lying just lateral to the optic radiations.

It deserves emphasis that in the phyletic development of primates, the convolutions in this part of the brain undergo a great expansion. Indeed, the fusiform gyrus lying next to the hippocampal gyrus represents a new convolution in the primate brain. There are some indications from the observations of Penfield (Penfield, 1954) and others that this region of the brain may be implicated in dreaming and imagination.

From the standpoint of paranoid symptoms, it is significant that some nerve cells in the parahippocampal retrosplenial cortex are photically activated only by the contralateral eye, suggesting that the impulses originate in the primitive temporal monocular crescent of the retina (MacLean, 1966a). As well known from personal experience, objects entering this part of our peripheral fields commonly cause emotional startle and alarm. This recalls the patient mentioned above who at the beginning of his seizure had a feeling that someone was standing behind him.

The parahippocampal cortex transmits impulses to the hippocampus which in turn projects to the hypothalamus and other structures of the brain stem involved in emotional, endocrine, and somato-visceral functions. Through such connections there is one possible mechanism by which the brain transforms the cold light with which we see into the warm light with which we feel.

We are currently exploring the limbic cortex of the claustral insula for evidence of inputs from the somatic and auditory systems (Reeves, *et al.*, 1968). Auditory and somatic stimulation evokes responses of a significant proportion of units in this area. No unit has responded to more than one modality. Two main types of auditory units are found, one of which responds with latencies as short as 10 msec. Somatic units are activated by pressure alone or by pressure and light touch. The receptive fields are usually large and bilateral. As the claustral insula has connections with the hippocampal formation, the findings suggest a pathway by which impulses of auditory and somatic origin may reach the hypothalamus and influence vegetative and emotional functions.

"Holonarchic" communication. On the basis of experiments such as these it is becoming evident that the limbic brain—just as the reptilian and neomammalian brains—receives specific information from the various sensory systems. Another crucial question in considering the elaboration of paranoid feelings is the one concerning the mechanism of communication among the three types of brains. The question promises to be the most difficult of all to answer. The limbic brain has many connections in series with the reptilian brain, but the indications are that its channels of communication with the neomammalian brain are largely indirect. This applies both to the horizontal lines of communication at the cortical level and the vertical lines of communication with the brain stem. Neuronographic studies would indicate that, as in the case of our business and governmental organizations, transactions between the new and old cortex depend largely on the relatively slow vertical lines of communication. The most probable loci for interaction are the reticulum of the midbrain and the intralaminar nuclei of the thalamus (MacLean, 1958). As was noted earlier, the opportunity exists in the antero-medial thalamus for reciprocity of action between the third subdivision of the limbic system and the prefrontal neocortex. Through these various vertical lines of communication, there is presumably the machinery for affects to arouse thoughts and for thoughts to generate affects. But whatever the mechanism, it may be inferred that the limbic cortex is too primitive in structure to allow communication in verbal terms. Hence, the neocortex with its problem of interpreting the feelings of the old brain might be imagined as being in a position analogous to that of a psychiatrist trying to analyse and explain the emotional feelings of his patient. As we shall see, in returning now to the

"paranoid streak", the same analogy would hold for human societies in which the medicine man, the religious reformer, the political leader, and the like, seek to interpret and give articulation to group feelings.

Final comments. In considering the genesis of paranoid feelings under normal and psychotic conditions let us begin with an illustration of the experience of a young man returning from Europe on the SS United States. One morning when the ship was off the Newfoundland Banks he walked out on deck and had a vivid feeling that it had been snowing. Presumably the appearance of sky and the temperature and the moisture of the air affecting his skin and respiratory passages added up to this snowy feeling. In communicating the nature of this feeling to his fellow passengers he found that they readily agreed with him. A short time later a news bulletin reported that it had been snowing in Newfoundland.

But let it be supposed that the young man had experienced and communicated similar feelings while cruising in the subtropical waters of the Caribbean. And suppose further, that the feeling persisted. Would he not perhaps become suspicious of fellow passengers who failed to agree with his inappropriate feeling or challenged his sanity? We may imagine that the schizophrenic patient with his persistent delusional symptoms finds himself in a comparable situation. Affective feelings provide the connecting bridge between our internal and external worlds and perhaps more than any form of psychic information assure us of the reality of ourselves and the environment around us: hence the statement that "a crazy man would be crazy not to believe in the reality of his crazy feelings" (MacLean, 1958). Beset by a chronic fearful feeling, the psychotic individual utilizes his preserved intellectual faculties to "explain away" his feelings. He may conclude that his fellow workers despise him and wish him to lose his job because they somehow learned that he had been involved in an unnatural sexual act. Less commonly in paranoid psychoses, there are chronic feelings of elation and delusions of grandeur which require some divine explanation.

The neurological condition that appears to distinguish the persisting affect of the epileptic from that of the schizophrenic is the presence in the former of a demonstrable irritative focus. What it is that accounts for the neural reverberation in the case of schizophrenia is entirely a matter of speculation. When a nagging, unpleasant feeling affects an entire social group, it may be concluded that it arises from something in the environment. Here the question of sanity does not arise because the group shares a feeling, which unlike that of the psychotic individual, lends itself to a collective belief-system.

What are some of the conditions that generate persisting unpleasant feelings among a society? Today, illustrations abound on every side. Perhaps the most generally prevailing disturbing factor is the pressure arising from over-population. Evidence is accumulating with respect to several animal species that aggressiveness increases with increasing density of population. It is often stated that man is uncommonly aggressive, and he is compared unfavourably with lower primates because of this and his propensity to kill. But the peaceful coexistence of groups of sub-human primates in the wild is possibly attributable to an abundance of living space and food; under conditions of captivity, crowding has been observed to result in violent and deathly struggle. Among animals, however, it is usually not death from combat that reduces population density to tolerable levels. Rather, nature appears to have ruled that aggression should take its toll indirectly through wasting diseases and loss of fertility. Surpassing other animals in intelligence and inventiveness, man has found an additional solution. The use of weapons, he discovered, works more quickly and just as effectively. The price of this discovery has been an enhancement of his natural paranoid tendencies, because since the slaying of Abel he has had to live with the realization that his fellow man holds over him the power of life or death. The story of the tree-of-the-knowledge-of-good-and-evil would appear to represent a paranoid delusion of early man, a delusion in which the weapon of procreation was mistaken for the weapon of destruction.

For the youth of today crowding means swimming in a school of mackerel and having the uncomfortable feeling of a loss of personal identity. In our teeming American universities it means the transformation of one's personality into the punched holes of an IBM card and the inability to stake out a little piece of intellectual territory of one's own. For the man on the street, it means a squeeze for living space, a squeeze for food, a squeeze for recreation, a squeeze in getting to work, a squeeze for one's job. For those in the establishment it means the threat of riot, revolution, and disestablishment. For the world it means the fear that some chance altercation will result in hydrogenic holocaust.

Conditions existing in Europe before World War II illustrate what can happen when there are widespread, uneasy feelings among whole nations. As feelings of this kind are commonly generated by situations that cannot be clearly identified, they are subject to a variety of explanations. At such times leaders step in and attempt to explain to individuals what they cannot see for themselves. The explanation that found appeal under the banner of the Swastika was that the Fatherland was threatened by one of its minority groups and that the resulting widespread sense of uneasiness could be relieved by the torture and extermination of this group. The explanation and recommendation were so psychotic in nature as to arouse wonder in the rest of the world how a civilized people could subscribe to such barbarism and carry out the dictates of its leaders. The lesson here seems to be that if a deranged leader rubs hard enough on the "paranoid streak" of his followers, it will flare into a pseudo-psychosis.

Just as puzzling is how a civilized people can be duped into the selection of a deranged leader. In discussing the reptilian brain, it was mentioned how dummies or fragments of dummies can trigger the sequential performance of various forms of instinctual behaviour. How influential, we might ask, is the reptilian counterpart of man's brain in selecting and following a leader? Is it possible that this brain in conjunction with the poorly discriminating limbic brain mistakes the caricature of a leader for a genuine leader. Particularly deceptive it seems, are the bold, aggressive qualities of the psychopath that make it possible for him to put on a big show and to talk louder and longer than anyone else.

With the increasing insights that are being obtained from the behavioural sciences, it is to be hoped that education through the "mass media" will make it possible for man to become increasingly thoughtful in his choice of leaders. This, short of Koestler's provocative prescription (Koestler, 1967), offers the most promise of eliminating those catastrophic dangers which Koestler sees lurking in the shadows of the "paranoid streak". Since the beginning of the humanitarian movement, the layman all on his own has been making significant strides in domesticating his emotions (MacLean, 1964). With the scientific knowledge now available to him, he should be able to harness them for progressively constructive purposes.

DISCUSSION

SMYTHIES   I think MacLean's concept of basing man's difficulties on the divorce between the two parts of the brain, the rational and emotional brains, is a very useful one. But could this not perhaps be used to explain a wider concept than paranoia: the general, emotional immaturity of human beings. The paranoid streak is one aspect of this.

KOESTLER   May I make a marginal comment? I incorporated a chapter on MacLean's work in my last book and have drawn some theoretical conclusions regarding the conflict between emotion and reason, and about the ways in which emotive belief-systems compel the neocortex to provide spurious rationalizations for these emotive beliefs. But I have rarely encountered such strong emotive resistance as on this subject; it was quite extraordinary, but at the same time rather gratifying, because it seems to be an ironic confirmation of the point that MacLean made.

KETY   Perhaps another way of expressing that point Paul was making is that the limbic system, these affects, may have played an important adaptive role in another era, when aggression, anger, could be said to have been adaptive at least under certain circumstances. One can demonstrate physiologically how in fact all of the semantic counterpoints of anger do play an important role in preserving function, accentuating function, mobilizing energy and terminating tissue damage and so forth. With the development of the neocortex and with the development of civilization, these kinds of adaptation are obsolete, but unfortunately have not become obsolete. And what is adaptive in the jungle is no longer adaptive in civilization. Unfortunately we still retain it and the combination of the neocortex and the reptilian brain is something that threatens to be a lethal one.

REFERENCES

BROCA, P. (1878) Anatomie comparée des circonvolutions cérébrales. Le grande lobe limbique et la scissure limbique dans la série des mamiféres. *Rev. Anthrop., 1*: pp. 385–498.

JACKSON, J. HUGHLINGS and COLMAN, W. S. (1898) Case of epilepsy with tasting movements and "dreamy state"—Very small patch of softening in the left uncinate gyrus. *Brain, 21*: pp. 580–590.

KOESTLER, A. (1967) *The Ghost in the Machine.* Hutchinson and Co., London, and Macmillan, New York.

KRAFFT-EBING, R. (1904) *Text-book of Insanity.* Trans by C. G. Chaddock, F. A. Davis Co., Philadelphia.

MacLEAN, P. D. (1949) Psychosomatic disease and the "visceral brain" Recent development bearing on the Papez theory of emotion. *Psychosom. Med., 11*: pp. 338–353

———— (1952) Some psychiatric implications of physiological studies on frontotemporal portion of limbic system (visceral brain). *Electroenceph. Clin. Neurophysiol., 4*: pp. 407–418.

———— (1954) The limbic system and its hippocampal formation. Studies in animals and their possible application to man. *J. Neurosurg., 11*: pp. 29–44.

———— (1958) Contrasting functions of limbic and neocortical systems of the brain and their relevance to psychophysiological aspects of medicine. *Amer. J. Med., 25*: pp. 611–626.

———— (1961) Psychomatics. In Field J. (ed.) *Handbook of Physiology, Neurophysiology III.* American Physiol. Soc., Washington, D.C., pp. 1723–1744.

———— (1962) New findings relevant to the evolution of psychosexual functions of the brain. *J. Nerv. Ment. Dis., 135*: pp. 289–301.

———— (1964) Man and his animal brains. *Mod. Med., 32*: pp. 95–106.

———— (1966a) The limbic and visual cortex in phylogeny: Further insights from anatomic and microelectrode studies. R. Hassler and H. Stephan (eds.) in: *Evolution of the Forebrain.* Georg Thieme, Stuttgart. pp. 443–453.

———— (1966b) Brain and vision in the evolution of emotional and sexual behavior. *Thomas William Salmon Lectures*, New York Academy of Medicine.

———— (1967) The brain in relation to empathy and medical education. *J. Nerv. Ment. Dis., 144*: pp. 374–382.

MALAMUD, N. (1966) The epileptogenic focus in temporal lobe epilepsy from a pathological standpoint. *Arch. Neurol., 14*: pp. 190–195.

MEYER, A. (1928) The evolution of the dementia praecox concept. *Res. Publ. Ass. nerv. ment. Dis., 5*: pp. 3–15.

PAPEZ, J. (1937) A proposed mechanism of emotion. *Arch. Neurol. Psychiat. (Chicago)*, *38*: pp. 725–743

PENFIELD, W. and JASPER, H. (1954) *Epilepsy and the Functional Anatomy of the Human Brain*. Little, Brown and Co., Boston.

REEVES, A. G., SUDAKOV, K. and MACLEAN, P. D. (1968) Exploratory unit analysis of exteroceptive inputs to the insular cortex in awake, sitting, squirrel monkeys. *Fed. Proc.*, *27*: pp. 388.

WIENER, N. (1948) *Cybernetics, or Control and Communication in the Animal and the Machine*. Wiley New York.

Empiricist and nativist theories of language: George Berkeley and Samuel Bailey in the 20th century

David McNeill

University of Chicago

For many years an empiricist theory of mind has dominated pschyological speculation concerning the acquisition of knowledge. This theory holds that the contents of the mind are derived from experience, and that the structure of the mind results from associating sensory ideas. Recently, and from a novel direction, the empiricist theory has come under attack. Chomsky (1959, 1965) and Katz (1966) argue that a native language cannot be acquired, given its levels of complexity and abstraction, unless children have innate knowledge of certain universal linguistic principles. Because such innate principles are available *a priori* languages everywhere have the same general form, as a reflection of the mind's intrinsic structure. My own work in the last several years has been devoted to developing a theory of language acquisition along these lines.

In the novelty of contemporary dispute it is easy to overlook the antiquity of the contending theories. The empiricist tradition is well known. For the opposing side, Chomsky (1966) has traced the history of rationalist speculation with which the nativist tradition is associated. Both theories are extremely well-knit collections of ideas. They contrast with each other in nearly all respects and have survived without essential modification down to the present day.

My aim is first to examine the nativist-empiricist difference where it has already been clearly developed—in the visual perception of depth—and then to relate the contrasting views as they appear there to modern views of language. To accomplish this translation I will attempt to express George Berkeley's "New Theory of Vision" (1709) as a theory of language. Doing so is not difficult, since to Berkeley vision *is* a language. The main problem is to express his theory abstractly enough so that it no longer has to do with vision; I am assisted in doing this by Berkeley's own use of language as a metaphor. The most important and least well-guided extension of Berkeley's views is into syntax. He said little about grammar that is explicit. However, Berkeley is so explicit on everything else that I am fairly confident about the extension I have made for him. For Samuel Bailey, Berkeley's 19th-century critic, I will take myself as a delegate. It is not that my views were developed to extend Bailey to language, but having independently developed my views and then having read Bailey I see that such is the relation.[1]

A number of "new" points arise when Berkeley and Bailey are extended into the domain of language, and certain "old" observations take on a new and surprising significance. Although some reshuffling of current ideas thus occurs, it is not because contemporary disputes are unrelated to the classical one. On the contrary, the contrast between contemporary nativist and empiricist positions emerges in full clarity when expressed as a contrast between Berkeley and Bailey. All the issues they originally raised are still present, implicitly or explicitly, today. That is the significance of my title: Berkeley and Bailey are alive and well in the 20th century.

My method is to outline Berkeley's views, presenting Bailey's criticisms as I go, and then to outline a Baileyan theory of language acquisition, which seems to me a better approximation of the truth.

Berkeley makes a number of interlocking assertions; Bailey makes a nearly equal number of interlocking counter-assertions. The table below is a summary of ideas that separate them.

Ideas separating Berkeley and Bailey

Berkeley	Bailey
No ideas are abstract	Some ideas are abstract
All ideas are derived from experience	Some ideas are innate
Justified assertion is correct assertion	Justified assertion is not correct assertion
Meaning is arbitrary	Meaning is not arbitrary

[1] I have been greatly aided in this project by being able to discuss many of the issues with Prof. R. J. Herrnstein of Harvard. It was he also who drew my attention to Bailey (1842), whose representative I have taken it upon myself to be.

1 The most fundamental of these differences arises from Berkeley's belief that there are no abstract ideas. That is, no ideas can be abstracted from perception. Berkeley cannot imagine, for example, a square without also imagining its colour, size and orientation. The word *square* for him refers to a set of perceptual ideas. Similarly in vision. Since geometrical distance is invisible but tangible distance can be felt, it must be *tangible* distance that we are aware of, even when we see. Bailey objected to Berkeley's argument on the ground that it is arbitrary to select touch as the basis of perceived distance. The puzzle exists as much for touch as for vision, according to Bailey, since tangible sensations give rise to ideas of depth no more readily than visual sensations do. This issue reflects a major difference between the empiricist and rationalist theories of mind.

2 Bailey's argument is intelligible only if we assume that he distinguished correct from justified assertion.[2] Berkeley's proof that abstract ideas do not exist, on the other hand, is intelligible only if we assume that he did not draw this distinction. The real division between them, as I reconstruct it, is this one. For Berkeley justified assertion is correct assertion. Hence he tested the concept of an abstract idea by looking for a *perceptual*-abstract idea. Could he imagine a black square, two inches on each side, which was none the less abstract? Berkeley is of course correct that such an idea is impossible. For Bailey, however, correct assertion is not the same as justified assertion. Justified assertion is about perceptual evidence; evidence suggests abstract ideas, which are the content of correct assertion; but as Berkeley showed abstract ideas are not perceptual ideas. To Bailey the word *square* refers to an abstract idea of squareness, and the only role of perceptual evidence is to suggest when it is justified to refer to this abstract idea by means of the word *square*. From Bailey's point of view, Berkeley's demonstration of the impossibility of an abstract idea takes for granted the controversial claim that justified assertion is correct assertion and does not touch the issue it was designed to settle—whether or not there are abstract ideas. I think Berkeley was quite unconscious of the circular reasoning behind his proof. More amazingly, I think Bailey was unconscious of it, too. Neither he nor Berkeley is explicit on the status of correct and justified assertion in their thinking, probably because neither thought it possible to differ over their two contradictory positions.

[2] This distinction was drawn to my attention by Chomsky, who uses it in distinguishing his own position from Wittgenstein's.

3. The difference between Berkeley and Bailey over correct and justified assertion coincides with an underlying difference between them over innate ideas. If there are no innate ideas and everything in the mind is acquired from experience, then justified assertion must be correct assertion. The correct use of words is precisely what the perceptual evidence suggests. If on the other hand there are innate ideas a distinction must be drawn between these ideas and the perceptual evidence that suggests them. For Berkeley nothing can exist in the mind unless it can be acquired from experience; procedures for the acquisition of knowledge are innate—in Berkeley's case, an ability to form associations—but there is nothing else. For Bailey there are in addition innate ideas, which must be related to percepts through experience.

Words suggest perceptual ideas, based on long experience. We can translate this claim of Berkeley into modern terms. A transformational grammar may correctly describe how sound and meaning are related, but it does not describe associations. But associations are the actual outcome of language acquisition. The relation between competence and performance in language is thus indirect. A transformational grammar describes competence, but a network of associations, of which a grammar is an idealization, describes performance.

Consider as an example certain strategies for recovering the deep structure of sentences. Fodor and Garrett (1967) have proposed that some verbs (e.g. *know*) suggest to a listener particular possibilities in the deep structure of sentences (e.g., that it contains a direct object or a complement). This proposal corresponds to one of Berkeley's analyses of vision, where he argued that a blurred, confused visual perception suggests very near objects. In both cases one must distinguish *seeing* depth or *hearing* meaning (in neither of which one thing suggests another) from being *reminded* of depth or meaning by virtue of an associative connection. In Berkeley's view a sign always functions to *suggest* to the mind a meaning. In Fodor and Garrett's terminology, the deep structure of a sentence is always *suggested* by a strategy applied to the surface structure. (Fodor and Garrett themselves have in mind a more restricted and plausible use of syntactic strategies.)

4 Any two perceptual ideas can be associated; these connections are arbitrary. Visual cues could conceivably have just the opposite significance from what they happen to have. So could words, and so

could sentences. A strategy or association is possible for any systematic relation, and hence in the case of language for any transformational grammar, since the procedure for acquiring knowledge is universal, innate, and applicable to any connection. In order for one idea to suggest another, therefore, it is not necessary to demonstrate the necessity of the coexistence of ideas. Indeed, such a demonstration is impossible to arrange for there are no necessary connections among perceptual ideas belonging to different senses (such as hearing and touch or sight).

Bailey would agree to all of this. However, he would add that arbitrary associations among perceptual ideas are peripheral, rare, and unimportant. Far more important and typical are connections between perceptual ideas and abstract ideas. Again the difference between Bailey and Berkeley is fundamental and related to their differences elsewhere. For Berkeley the meaning of a word is the perceptual evidence that justifies its use. Since associations are established through long experience between arbitrarily connected perceptual ideas, *meaning* is arbitrary. For Bailey on the other hand the meaning of a word is the abstract idea with which the word is associated. Perceptual ideas are arbitrarily associated with abstract ideas, so words and the justification of their use are arbitrary, but *meaning* is not.[3]

As an example of this difference between Bailey and Berkeley, consider that for Bailey it is perfectly intelligible if an expression appears in the language of children before the "intellectual" equivalent appears outside language. Words or sentences can be associated with abstract ideas before the same abstract ideas are associated with non-linguistic perception; the opposite, of course, also can happen. For Berkeley, however, such discrepancies are perfectly unintelligible. It is possible in Bailey's theory but not in Berkeley's for children to use words or sentences meaningfully without at the same time knowing what they refer to in the external world. Children can have in mind the abstract idea of causation, for example, without understanding anything of physical or psychological causation. In this case

---

[3] Actually Bailey is unclear about whether or not he thought the connections between abstract ideas and perceptual ideas are arbitrary. In discussing vision he seems to maintain they are not. But such a claim is separate from the claim that the mind has available innate abstract ideas. To emphasize the distinction, I have dropped in what follows Bailey's apparent assumption of non-arbitrariness while keeping his assumption of innateness. None the less Bailey may be right about arbitrariness, at least in some cases.

they utter such sentences as *why not me careful?* or *why me bend that game?*—questions about causation that simultaneously betray confusion over actual causes. Berkeley would say that such questions are not about causation at all, but rather about other external relations among events.

Berkeley argued that suggestion operates with such great speed and certainty that sound and meaning enter the mind in the same moment. Moreover, the mind pays little attention to signs. Attention falls instead on meaning, and for this reason one believes he *hears* meaning, although actually meaning is *suggested* by sound.

Bailey's main counter-argument is raised against this claim.[4] Associations only guarantee that one thing reminds you of another. In the case of fuzzy vision, when an association suggests a nearby object, two things are present in the mind: a blurred image and the idea of an object very close before the face. It is plausible that one reminds you of the other. Similarly, the verb *know* suggests two different deep structures, whereas other verbs, say *hit*, suggest only one.

However, effects of this kind are not the phenomena to be explained. The real phenomena in Berkeley's view are that you *think* you hear meaning or see depth. But, Bailey points out, such effects are ". . . altogether at variance with the acknowledged laws of the human mind", namely, the laws of association, and Berkeley suggests no other laws. In fact, Berkeley's theory merely restates the phenomena it sets out to explain.

Berkeley gave two explanations of the common belief that meaning is heard and depth seen. One is that sound and meaning enter the mind at the same moment. The other is that attention falls entirely on meaning. Because of these two processes we think we hear meaning and see depth. But the speed of association is irrelevant. Even if sound and meaning enter the mind at precisely the same instant, there is no reason not to be conscious of both. The simultaneous perception of different sensory ideas is clearly possible. The second process—that attention falls entirely on meaning—is *ad hoc* since it has nothing to do with association. An association brings to mind

[4] Bailey, it should be noted, did not regard the following argument as his main point. As Herrnstein has pointed out, Bailey misunderstood Berkeley to say that we learn to *see* depth by associating visual ideas with tangible ideas of distance. Had Berkeley said such a thing, his theory of vision would be based on a non-sequitur. But Berkeley did not say it, and what Bailey thought of as his main point—that Berkeley's theory is based on a non-sequitur—turns out to be Bailey's categorical error.

one idea given the presence of another; it does not make one idea disappear. Moreover, since by "attention" Berkeley means what the mind is conscious of, his explanation merely restates the original problem—that for some reason attention falls entirely on meaning. The basic flaw in Berkeley's argument exists in this very formulation of the problem. Contrary to Berkeley's assumption about the facts, meaning *is* heard; it is not in the normal course of events suggested. Bailey asserts the same about the visual perception of depth.

Although Bailey did not notice it, Berkeley himself unintentionally provides an example counter to his theory. The case under consideration is reading, not speech, but the structure of the argument is the same. Berkeley writes: "It is customary to call written words and the things they signify by the same name; for, words not being regarded in their own nature, or otherwise than as they are marks of things, it had been superfluous and beside the design of language to have given them names distinct from those of the things marked by them" (NTV. p. 88). In other words, printed signs are ignored in favour of meanings, and because of this printed words do not have their own names but take the names of things.

It is unquestionably true of printed words that attention is focused on meaning. But Berkeley is wrong in supposing that printed words do not have independent names. They do. It is just that printed words are *read* with the names of things. "s" "q" "u" "a" "r" "e" is the name of a particular printed word. It is read/skwær/, the name of a □. If attention were deflected from names to the things they suggest, as Berkeley argues, in the case of print the mind would be conscious of what Berkeley says it is unconscious—it would become aware of "square" when it perceived "s" "q" "u" "a" "r" "e". This does not happen, of course, because the mind does what Berkeley assumes is impossible: it perceives meaning directly from the printed page. That proof-readers do not easily perceive meaning and meaning-readers do not easily perceive proof errors is perfectly consistent with this analysis. Berkeley's theory describes proof-readers only. Bower (in press) has recently argued that in reading meaning is perceived directly from print.

For Bailey Berkeley's problem does not exist. The ability to obtain meaning directly from a printed page (or from speech) is not mysterious; in fact, it is inevitable. What is named by the word *square* is the abstract concept of squareness. Many different visual, auditory, or tangible ideas may suggest the concept of squareness, for any perceptual idea can become associated with an abstract idea. Indeed, outside of special situations, such as some psychological experiments, perceptual ideas are associated only with abstract ideas. The "name" of a printed word therefore suggests an abstract idea directly. It is connected with nothing else. Bailey's explanation is not available to Berkeley, since to escape along this avenue is to introduce abstract ideas and to distinguish correct and justified assertion.

Some modern observations bear on the question of whether the mind perceives meaning directly or first perceives sound and then is reminded of meaning. One conclusion to be drawn from G. A. Miller's early research on the perception of speech is that a listener can hear sound imperfectly but meaning perfectly. Over a wide range of variation subjects correctly "shadow" speech despite the presence of a white noise—a noise that sounds like escaping steam, in which energy is equally probable at all frequencies. Such observations are significant because they show that meaning can be recovered even though speech is disrupted at random. Since the disruption is random and thus unpredictable it is not possible that different associations are involved in the recovery of meaning. Most of the mutilated speech sounds are encountered for the first time in the experiment. It cannot be "generalization" among sounds either— that is, a single association from a class of sensory ideas—because the perceived meaning is unchanging while the perceived sound varies. The sounds are novel, so the relevant class of sounds cannot be learned. There are no innate abstractions, so the class of sounds cannot be innate. In fact, the required basis of generalization is in terms of what is recovered, not what is given, and this is logically inconsistent with Berkeley's theory. The theory predicts the opposite of actual fact—that the suggested meaning would be generalized along with the sound. The comprehensibility of a foreign accent raises the same problem. In none of these cases can meaning be suggested to the mind by the perception of sound. Meaning must be heard directly.

Such observations pose no difficulty for Bailey's theory. Meaning according to Bailey is not arbitrary, so if a listener perceives any speech sounds at all, they must lead him to some definite abstract idea. The sounds may be so distorted, of course, that they suggest several abstract ideas or the wrong abstract idea or no abstract idea at all—all this can happen—but there are not gradations of meaning with gradations of sound, as in Berkeley's theory.

Berkeley holds that there is true meaning (as there is fixed magnitude). Berkeley never says what he thinks true meaning actually is, except that words are arbitrarily connected with things—i.e., with the perception of things. He does comment that "Whenever . . . we speak of the magnitude of anything . . . we must mean the tangible magnitude" (NTV, p. 44), and that the meaning of, for example, the word "erect" is " . . . nothing more . . . than that perpendicular position of [say] a man wherein his feet are nearest to the earth . . ." (NTV, p. 67).

Bailey's theory of meaning has already been stated. The meaning of a word is the abstract idea to which it is associated. Unfortunately this theory is no more explicit than Berkeley's, so although there are facts inconsistent with Berkeley's theory we do not know if the same facts are consistent with Bailey's.

In this section I look briefly into the question of word meaning. The question of syntax is taken up in the next. Let us look in particular at Berkeley's one claim about word meaning—that thing-words are associated with perceptual ideas of things. This view is incomplete, if not actually wrong, and there is no way to extend it without violating the hypothesis that the meaning of a word is the set of perceptual ideas it suggests. Consider the concrete noun *collie*. It refers to a certain class of hairy quadrupeds and we suppose it to be associated with certain perceptual ideas—visual, tangible, auditory, and (in some cases) olfactory ideas. But we also know of this word that: a collie is a dog; a dog is an animal; an animal is a living thing; and a living thing is a physical object. *Collie* implies the entire series. If we say "my collie blooms in late May", for example, we violate one member. We might think that *collie* is associated with the perceptual ideas suggested by the various members of the series. "Physical object", for example, may suggest a vague, hazy visual idea. If this idea is part of the meaning of *collie* then it could be labelled "collie", for it is part of what the word suggests. But perceptual evidence cannot be all there is to the meaning of *collie*. If it were, we should be able to say "a physical object is a collie". However, we cannot. Indeed, none of the following is admissible: a dog is a collie, an animal is a dog, a living thing is an animal, and a physical object is a living thing. *Collie* is in fact asymmetrically related to the ideas of "dog", "animal", "living thing", and "physical object" by means of a certain predicate, "is a". Predicative relations are inherent in the notion of a "sentence" and for this reason sentences are more than associations of words. If one *asks* people for associations, the result is the first (predicative) series, not the second (associative) series. *Dog* suggests *animal* far more readily than *animal* suggests *dog*. Miller (1967), from whom the above examples come, proposed that nouns are mentally organized around such predicates in order to facilitate the understanding of sentences.

One of Berkeley's main points is that new ideas are named in a certain way when they have something in common with old ideas, but if there is nothing in common a name cannot be given (as in the case of a newly sighted blind man). As noted before, Berkeley did not discuss syntax but it is possible to guess what his views were. Because of syntax names need not be identical if ideas are not. It is consistent with Berkeley's thinking to assume that sentences accomplish this kind of flexibility by limiting the ideas suggested by individual words. He writes ". . . a word pronounced . . . in a certain context with other words has not always the same import and signification that it has when pronounced in . . . different contexts of words" (NTV, p. 52). A plausible extension of this remark is that sentences are special cases of names. Names have more or less fixed significance; sentences are impromptu names that suggest impromptu meanings. That is what a child acquires.

For Berkeley the question of language acquisition is how to get from simple names to complex names, that is, from correlations between simple perceptual ideas to correlations between complex perceptual ideas. In a transformational grammar every sentence has a subject and predicate. The predicate has a verb, and sometimes an object or prepositional phrase, either of which may include another sentence. Even more complex variations occur. These deep forms cannot be abstract ideas in Berkeley's view. They must be perceptual ideas, and probably always tangible ideas, which come from relations experienced in the external world. Several contemporary psychologists have suggested how this is done (e.g., Slobin, 1966; Schlesinger, in press). The relation of subject and predicate in a sentence is associated with actors and actions in perceptual experience. Among all real actions there are some that act on things, others that act *in, near, by,* or *after* things, and still others that act on things that themselves include an actor and an action. In this theory, every sentence, however complex, suggests some such combination of perceptual (and probably tangible) ideas.

If Berkeley's theory is correct one would certainly expect children's first uses of the grammatical relations of subject and predicate to be relations among perceptual ideas. Indeed, it is difficult to understand in Berkeley's theory how these relations could ever be anything else. A sentence like *McGregor admires sincerity* seems beyond the scope of the theory, although Berkeley's views might possibly be extended to cover such cases (e.g., "admiration" is a certain tangible idea, "sincerity" is a certain visual idea, etc.).

If we examine the speech of young children, however, these problems of extension do not arise. Child speech is almost always about perceptual ideas; indeed, it is almost always about tangible perceptual ideas. For child speech, therefore, Berkeley is strikingly well confirmed. The following utterances are selected at random from the tape-recorded speech of a two-year-old:

> where birdie go?
> birdie up dere
> hit hammer, mommy
> screw did happen
> yep it fit.

In each one concrete perceptual idea is related to another, and in several (if not all) the relations are tangible. It is just as Berkeley says.

However, there are other more subtle characteristics of child speech, which suggest that Berkeley's theory is not correct; that, in fact, a totally different theory is correct.

Certain perceptual relations do not appear in early child speech even though children experience them. Children surely experience tangible sequences of acts—running and picking up a toy, for example. But such sequences do not appear in the speech of young children; there are no sentences with pairs of verbs (e.g., *run (and) get (the) toy*). Such is the situation even though sentences with pairs of verbs are spoken to children and children commonly utter sentences with single verbs (e.g., *get toy*). Children do not associate sentences with every tangible idea they experience, therefore, only with some tangible ideas. There is no way to accommodate such systematic selectivity into Berkeley's theory; it contradicts the claim that meaning is arbitrary.

The opposite phenomenon also occurs. Relations are expressed in child speech that cannot be perceived in the external world. Other people's habitual activities, for example, cannot be perceived. They can only be remembered. But predicates expressing such habitual activities are among the first to appear in child speech. Some examples are the following: in answer to the question "where is daddy?" one (Japanese) child said "at the office"; another child (American), in answer to the question "when is Cromer coming?" said "on Wednesday". In both cases the child replied with a predicate that describes someone else's habitual action. Berkeley could reply that children associate answers with questions, not with habitual activities, so the examples are not what they seem. But this is a band-aid for a decapitation.

Consider the problem that arises for Berkeley when associations manifestly do not exist. Take the sentence *that no fish school*. It was recorded in the speech of one of the children studied by Roger Brown, and is characteristic of child speech at a certain stage of development. If this sentence is meaningful at all, its meaning must be different (in Berkeley's theory) from "that's not a fish school", "that's no fish school", "that's a school for non-fishes", or any other expansion into adult English. The sentence could possibly have an intermediate meaning, but it could have no meaning in adult English. *That no fish school* cannot suggest to an adult an adult meaning, since adults have had no chance to learn an association to it; they know only adult forms. It cannot suggest to a child an adult meaning either since children also have had no chance to learn an association to it; they know only child forms. In Berkeley's theory children must express different meanings from adults when they produce sentences that differ. I cannot say if this is right or wrong—children might express esoteric meanings in their speech. The possibility cannot be investigated since doing so depends on the interpretation of an adult investigator.

However, Berkeley's argument leads to another conclusion. Not only should children express different meanings from adults, but also adults should understand child speech as having different meanings. In this case the implication is demonstrably false. Adult expansions are invariably syntactically well-formed and meaningful adult sentences. Unless adults actively confabulate, this fact is inexplicable in Berkeley's theory. Adults are former children. There is no way for them to acquire the associations that produce expansions unless as children *they* expressed adult meanings through child sentences. But if adults did it as children why do present children not do it, particularly in view of the inevitable fact that they, too, will some day expand child speech? Berkeley, to escape one horn of this dilemma, thus falls on the other. The difficulty here is the same as with sentences heard in noise. If meaning is arbitrary, as Berkeley maintains, an esoteric sound must suggest an esoteric meaning. This proposition is a consistently false implication of the theory. I see no way of avoiding Berkeley's dilemma without untying the entire knot of ideas that makes up his empiricism; without, in fact retying it in Bailey's manner.

Bailey's theory avoids Berkeley's dilemma altogether. Meaning is not arbitrary; it is only suggested by arbitrary signs. If sounds are distorted, the correct meaning stands a poor chance of being suggested, but the meaning suggested will not be esoteric. It can only be some other abstract idea.

Let me now try to extend Bailey's views in the direction of language acquisition. I offered myself above as his delegate; I do so even though other theories of language acquisition can be imagined which also are consistent in the main with Bailey's position. The differences are not always minor. For example, depending on whether or not we take Bailey to say that necessary connections exist between sensory and abstract ideas, we obtain quite different theories. The one sketched below assumes there are no such connections.

In contrast to Berkeley's theory, where sentences are special cases of names, one can argue in exactly the opposite direction that names are special cases of sentences. Such an argument is possible if a sentence is an innate abstract idea. If sentences are innate ideas, the structure of a sentence is predetermined, sentences are not arbitrary, and children's linguistic concepts inevitably will include them. In this theory a child can no more avoid speaking sentences than, according to Bailey, he can avoid seeing three dimensions. If sentences are in addition abstract ideas they cannot be identified with any organization of perceptual ideas. They are, rather, associated with perceptual ideas.

The problem now facing us is to determine what the intrinsic form of a sentence is, and then to see if children show evidence of honouring it at an early age. As I will try briefly to show, not only is there such evidence, but all the difficulties and absurdities that arise from Berkeley's theory vanish if we assume that a sentence is an innate abstract idea.

In linguistic notation a sentence is a noun phrase and verb phrase bracketed together, with no constituents intervening. We can write it (NP, VP)$_s$. Between the NP and VP we can define a relation of *predication*. The deep structure of every language and of every sentence in every language is built upon this relation, apparently without exception.

Intuitively, predication consists of a comment on a topic. Examples abound. Holding up an empty cocktail glass in one hand and two fingers on the other means "two more of the same"—a comment and topic. The word in my dialect pronounced / meriy / alone and out of context is neither a topic nor a comment; it is nothing grammatical. Likewise the word *sings* alone and out of context is nothing grammatical. But when the two are put together something more than a list or a paired-associate emerges—one word comments on the other. We have *sings marry*—a sentence. That we have a sentence in even this simple and (I assume) unexpected case reflects the fact that predication is one intrinsic feature of sentences.

When children begin to speak, utterances are limited to single words—actually to the phonological mutilations of single words. Berkeley and Bailey would agree that such isolated words are the beginning of grammar. But they differ on the significance of this beginning. In Berkeley's theory sentences are developed from an initial base of isolated names; names are what children first utter. In the Baileyan theory I am sketching, the isolated words of young children are already sentences. Single words are used predicatively, as comments on the extra-linguistic context. Names come later as special cases, as the topics of comments, and thus depend on the prior existence of predication. Very little is known of the earliest phases of language development, but such information as exists supports a Baileyan view. The best examples come from O. Bloch (1921). One of his children, a little girl, pointed to her older brother's slippers and said his name, "Raymond". Unless the little girl was confused over what exactly her brother *was* and was actually labelling a brother-part, she spoke predicatively. Another time she carried in a newspaper and said "Papa". Again, apparently, she spoke predicatively. Bloch describes a number of such cases.

If, when children begin to speak, they do so because they associate single words with the abstract idea of a sentence, it is easy to see why Bloch's result should arise and why adults can understand even this most distorted of child speech. In fact adults cannot avoid understanding holophrastic speech. All an adult must do is believe that he is engaged in the socially universal act of conversation; then he cannot avoid being correct, since the processes underlying both child and adult speech are guaranteed in advance to be the same. Both have in mind the abstract idea of a sentence. The extreme distortion of child speech makes uncertain the particular adult sentence that corresponds to what a child says; but distortion does not make uncertain the method of combining meanings, which is fixed and not arbitrary.

The adjustment of adults to children thus depends only on meeting the most minimal condition—that an adult believes he is conversing. Such a modest requirement can exist because the initial ideas of children coincide with the entire structure of language. Predication is the first to appear of several so-called basic grammatical relations. The others—subject, verb and object, modifier—together with predicates comprise a major part of the deep structure of sentences. It is not known how the other basic grammatical relations arise, though possibly it is through differentiation of primitive predication. However they arise, the basic grammatical relations appear in full complement by the second birthday. I will mention some of the evidence for this in a moment, but first I must say something about the design and structure of language.

A traditional view holds that language is a systematic relation between sound and meaning. Both Bailey and Berkeley accept this view. A transformational grammar likewise accepts it, and goes on to resolve the relation of sound to meaning into three major (and again traditional) parts—phonology, syntax and semantics. I will concentrate on the middle component of syntax. Every sentence includes a superficial, manifest structure, where words are organized and arranged as we actually perceive them; the phonological aspect of a sentence naturally depends on the surface structure. Sentences also have a deeper structure, underlying the surface and generally quite different from it in form. *John is easy to please* and *John is eager to please* have the same surface structure but obviously have different deep structures—*John* is subject in one and object in the other. The deep structure of a sentence includes syntactic information necessary for a semantic interpretation—the basic grammatical relations, for example, and the constituent membership of the words. The exciting discovery made by Chomsky fifteen years ago is that deep and surface structure, although different, always stand in a precise relation to one another, a relation that Chomsky and his colleagues have been able to describe exactly in many cases. These relations are the transformations.

Transformations explicitly state the relation between sound and meaning in each language. It has been observed in work on transformational grammars that although different languages have idiosyncratic transformations, many aspects of the deep structure are linguistically universal. Bailey's theory explains why this empirical generalization should hold. If the idea of a sentence is given *a priori*, then to learn a language a child must discover how to relate the abstract idea of a sentence to the surface structure of his language; that is, he must learn precisely what are described as transformations. Since associations between perceptual ideas and abstract ideas are arbitrary, transformations are idiosyncratic among languages; since innate ideas are not arbitrary, deep structures are universal among languages. In Bailey's theory child and adult speech converge very beautifully at the most crucial level—at the level of deep structure where meaning is organized—and diverge elsewhere, at the level of sound.

Bailey's theory also explains why certain patterns appear in child speech while other patterns do not. Recall that children do not utter sentences with pairs of verbs even though such sentences are exposed to them and children themselves utter sentences with single verbs. Other combinations of grammatical classes also do not occur. At an early point in development one child had evolved three grammatical classes. Of the $(3)^2 = 9$ two-word combinations thus available to him, the child used four; of the $(3)^3 = 27$ three-word combinations available to him, he used eight. Five and nineteen combinations did not occur. We can find a systematic difference between the two sets—the occurring and the non-occurring—which suggests an explanation of what the child was doing. The sentences that did occur are direct expressions of the basic grammatical relations. Those that did not—such as verb-verb-noun—are expressions of the basic grammatical relations only through the intervention of various transformations not yet known to the child. The dichotomy between sentences that honour the basic grammatical relations and ones that do not held without exception in the child's speech at two years. The principle of selection, lacking in Berkeley's theory but present in Bailey's, is the abstract idea of a sentence.

If the abstract idea of a sentence is innate, and children have to discover arbitrary associations between it and sound, children everywhere should employ the same initial grammatical system, even though they are exposed to languages of very different structure. Children should speak a universal language, in which only the words differ at first, but the underlying abstractions are the same. I will mention three examples out of many. Children exposed to a wide variety of languages (English, Japanese, German, French, Hungarian, Finnish, Russian and Chinese) always pass through a holophrastic stage, in which single words are used predicatively. Later, children exposed to all these languages construct sentences that express the basic grammatical relations but are not yet complicated with transformations (the so-called Pivot-Open grammars). Finally, all children first negate sentences in the same way, though their languages differ considerably on how negation is expressed—for young children it is always Neg+S or the reverse, as in *no drop mitten, non yeux ouverts*, or *tebakuro hame nai*. Negation at the outset is a sentence with a minus sign.

I have discussed several of Berkeley's propositions—that justified assertion is correct assertion, that there are no innate abstract ideas, and that meaning is therefore arbitrary—and have tried to show how they yield consequences fatal to themselves and to Berkeley's theory. When taken seriously, they lead to absurdities, for Berkeley's propositions are in fact mirror images of the truth. If we replace each proposition with its opposite—that justified assertion is distinct from correct assertion, that innate abstract ideas exist, and that meaning is therefore not arbitrary—we obtain a natural explanation of the facts of language and its acquisition.

DISCUSSION

SMYTHIES I would like to make a few naïve comments from the evidence in the experiments of psychology and neurology on the relation of meaning and function, and I don't know whether really this is relevant to what you've been saying. The first is to recall the experiments of Michotte who showed that causality can be an immediate perceptual experience. You observe a number of lines moving behind a slit and if the sequence is just right you experience an immediate impact of causality, which is not quite really within the concept of perception as such.

There might be some analogy to the effect that meaning and sound come together.

The second one is the problem of agnosia, of course, which merely indicates that we hear the sounds without any meaning at all. This merely suggests that the cerebral mechanisms supplying meaning and those supplying reception separate very low down in the organizational structure of the nervous system. Whether that is relevant or not, I don't know.

MCNEILL What is the evidence in support of your statement (that it is low down in the nervous system)?

SMYTHIES By that I mean it's subconscious; all the sounds appear in consciousness bereft of meaning; so the mechanisms must separate lower down.

Another point is the visual depth argument. You can't see visual depth unless you have experienced tactual objects. There is one very nice case of someone who was brought up in an extraordinary situation. Since a very early age this child had been confined to bed. This was because his father had a small pawn shop and all the objects in the pawn shop were strewn on the floor and all the tickets were tied to objects. So he had never been allowed to climb out of his bed, and thus had had a grossly defective tactual experience. In later years he was examined by a psychologist and his visual depth perception was found to be extremely inadequate.

MCNEILL This, of course, supports Berkeley's theory of vision. There are other indications that Berkeley was wrong about vision; Bower's experiments, for instance. In any case, it's the form of the argument that counts outside of vision—in language, say.

SMYTHIES Perhaps a case of getting to the right place by the wrong route.

HYDÉN If I understood correctly, you began with the picture of a tabula rasa and the development of language. In other words, is there an instructional brain mechanism to transform the tabula rasa into an integrating part of the brain? It seems that the sentences of the human language are handled by the brain mechanism as packages of sub-units rather than as discrete units in a digital form. Therefore it seems unlikely that an instructional mechanism would be used based on units ordered in a linear way to fill up the tabula rasa intended for languages, if such existed.

MCNEILL I don't quite know what you mean by units in this case. Are you raising the question of how linguistic knowledge can be expressed in the form of actually uttered speech or actually comprehended speech?

HYDÉN I am raising the question whether language could emerge by an instructional mechanism from units ordered in a linear form to express ideas.

MCNEILL By linear form you mean a noun phrase, verb phrase— that kind of linearity?

HYDÉN Yes.

MCNEILL I am not sure I can answer your question. I think it's almost an historical accident in the development of linguistic theory that the deep structures of sentences are ordered according to the elements of grammar. The same relations of subject and predicate, for instance, can be defined as well for an unordered set as an ordered set. In fact that would be even a stronger argument against any explanation of these relations which attempts to derive them from motor acts, as Berkeley's theory presumably would have done. These theories would necessarily consider the deep structure to be sequential. I would imagine that unordered representations of the deep structure of sentences would have to a certain degree a more obvious psychological interpretation than the ordered arrangement.

Perhaps none of us is persuaded by these arguments concerning order because they have to fit a distinction very difficult to maintain in practice. A linguistic description is a reconstruction of the knowledge that any fluent speaker of a language must have if he is to perform linguistically; it makes no claim whatsoever about the mechanism of performance. But these arguments really have to do with performance.

KETY I am not sure that I understood all McNeill was saying, but if the tabula rasa idea is not supported by the evidence, and if language and concepts are ordered in some way, are we sure that this ordering is something innate, that is, in the brain? And if so, what progress has been made in defining this structure? Or is it possible that the built-in components are built into our culture, or to our being human beings and are nevertheless part of experience except that they are common to all the cultures that we are aware of or all the human communities that we are aware of?

MCNEILL   You are quite right in questioning the identification of innate ideas and linguistic universals. The inference from observed universals to innate ideas is completely illegitimate. The converse, however, is not. If something is an innate idea it must be universal, but not everything universal must be innate. For example, every language includes kinship terms, and certain kinship terms are present in every language. For example, a word for maternal parent. While that's universal, no one claims to my knowledge that "Mother" represents an innate idea. The status of innate ideas, I think, has to be handled very cautiously because it's so easy to mistake the level of discourse on which this kind of statement makes sense. It does not make sense as a description of all cross-cultural similarities.

INHELDER   You have already answered part of my question. I think you have never accused Piaget and myself of being crypto-empiricists. Still, I want to come back to the question: does the nativistic point of view—defended, not so much by yourself, I think, but by many of Chomsky's followers—does this theory have an explanatory value or not? And wherefrom comes the resistance from people in this group against thinking of the development of language in the sense we do, as the development of cognitive structures—that knowledge is the result of self-regulating processes in the Waddington sense, building up structures.

Let me stress again that if my colleague, H. Sinclair, approaches psycholinguistic problems from this angle, it is because we are thinking of acquisition mechanisms. That is to say, we should like to determine by which mechanism the non-speaking child is able to build his interiorized frames from what he hears around him. Moreover, we always like to stress that what he learns around him relates in some way or other to reality, as a symbolization, and, moreover, that his linguistic experience takes place in a context of communications.

Where we see the role of self-regulatory processes and the parallel with cognitive growth is in the mechanisms which *allow* the child to build up a language from what Chomsky calls the degenerate quality and limited extent of data. From this point of view it is immaterial whether predication is universal, or whether elision is a universal transformation; what matters is which mechanisms make it possible for the child to infer these rules from what he hears.

We find our position clashing with people from Chomsky's group, and perhaps you could give us some more information about this.

MCNEILL I'm not sure I can say why you encounter resistance. Some of it, I think, lies in the specific form of some linguistic universals. Certain of them are quite exotic. Maybe the best thing for me to do is to talk about the explanatory value of universal grammar. Within linguistics its role is clear enough. It's the sole source of explanatory power. There's no inconsistency with Piagetian theory here, no more than there's an inconsistency with quantum mechanics. The domains of discourse are simply different and conflicts can't arise. A possibility of conflict arises when universal grammar is given some kind of psychological interpretation. In this case there are many possible conflicts among the various interpretations. Certainly there would be conflicts between Piagetian and S-R theory, for example. As I mentioned before, there is no reason to assume in advance that all (or any) universals must be explained in terms of a standard psychological theory of development, be it cognitive or behavioural. In case some (or all) universals cannot be so explained, then interpreting them as reflecting innate abilities is the *only* theory with explanatory value. I fail to understand the argument, that describing innate linguistic abilities as linguistic universals, somehow makes explanation of the behaviour resting on these abilities impossible. It seems to express an *a priori* conviction that such abilities don't exist; therefore that mentioning them can't explain anything.

INHELDER Still it's an open question for you.

MCNEILL Definitely. I think it's an open question to everyone.

KOESTLER The discussion centred on one very essential problem, of nativism versus empiricism. But I would like to go back to reductionism, or rather the rejection of naïve reductionism, which is our common ground. Now, language in the light of modern psycholinguistics is of course one of the strongest arguments against reductionism. There have been endless discussions whether, when you know the properties of oxygen and the properties of hydrogen, the wetness of water and the other properties of water can be predicted. Now if you substitute phonemes or morphemes for chemical substances, then the unpredictability of what happens when you combine them becomes very obvious. If you regard phonemes or morphemes or words as atoms of behaviour, operant responses, then evidently it is impossible to predict the relations that will emerge on the higher level, or to reduce semantics to acoustics. But the trouble is that we write from left to right in a linear sequence, and we speak in a linear sequence in time, so language gives the misleading impression of being a linear affair, a Skinnerian chain, because the hierarchic tree behind it, Chomsky's tree, with its ramifications and transformations, is invisible. The phrase-generating machinery is invisible, and the meaning-extracting machinery in the listener's head is also invisible. And the rules of grammar and syntax operate unconsciously in the gaps between the words. So there was a superficial plausibility in Skinner's view that the emission of speech sounds is like the emission of any other operant behaviour, and that the linear chaining of these operants is what constitutes language. But there is no need to go into this since Chomsky's devastating critique—or execution—of the behaviourists' attitude to language.

Now to a different point. I would suggest that all the canonical rules and properties of SOHO—Self-regulating Open Hierarchies—which I enumerated yesterday, are beautifully epitomized in language, and can be demonstrated by the linguist. Language has its "fixed rules and flexible strategies", its "arborizing tree" and "reticular networks". A word is a "whole" relative to its parts, but a "part" relative to the sentence—a Janus-faced holon. Speech also shows the scale of gradations from "mechanization" to "freedom", for which I gave yesterday the example of driving a car "mechanically" or "mindfully". At one extreme of the scale, you may remember Lashley quoting his behaviourist colleague who said: "When I give a lecture I turn my mouth loose and go to sleep." That is one extreme end of the scale, the mechanized end. At the other end of the scale you have the situation of groping for words to express your meaning, or you ask a person a question and you see him hesitate, search for the answer or consider what to say. You can see that he is at least under the illusion of freedom. And he has indeed an enormous number of choices before him because, according to the canon of the open hierarchy, there are more degrees of freedom on the higher, more complex levels than on the lower levels. I will stop now; I just wanted to point out how neatly all this fits into the scheme.

WEISS I am going to be daring and link up in accordance with Arthur's comments with the basic problem we discussed yesterday. Let's see whether it can be extrapolated.

I think I emphasized that things in an organic system do not happen as isolated events. (Draws seven dots on the blackboard.) Basically a mathematical point is an isolated datum. Now, if we look at brain processes, for example, the production of an individual word, an individual letter even, these cannot be conceived of as isolated events. The event has no static existence but is the centre of a *process*. If you have a single process in a homogeneous environment, then you will get a circular field of activity in the surrounding matrix around the active point. But if you have two points within a distance allowing interaction, you never get two overlapping concentric patterns, but you get instead a mutual distortion between the two domains as a result of their interaction. The "Liesegang rings" phenomenon, which I use to demonstrate this effect, is a rhythmic chemical interaction in a jelly. The jelly represents the field continuum in which particular point sources of other chemical emanations are set into action. As a result of their interaction, the circular patterns are deformed along the line connecting the two centres, as if by reciprocal "attraction". In other words, the dynamic pattern of the whole limits the possibility of expression. This is an emergent geometric novelty, over and above what could be predicted from the algebraic summation of the "isolated" point effects. The reason why I drew these dots on the board in the shape of the stellar constellation of Septemtriones is that the chemical connections in the jelly, as you can see, have automatically connected the points in the same configuration as man's imagery has linked them in the sky. This I ascribe to the unequivocality of dynamic interactions of "point representations" in the matrix of the brain. Theoretically, in your mind you can have any number of connections between the points, and that, I suppose, would be the legitimate way to look at it in the Berkeleyan scheme of the brain as a computer: that is, you get them up as independent variables which you can shuffle and re-shuffle in any way you want to. In reality, however, all peoples have, as far as I know, always united them in this unequivocal way, whether they compared the configuration to a Great Dipper or to a Great Bear. So the point to remember is that if a point receptor sets into action all over a cortex an activity similar to our model, then two points will similarly interact and produce a unique connection between the two points, and three points will form a triangle and so forth. And now my question is whether two words, or two concepts like a noun and a verb, do not produce the same kind of connective process in the brain so that there would be a *physiological* foundation for the idea that the whole in this case is more than the sum of the parts, and that not "anything goes" in the way of algebraic permutation and combination. In short I postulate dynamic singularities in brain action to be the physiological correlates of preferential or consistent linguistic patterns.

MCNEILL   How do we get together? I think I described predication as a way of combining ideas. It is the name of a process for combining ideas and invariably takes a particular form. What you would want to say, I think, in following up this line of argument is that the human mind is so constructed that it inevitably follows this particular path, in this particular case, and so assumes a predicative configuration whenever it combines ideas. One thing in my paper might be relevant here. "Thing" words are at least in part, if not exclusively, predicates. For Berkeley, thing words should be the clearest case of what he had in mind—words associated with the perception of things. But at best that would be an incomplete view of the matter. For much of what we know about thing words is not connected with things at all. The example of the concrete noun, collie, is a case in point. It might be associated with certain perceptual ideas, but we also know that a collie is a dog and a dog is not a collie, and so forth.

WEISS   What I'm trying to do is simply ask a question: whether an intrinsic dynamics of the activity of the brain wouldn't explain the universality of all your primitive language structures?

MCNEILL   I assume that must be the case. But the hypothesis about brain dynamics has to be very specific to account for the specific kinds of linguistic universals that exist. What in the brain explains the special method of combining ideas we refer to as predication? It is asymmetrical; it *can* be negated; it *cannot* be doubled up (you can't say "this figure is round is big", for example).

KOESTLER   It's no good asking him for a more precise answer because between the two end-points there is nothing but missing links—gaps.

WEISS   But their postulates from the two ends agree. Assume it's a hypothesis that they must meet, then you proceed as in building a tunnel. You look at the other end when you are proceeding into the unknown, or else the two dividings will never connect.

MCNEILL I would like to know if there is anything in the physiology of the brain that would even suggest the sorts of asymmetries I outlined with "collie".

WADDINGTON [who came in late]: I don't really know what you have been discussing before, but concerning this asymmetry of "collie" with respect to the larger classes it's related to. How is this related to the asymmetry in the act of perception itself? When you perceive a collie, you see certain bits of fur, a collection of patches of colour, and you associate this with an idea of "collie" which you have already got internally in the mind. If you listen to people talking about perception nowadays this is the kind of scheme you find. The title of the last paper I heard was "How does so little information control so much behaviour". The point is that perception involves something more than being aware of what is on the retina. The image which comes into the eye latches on to an already existing structure.

MCNEILL In both the Baileyan and Berkeleyan theories, which you haven't heard me outline, this is completely intelligible. For Bailey both the word "collie", and the real perception of a collie are associated with the same abstract idea. This idea in turn is associated with others, but the association runs in one direction only. Berkeley's theory also applies, but it cannot account for the asymmetries—the perception of physical-object-bits doesn't suggest the idea of collie-bits in the same way as the perception of collie-bits suggests the idea of physical-object-bits. But it should do so if the retinal image is enriched by association, as Berkeley maintained.

HAYEK I've got only a very brief point. It's going back to the question of nativism and empiricism. I wonder why you connected this particular controversy at all with nativism, because the point at issue between Berkeley and Bailey was really whether the growth of knowledge proceeds from the particular to the general, or whether it proceeds from the general to the particular. Now this is not necessarily connected with the question of nativism versus empiricism. It is true that at one time it seemed easier to explain the growth of knowledge by starting with the particular and assuming that by a process of abstraction we arrived at the general. We haven't any similarly plausible theorems beginning with knowledge of the general, but it is by no means necessary that such a primary capacity to perceive generals must be innate. There are many other ways in which generalization may come first, and then is tested on the particulars. I would like to separate these two things.

MCNEILL   I quite agree with you. This is why I wrote the paper the way I did, to express the underlying differences between Bailey and Berkeley.

HEYEK   It may have been the atmosphere of their time. They had to talk in terms of nativism or empiricism.

MCNEILL   I agree with you on this point. That is, the direction of inference is not from abstraction to nativism. It is possible to maintain that we have abstract ideas and not be committed to assume they are innate. On the other hand, if you believe there are innate ideas, then you're committed to assume they are abstract. This is why I made the paper revolve around the nativist controversy, because it implies the entire set of ideas that makes up the two positions. But you're quite right, you can break this down and not be logically committed to nativism or empiricism by accepting abstract ideas. A further motivation for presenting the arguments in this form is that the contemporary dispute, in the context of language acquisition, is over nativism, I wanted to elucidate the knot of ideas this dispute drags along with it.

HAYEK   I especially wanted to safeguard against seeming to accept innate ideas because I agree so much with most of what you have said.

THORPE   As one involved in the study of bird vocalization, I would like to say how intrigued I am by Chomsky's idea of deep structures. It would, I think, be unjustifiable to pass from bird vocalizations to deep structures in a single leap. But in the organization of bird song we find innate propensities which lead to a song of a certain type, or structure in general, characteristic of the species. For instance, the chaffinch (*Fringilla coelebs*) has a song which is usually in three sections: a, b, c, descending stepwise in overall pitch with a "flourish" at the end. It is possible by playing a tape to it, to make an uninstructed chaffinch put the end in the middle. A chaffinch so treated in the critical period will, therefore, come out with a song: a, c, b. It is still more difficult to make it say b, c, a, but the interesting point is that the last phrase terminates in what I have called a "flourish", and this "flourish" is, I think, particularly useful for recognizing individuals.

You can teach a chaffinch to sing without a "flourish" at the end by keeping it in a soundproof room and making sure that it never hears any sound patterns resembling the "flourish" in structure. If after this you then allow it to hear a "flourish" it will show a strong tendency to copy this and attach it to the end of its song rather than insert it in the middle or put it at the beginning. In other words there is a very considerable resistance to producing a song which has the stages in the wrong order. Similarly, I think you could never teach a cuckoo to say "coo-kook". And there are many similar instances where at least the elementary order of the phrases seems to be genetically determined in some degree. I don't know whether it is too wild an idea to suggest that to account for these facts we have to postulate in birds something like these deep structures which are the basis of human language.

MCNEILL I told you the other day that I cribbed rather extensively from your book *Bird Song*. The reason is that I took your description of how chaffinches develop full song as a very good example of how an innate capacity to produce a certain type of structure is elaborated in the light of experience. I tried to point out the analogy between the development of the chaffinch song and the development of language. I think there is such an analogy, and it suggests that whenever there is a problem of organization in a communication system, whatever the species, it tends to be solved in the same way; this is captured by your metaphor of crystallization. However, as far as I can tell, bird songs include nothing like deep structure. That is, nothing like an abstract structure which carries meaning. Bird songs seem to be reacted to superficially, the way we recognize a landmark.

THORPE Another very obvious example is provided by a common European bird, the yellow bunting (*Emberiza citrinella*). In England this bird's song is thought to suggest the phrase "A little bit of bread and no cheese". Now, with immature or inexperienced birds you may get the "A little bit of bread" phrase without the "and no cheese". If, at a later stage, you then give such birds repeatedly the sound "no cheese" on tape, they are likely to attach it almost at once to the end of the song. But I doubt if you could ever get the yellow bunting to say "no cheese—a little bit of bread and", except by using a hand-reared "auditory isolate" bird and giving it prolonged indoctrination.

KOESTLER I would like to ask a question. I read some years ago in Kretschmer—and I wonder whether his remarks are still valid—that the language of the Bushman is much like the language of the child at an early age, that it expresses concrete situations like a picture-strip without abstract formulations. One of his examples runs like this: Tom-he-take-tobacco, white-man-angry-take-stick, Tom-he-run, white-man-run, white-man-beat-Tom, and so on, like a comic strip.

MCNEILL This reminds me very much of Whorf's method of showing the alleged difference between SAE (Standard, Average European) languages and American Indian languages, for example, Apache. His method was to translate. One example is this:

Tó-nō-ga in Apache loosely translated means: "it is a dripping spring".

But Whorf translated it literally with: "ga, whiteness, nō, downward motion, tó water", which he then "synthesized" into "as water, whiteness moves downward". Then he commented "How utterly unlike our way of thinking!" It is like Kretschmer, I think, It's obvious that he has omitted all of Bushman syntax. But this is what Whorf did to Apache. These Pearl Buck translations tell you nothing.

WADDINGTON One question going back to Thorpe's birds. You have a lot of different kinds—the cuckoo, the chaffinch, the yellowhammer, with different song patterns. The species are different, and have presumably been produced by some sort of selection process or other. If you are going to say a man has a certain pattern in his language, are you saying that this is just the pattern that the human species happens to have got, and that it is pure chance it's got that pattern? That it is no more fundamental than the pattern of the cuckoo and so on? Or, are you saying that the structure of human languages has got a deeper meaning than just what happened to be there when natural selection settled on it, when evolution was first developing a speaking ape?

THORPE I would regard the deep structures for languages in the human stock as certainly adaptive in the sense that they are such as to provide the basis for a language which is suitable for communicating information about action in the real world as perceived by the human stock. Similarly the deep structures in birds are such as to ensure effective species-characteristic sound signals which can be used as territorial proclamations and so forth. In connection with this it is relevant to consider the results, so far announced in preliminary form only, of R. A. and B. T. Gardner, of their attempts to teach American "sign language", the gestural system of communication used by deaf persons, to an infant female chimpanzee. Previous attempts to teach chimpanzees to talk have been almost completely unsuccessful and it was a highly relevant and sensible proposal to test their abilities to acquire "sign language". The experimental animal was not allowed to hear any human speech, and various methods were used to secure imitation of gestures. By the end of the twenty-second month of the project the animal was able to use over 30 separate signs spontaneously and appropriately. The first were simple demands; later signs included an increasing proportion of names for objects such as flower, blanket and dog. These names have been used both as demands and as answers to questions. From the time that the animal had 8 or 10 signs in her repertoire she began to use them in strings of two or more although no deliberate effort was made to elicit specific combinations. Ultimately, signs for certain pronouns: "I-me" and "you" have been acquired and when these occur in combinations the result resembles a short sentence. The published accounts of the work,— which is still in an early phase—are not enough to say exactly what the implications are in relation to the existence in chimpanzees of deep structures. But obviously the work is of great promise and importance.

WADDINGTON You think it was just what natural selection happened to strike on, when it became advantageous for apes to develop speech, and it's of no particular significance, and it's of relatively little interest?

MCNEILL It would certainly be of interest, but I can't possibly answer your question. The Gardners' ape is fascinating, but the few examples of her "speech" I have seen appear to be quite different from the speech of young children. The structural arrangement, if there is any, looks unlike anything that occurs in the development of language. Of course, what would be very exciting would be to discover that the chimp arrived at some kind of structure, but *not* the structure of language. That, along with the possibility of a very rich capacity for language in humans might suggest some kind of canalization of evolution. If chimps have advanced a small distance in one direction and we have advanced a great distance in another, it might mean that our direction was more strongly canalized.

REFERENCES

BAILEY, S. (1842) *A review of Berkeley's theory of vision, designed to show the unsoundness of that celebrated speculation.* Ridgway, London.

BERKELEY, G. (1709) *An essay toward a new theory of vision* (Citations to NTV are from *Works on vision*, C. M. Turbayne (ed.) Bobbs-Merrill, New York, 1963).

BLOCH, O. (1921) *"Les premiers stades du langage de l'enfant"*, *J. de psychol.*, *18*, pp. 693–712.

BOWER, T. (in press) "Reading by ear, reading by eye", in: H. Levin & J. Williams (eds.) *Basic studies in reading.* Harper and Row, New York.

CHOMSKY, N. (1965) *Aspects of the theory of syntax.* MIT Press, Cambridge, Mass.

———— (1959) "Review of B. F. Skinner, 'Verbal behavior'", *Language*, *35*, pp. 26–58.

———— (1966) *Cartesian linguistics.* Harper and Row, New York.

FODOR, J. A. and GARRETT, M. (1967) "Some syntactic determinants of syntactic complexity", *Perception Psychophys.*, *2*, pp. 289–296.

KATZ, J. J. (1966) *Philosophy of language.* Harper and Row, New York.

MILLER, G. A. (1967) "Psycholinguistic approaches to the study of communication", in: D. L. Arm (ed.). *Journeys in science: small steps—great strides.* U. New Mex., Albuquerque.

SCHLESINGER, I. M. (in press) "Production of utterances and language acquisition", in: D. I. Slobin (ed.), *The ontogenesis of language.* Academic Press, New York.

SLOBIN, D. I. (1966) "Comments on 'Developmental psycholinguistics',", in: F. Smith and G. A. Miller (eds.), *The genesis of languages: a psycholinguistic approach.* MIT Press, Cambridge, Mass.

The primacy of the abstract[1]

F. A. Hayek

University of Freiburg

I did not bring a written paper as I preferred to leave it to the course of the discussion to determine in what direction I could best supplement it. Perhaps, however, it was quite as much a tacit hope that the discussion would provide me with an excuse to talk about a problem in which at the moment I am much interested but on which my ideas have not yet reached the clarity required for writing a formal paper. As I was listening I have indeed come to the conclusion that this is the most useful thing I can attempt to do and I am now taking my courage in both hands to present to you, as well as I can from a few notes,[2] some half-baked ideas about what I have called "The Primacy of the Abstract".[3]

[1]

What I shall try to explain under this paradoxical heading seems to me in some ways merely a final step in a long development, which

---

[1] On the original programme of the conference, Professor Hayek was scheduled to give a paper on "Group Behaviour and Individual Behaviour" and the applications of the hierarchic concept in the social sciences. In the preamble to his presentation, he explained why he preferred to talk on "The Primacy of the Abstract" instead. This also explains why, except for references in passing, social science in the narrower sense was not discussed at the Symposium. *Eds.*

[2] The numbered paragraphs in the present written paper correspond to the headings of the notes from which I spoke. Beyond this I have followed only partly the transcript of the tape recording. Not everything as now written was contained or came out clearly in the oral presentation.

[3] I could, of course, instead have spoken of "the primacy of the general", but this would not have had the shock effect which is the merit of the phrase chosen.

would probably have been explicitly formulated some time ago had it not required the overcoming of a barrier built into the language which we have to employ. This is shown by the necessity in which I found myself of describing my subject by an apparent contradiction in terms. We simply have no other suitable term to describe what we call "abstract" than this expression which implies that we deal with something "abstracted" or derived from some other previously existing mental entity or entities which in some respect are richer or "more concrete". The contention which I want to expound and defend here is that, on the contrary, all the conscious experience that we regard as relatively concrete and primary, in particular all sensations, perceptions and images, are the product of a super-imposition of many "classifications"[4] of the events perceived according to their significance in many respects. These classifications are to us difficult or impossible to disentangle because they happen simultaneously, but are nevertheless the constituents of the richer experiences which are built up from these abstract elements.

My main concern in all this will not be to argue the truth of my contention but to ask what is its significance if true. I shall in a moment try to show that the phrase of the title merely brings under one heading several conceptions which have emerged independently in different fields. They will not be quoted as conclusive evidence for the truth of my thesis, but merely as a justification for examining the consequences that would follow if it were true. Without entering into a detailed account of the different theories in question, these references must remain very summary and incomplete. But I want to leave as much time as possible for showing in what way the conception suggested might provide a clue to a number of interesting questions, and have a liberating effect on one's thinking.

[2]

First I want to explain more fully what I mean by the "primacy" of the abstract. I do not mean by this primarily a genetic sequence, although an evolutionary movement from the perception of abstract patterns to that of particular objects is also involved. The primacy

---

[4] For a justification of this, and a few related terms I shall occasionally use, see my earlier book *The Sensory Order* (London and Chicago 1952), in which, as it now seems to me, much of what I shall have to say was already implicitly contained.

with which I am mainly concerned is a causal one, that is, it refers to what, in the explanation of mental phenomena, must come first and can be used to explain the other. I do not wish to deny that in our conscious experience, or introspectively, concrete particulars occupy the central place and the abstractions appear to be derived from them. But this subjective experience appears to me to be the source of the error with which I am concerned, the appearance which prevents us from recognizing that these concrete particulars are the product of abstractions which the mind must possess in order that it should be able to experience particular sensations, perceptions, or images. If, indeed, all we are aware of are concrete particulars, this does not preclude our being aware of them only because the mind is capable of operating in accordance with abstract rules which we can discover in that mind, but which it must have possessed before we were able to perceive the particulars from which we believe the abstractions to be derived. What I contend, in short, is that the mind must be capable of performing abstracting operations in order to be able to perceive particulars, and that this capacity appears long before we can speak of a conscious awareness of particulars. Subjectively, we live in a concrete world and may have the greatest difficulty in discovering even a few of the abstract relations which enable us to discriminate between different things and to respond to them differentially. But when we want to explain what makes us tick, we must start with the abstract relations governing the order which, as a whole, gives particulars their distinct place.

So far this may sound pretty obvious, but when we reflect on the implications, they would mean little less than that psychology and the theory of knowledge frequently start at the wrong end. From the assertion that the abstract presupposes the concrete rather than the concrete the abstract (in the sense that in the mind the abstract can exist without the concrete, but not the concrete without the abstract) a wholly erroneous approach results which treats as given what most requires explanation.

[3]

Let me now remind you briefly of the chief developments in the various disciplines concerned, which seem to me instances of my general proposition. The chief support comes, of course, from ethology, and especially from the dummy experiments with fishes and birds that show that they respond in the same manner to a great variety of shapes which have only some very abstract features in common. It seems to follow that probably most animals recognize, not what we would regard as concrete particulars, or particular individuals, but abstract features long before they can identify particulars. This is indicated most clearly by the theoretical framework developed by ethology, which distinguishes between the "innate releasing patterns" and the mechanism through which these evoke certain "action patterns", where both concepts refer not to particular events, but to classes of combinations of stimuli and their effects in inducing a preparedness for one of a class of actions, which are both definable only in abstract terms.[5]

Similar insights have been gained by human sensory psychology in the course of its gradual emancipation from the conception of simple elementary sensations from which, in a mosaic fashion, the representations of the environment were supposed to be built up.[6] From H. von Helmholtz's still insufficiently-appreciated conception of "unconscious inference" and the similar ideas of C. S. Peirce[7] to F. Bartlett's interpretation of perceptions as "inferential constructs", of which Koestler has reminded us, and culminating in the Gestalt school, which now proves to have emphasized only one aspect of a much wider phenomenon,[8] they all stress in one way or another that our perception of the external world is made possible by the mind possessing an organizing capacity; and that what used to be called elementary qualities are its product rather than its material.[9]

Another important development in a similar direction is the increasing awareness that all our actions must be conceived of as being guided by rules of which we are not conscious but which in

[5] Cf. e.g., W. H. Thorpe, *Learning and Instinct in Animals*, 2nd ed., London 1963, p. 130.

[6] See on what immediately follows my *The Sensory Order* quoted before.

[7] C. S. Peirce, *Collected Papers*, I, 38.

[8] In a paper which I have come to know only since delivering this talk, M. Merleau-Ponty discusses under a heading very similar to that of this paper, the "primacy of perception" over sensation. See his volume, *The Primacy of Perception*, ed. by J. M. Edie, North-western University Press 1964, pp. 12 et. seq.

[9] Cf. further J. C. Gibson, *The Perception of the Visual World*, Boston 1950, W. H. Thorpe, *op. cit.*, p. 129, and particularly Ivo Kohler, "Experiments with Goggles" *Scientific American*, May 1962, who speaks of the "general rules" by which the visual system learns to correct exceedingly complex and variable distortions produced by prismatic spectacles.

their joint influence enable us to exercise extremely complicated skills without having any idea of the particular sequence of movements involved. (This capacity is often inadequately described as "intuitive knowledge".) From Gilbert Ryle's now familiar distinction between the "knowledge how" to do a thing and the "knowledge that" it is so and so,[10] through Michael Polanyi's analysis of skills (and the closely connected concept of "physiognomy perception"),[11] to R. S. Peters' highly important discussion of the significance of non-articulated rules in determining action, there has been an increasing stress on mental factors which govern all our acting and thinking without being known to us, and which can be described only as abstract rules guiding us without our knowledge.

The field, however, in which it has come out most clearly that our mental activities are not guided solely or even chiefly by the particulars at which they are consciously directed, or of which the acting mind is aware, but by abstract rules which it cannot be said to know yet which nevertheless guide it, is modern linguistics. I do not know enough about it to discuss it at any length, but the chief point has indeed been brought out as long as two-hundred years ago by Adam Ferguson in one of my favourite passages of his great work which I cannot refrain from quoting:

> The peasant, or the child, can reason and judge, and speak his language, with a discernment, a consistency, and a regard to analogy, which perplex the logician, the moralist, and the grammarian, when they would find the principles upon which the proceeding is grounded, or when they would bring to general rule, what is so familiar, and so well sustained in particular cases.[12]

You all know how far this conception of the elaborate theory of the grammar of his language which the small child can observe without having any conscious idea of its existence has been carried by Noam Chomsky[13] and his school of transformational-generative grammar.

[10] G. Ryle, "Knowing How and Knowing That", Proc. Artist. Soc. 1945–6 and *The Concept of Mind*, London 1949.

[11] M. Polanyi, *Personal Knowledge*, London 1959.

[12] Adam Ferguson, *An Essay on the History of Civil Society*, p. 50, London 1767.

[13] N. Chomsky, *Syntactic Structures*, s'Gravenhage 1957. I notice in R. H. Robins, *A Short History of Linguistics* (London 1967), p. 126, that L. Hjelmslev in his early *Principes de grammaire générale* (Copenhagen 1928), pp. 15,

[4]

When I now turn to the substance of my thesis it will be expedient to begin by considering, not how we interpret the external world, but how this interpretation governs our actions. It is easier first to show how particular actions are determined by the superimposition of various instructions concerning the several attributes of the action to be taken, and only afterwards to consider in what sense the perception of events can also be regarded as a subsummation of particular stimuli, or groups of stimuli, as elements of an abstract class to which a response possessing certain characteristics is appropriate.

The most convenient starting point is the conception of a disposition (or "set", or propensity, or state) which makes an organism inclined to respond to stimuli of a certain class, not by a particular response, but by a response of a certain kind. What I mean to show in this connection is that what I have called an abstraction is primarily such a disposition towards certain ranges of actions, that the various "qualities" which we attribute to our sensations and perceptions are these dispositions which they evoke, and that, both the specification of a particular experienced event and the specification of a particular response to it, are the result of a superimposition of many such dispositions to kinds of actions, which result in the connection of particular stimuli with particular actions.

I need not enter here into the detail of the physiological processes involved through which, by raising the threshold of excitation of a great many other neurons, the stream of impulses issuing from one will put a great many others in a state of preparedness to act. The important point is that only very rarely if ever will a single signal sent out from the highest levels of the nervous system evoke an invariable action pattern, and that normally the particular sequence of movements of particular muscles will be the joint result of many superimposed dispositions. A disposition will thus, strictly speaking, not be directed towards a particular action, but towards an action possessing certain properties, and it will be the concurrent effect of many such dispositions which will determine the various attributes of a particular action. A disposition to act will be directed towards

---

268, demanded a universal *état abstrait* comprising the possibilities available to language and differently realized in *états concrets* for each particular one which I quote for the interesting use of "abstract" and "concrete" in this connection.

a particular pattern of movement only in the abstract sense of pattern, and the execution of the movement will take one of many different possible concrete forms adjusted to the situation taken into account by the joint effect of many other dispositions existing at the moment. The particular movements of, say, a lion jumping on the neck of his prey, will be one of a range of movements in the determination of which account will be taken not only of direction, distance and speed of movement of the prey, but also of the state of the ground (whether smooth or rough, hard or soft), the fact whether it is covered or open territory, the state of fitness of its various limbs—all being present as dispositions together with its disposition to jump. Every one of these dispositions will refer not to a particular action but to attributes of any action to be taken while the dispositions in question last. It will equally govern the lion's action if it decides to slink away instead of jumping.

The difference between such a determination of an action and the unique response of what we usually call a mechanism when we pull a trigger or press a button, is that each of the various signals ultimately determining the action of the organism at first activates merely a tendency towards one of a range of in some respect equivalent movements; and it will be the overlapping of many generic instructions (corresponding to different "considerations") which will select a particular movement.

These several dispositions towards *kinds* of movements can be regarded as adaptations to typical features of the environment, and the "recognition" of such features as the activation of the kind of disposition adapted to them. The perception of something as "round", e.g. would thus consist essentially in the arousal of a disposition towards a class of movements of the limbs or the whole body which have in common only that they consist of a succession of movements of the several muscles which in different scales, dimensions and directions lead to what we call a round movement. It will be these capacities to act in a kind of manner, or of imposing upon the movements certain general characteristics adapted to certain attributes of the environment, which operate as the classifiers identifying certain combinations of stimuli as being of the same kind. The action patterns of a very general character which the organism is capable of imposing upon its movements operate thus as moulds into which the various effects upon it of the external world are fitted.

What this amounts to is that all the "knowledge" of the external world which such an organism possesses consists in the action patterns which the stimuli tend to evoke, or, with special reference to the human mind, that what we call knowledge is primarily a system of rules of action assisted and modified by rules indicating equivalences or differences or various combinations of stimuli. This, I believe, is the limited truth contained in behaviourism[14]: that in the last resort all sensory experience, perceptions, images, concepts, etc., derive their particular qualitative properties from the rules of action which they put into operation, and that it is meaningless to speak of perceiving or thinking except as a function of an acting organism in which the differentiation of the stimuli manifests itself in the differences of the dispositions to act which they evoke.

The chief points I want to drive home here are that the primary characteristic of an organism is a capacity to govern its actions by rules which determine the properties of its particular movements; that in this sense its actions must be governed by abstract categories long before it experiences conscious mental processes, and that what we call mind is essentially a system of such rules conjointly determining particular actions. In the sphere of action what I have called "the primacy of the abstract" would then merely mean that the dispositions for a kind of action possessing certain properties comes first and the particular action is determined by the superimposition of many such dispositions.

[5]

There is still one special point to which I must draw your attention in connection with these action patterns by which the organism responds to—and thereby, as I like to call it, "classifies"—the various effects on it of events in the external world. This is the limited extent in which it can be said that these action patterns are built up by "experience". It seems to me that the organism first develops new potentialities for actions and that only afterwards does experience select and confirm those which are useful as adaptations to typical

---

[14] A truth, however, often much more clearly expressed by authors who were very far from being behaviourists: cf. e.g. E. Cassirer, *Philosophie der symbolischen Formen II* (Berlin 1925), p. 193: Nicht das blosse Betrachten, sondern das Tun bildet den Mittelpunkt, von dem für uns Menschen die geistige Organisation der Wirklichkeit ihren Ausgangspunkt nimmt".

characteristics of its environment. There will thus be gradually developed by natural selection a repertory of action types adapted to standard features of the environment. Organisms become capable of ever greater varieties of actions, and learn to select among them, as a result of some assisting the preservation of the individual or the species, while other possible actions come to be similarly inhibited or confined to some special constellations of external conditions.

Perhaps I should add, in view of what we have discussed earlier, that nothing in this commits us to a choice between nativism and empiricism, although it makes it seem probable that most of the action patterns by which the organism responds will be innate. The important point is that the action patterns are not built up by the mind, but that it is by a selection among mechanisms producing different action patterns that the system of rules of action is built up on which rests what we regard as an interpretation of the external world by the mind.

You may already have noticed that what I have been arguing is in some way related to certain developments in the modern theory of knowledge, especially Karl Popper's argument against "inductivism" —i.e., the argument that we cannot logically derive generalizations from particular experiences, but that the capacity to generalize comes first and the hypotheses are then tested and confirmed or refuted according to their effectiveness as guides to actions. As the organism plays with a great many action patterns of which some are confirmed and retained as conducive to the preservation of the species, corresponding structures of the nervous system producing appropriate dispositions will first appear experimentally and then either be retained or abandoned.

I cannot here more than just mention that this approach evidently also sheds important light on the significance of purely playful activities in the development both of animal and of human intelligence.

[6]

While my chief contention is the primacy of the rules of action (or dispositions), which are abstract in the sense that they merely impose certain attributes on particular actions (which constitute the "responses" by which the stimuli or combinations of stimuli are classified) I will now turn to the significance of this for the cognitive processes. I will put my main point in front by stating that the formation of abstractions ought to be regarded not as actions of the human mind but rather as something which happens to the mind, or that alters that structure of relationships which we call the mind, and which consists of the system of abstract rules which govern its operation. In other words, we ought to regard what we call mind as a system of abstract rules of action (each "rule" defining a class of actions) which determines each action by a combination of several such rules; while every appearance of a new rule (or abstraction) constitutes a change in that system, something which its own operations cannot produce but which is brought about by extraneous factors.

This implies that the richness of the sensory world in which we live, and which defies exhaustive analysis by our mind, is not the starting point from which the mind derives abstractions, but the product of a great range of abstractions which the mind must possess in order to be capable of experiencing that richness of the particular. The difference between this approach and the still predominant one is perhaps best illustrated by an oft-quoted phrase of William James which is very characteristic of the idea that the primitive mind of a higher animal or a small child perceives concrete particulars but lacks abstract relations. James speaks of the "blooming, buzzing confusion" of the baby's sensory experience of his environment. This presumably means that the baby can fully perceive such particulars as coloured spots, particular sounds, etc., but that for him these particulars are unordered. I am inclined to believe that, in the case of the baby as well as in that of higher animals, almost the exact opposite is true, namely that they experience a structured world in which the particulars are very indistinct. The baby and the animal certainly do not live in the same sensory world in which we live. But this is so, not because, though their "sense data" are the same, they have not yet been able to derive from them as many abstractions as we have done, but because of the much thinner net of ordering relations which they possess—because the much smaller number of abstract classes under which they can subsume their impressions makes the qualities which their supposedly elementary sensations possess much less rich. Our experience is so much richer than theirs as a consequence of our mind being equipped, not with relations which are more abstract, but with a greater number of abstract relations not derived from given attributes of the elements. It rather confers these attributes on the elements.

[7]

Some people are likely to object to this analysis on the ground that the term "abstract" is properly attributed only to results of conscious thought. I shall later return to this point and in fact question whether we can ever in the same sense be conscious of an abstraction in which we are conscious of the intuitive perceptions of particular events or of images. But before I turn to this question I want to examine a tacit assumption which seems to be uncritically accepted in most discussions of these problems.

It is generally taken for granted that in some sense conscious experience constitutes the "highest" level in the hierarchy of mental events, and that what is not conscious has remained "sub-conscious" because it has not yet risen to that level. There can of course be no doubt that many neural processes through which stimuli evoke actions do not become conscious because they proceed on literally too low a level of the central nervous system. But this is no justification for assuming that all the neural events determining action to which no distinct conscious experience corresponds are in this sense sub-conscious. If my conception is correct that abstract rules of which we are not aware determine the sensory (and other) "qualities" which we consciously experience, this would mean that of much that happens in our mind we are not aware, not because it proceeds at too low a level but because it proceeds at too high a level. It would seem more appropriate to call such processes not "sub-conscious" but "super-conscious", because they govern the conscious processes without appearing in them.[15] This would mean that what we consciously experience is only a part, or the result, of processes of which we cannot be conscious, because it is only the multiple classification by the super-structure which assigns to a particular event that determined place in a comprehensive order which makes it a conscious event.

This brings me back to the question of whether we can ever be conscious of all the higher abstractions which govern our thinking. It is rather significant in this connection that we seem to be unable to use such abstractions without resort to concrete symbols which appear to have the capacity of evoking the abstract operations that

[15] I did not mention in my oral exposition, and therefore will not enlarge here on, the obvious relation of all this to Kant's conception of the categories that govern our thinking—which I took rather for granted.

the mind is capable of performing, but of which we cannot form an intuitive "picture", and of which, in this sense, we are not conscious. It seems to me that if we ask whether we can ever strictly be conscious of an abstraction in the same sense in which we are conscious of something that we perceive with our senses, the answer is at least uncertain. Is what we call an abstraction perhaps something which had better be described as an operation of the mind and which it can be induced to perform by the perception of appropriate symbols, but which can never "figure" in conscious experience? I would suggest that at least those abstractions of which it can in some sense be said that we are aware of them, and can communicate them, are a secondary phenomenon, late discoveries by our mind reflecting on itself, and to be distinguished from their primary significances as guides to our acting and thinking.

[8]

The point in all this which I find most difficult to bring out clearly is that the formation of a new abstraction seems *never* to be the outcome of a conscious process, not something at which the mind can deliberately aim, but always a discovery of something which *already* guides its operation. This is closely connected with the fact that the capacity for abstraction manifests itself already in the actions of organisms to which we surely have no reason to attribute anything like consciousness, and that our own actions certainly provide ample evidence of being governed by abstract rules of which we are not aware.

I may perhaps mention here my interest in two apparently wholly different problems, namely the problem of what makes the observed action of other persons intelligible to us, and the problem of what we mean by the expression "sense of justice".[16] In this connection I was driven to the conclusions that both our capacity to recognize other peoples' actions as meaningful, and the capacity to judge actions of our own or of others as just or unjust, must be based on the possession of highly abstract rules governing our actions, although we are not aware of their existence and even less capable of articulating them in words. Recent developments in the theory of linguistics

[16] Cf. chapters 3, 4, and 11 in my *Studies in Philosophy, Politics and Economics* and section III of my pamphlet *The Confusion of Language in Political Thought*, London (Institute of Economic Affairs) 1968.

at last make explicit those rules to which older linguists used to refer as the *Sprachgefühl*[17]—which is clearly a phenomenon of the same sort as the sense of justice (*Rechtsgefühl*). Once more the jurists, as they did in ancient Rome,[18] could probably learn a great deal from the "grammarians". The point which the lawyers have yet to learn is that what is "felt but not reasoned" is not, as the word "feel" might suggest, a matter of emotion, but is determined by processes which, though not conscious, have much more in common with intellectual than with emotional processes.

There is still another problem of language on which I must briefly touch. It is probably because in the development of language concrete terms seem to precede abstract terms that it is generally believed that the concrete precedes the abstract. I suspect that even the terms "concrete" and "abstract" were introduced by some ancient Latin grammarian and then taken over by the logicians and philosophers. But even if the evolution of words should proceed from concrete to abstract terms, this would not disprove that mental development proceeds in the opposite direction. Once we realize that the capacity to act in accordance with very abstract rules is much older than language, and that man developing language was already guided by a great many abstract rules of action, the fact (if it is a fact) that language begins with names for relatively concrete things would mean no more than that in the development of language the sequence characteristic of the development of mind is reversed.

Even that may be true, however, only if we mean by language the words of which it is made up and not also the manner in which we handle the words. We do, of course, not know whether vocal signs for such abstract concepts as "danger" or "food" did not actually appear earlier than names for particular things. But if they did not, this is probably due to the fact, already mentioned, that of such abstractions no conscious image can be formed but that they are represented directly by dispositions to certain kinds of actions, while words were developed largely to evoke images of absent things. However that may be, it does not seem to me to mean that if in language abstract terms appear relatively late, we can draw from

[17] Cf. F. Kainz, *Psychologie der Sprache*, vol. IV (Stuttgart 1956), p. 343: "Die Normen, die das Sprachverwenden steuern, das Richtige vom Falschen sondern, bilden in ihrer Gesamtheit das Sprachgefühl."
[18] Peter Stein, *Regulae Iuris*, Edinburgh 1966.

this any conclusions concerning the development of the mental faculties which govern all action (including speaking).

To identify and name the regularities which govern our own actions may be a much more difficult task than to identify objects of the external world, even though the existence of the former be the condition which makes the latter possible. If, as I suggested, abstractions are something that the conscious mind cannot make but only discover in itself, or something the existence of which constitutes that mind, to become aware of their existence and to be able to give them names may indeed be possible only at a very late stage of intellectual development.

[9]

Before I attempt briefly to sum up I should like at least to mention, although I cannot pursue this point at any length, that only the recognition of the primacy of the abstract in the production of mental phenomena can enable us to integrate our knowledge of mind with our knowledge of the physical world. Science can deal only with the abstract. The processes of classification and specification by superimposition of many classes, which would turn out to be the determinants of what we experience subjectively as events in our consciousness, appear then as processes of the same general kind as those with which we are familiar in the physical sciences. And although, as I have argued at length in other places,[19] a complete reduction of the subjectively-experienced mental qualities to exhaustively-defined places in a network of physical relations is in principle impossible for us, because, as I would now like to put it, we can never become consciously aware of all the abstract relations which govern our mental processes, we can at least arrive at an understanding of what ranges of events lie within the power of those physical forces to produce—even if we cannot aspire to more than what I like to call a limited "explanation of the principles" involved.

[10]

In the course of this sketch I have repeatedly used the phrase "specification by superimposition", meaning that particular actions

[19] See *The Sensory Order*, chapter VII and *Studies in Philosophy, Politics and Economics*, pp. 39 and 60–63.

are selected from fields of in some respect equivalent action patterns for which the threshold of activation is lowered, by those being reinforced which also belong to families of action patterns which are equivalent in other respects. This phrase "specification by super-imposition" seems to me to be the best description of the mechanism for the operation of which I have claimed the "primacy of the abstract", because each of the causal determinants decides only one of the attributes of the resulting action.

It is this determination of particular actions by various combinations of abstract propensities which makes it possible for a causally determined structure of actions to produce ever new actions it has never produced before, and therefore to produce altogether new behaviour such as we do not expect from what we usually describe as a mechanism. Even a relatively limited repertory of abstract rules that can thus be combined into particular actions will be capable of "creating" an almost infinite variety of particular actions.

I do not know how far Koestler would be prepared to accept this as a generalization of his account of creation by "bisociation". To me it seems to describe much the same process he had in mind in coining that term, except that under my scheme the new may be the result of a combination of any number of separately existing features. However, I am concerned with the appearance of the new in a much wider—and more modest—sense than he was in *The Act of Creation*. I am concerned with the fact that almost every action of a complex organism guided by what we call mind is in some respect something new.

I know that we both have in this connection been vainly endeavouring to find a really appropriate name for that stratification or layering of the structures involved which we are all tempted to describe as "hierarchies". I have throughout disregarded the fact that the processes I have been considering occur not just on two but on many superimposed layers, that therefore, for instance, I ought to have talked not only of changes in the dispositions to act, but also of changes in the dispositions to change dispositions, and so on. We need a conception of tiers of networks with the highest tier as complex as the lower ones. What I have called abstraction is after all nothing but such a mechanism which designates a large class of events from which particular events are then selected according as they belong also to various other "abstract" classes.

DISCUSSION

KETY   What Dr. Hayek said makes a very beautiful substrate for the talk on pharmacology, of all things, that I shall be trying to give tomorrow. I would really like to discuss this paper tomorrow, but I cannot resist one comment on the essence of his talk, the primacy of the abstract. He presented the novel idea that the abstract comes first and then raised the question how can the child start with the abstract when he has very little experience to draw on. But we must not forget the child has the experience of the species to draw on, and in a way it makes a very useful kind of concept to think that the child starts up with abstracts and in our later intellectual development what we are doing is adding concrete elaborations on them. Well, one example of an abstract the child starts out with is that things which cause pain are bad and should be avoided. Now this is an abstract of a million years of evolution, and yet the child starts out with that notion and we developed and elaborated upon it in terms of what constitutes pain and how it should be avoided.

HAYEK   It becomes an abstraction by being linked up with families of action patterns which may be innate or acquired by the organism.

BRUNER   I would like to mention one example of a very young baby forming an abstraction. A four-week-old baby in the laboratory was shown his mother's face as a reward for doing some task, such as turning his head to the right. Contrary to the experimenter's expectations, the baby cried when he saw his mother's face, and it was soon evident that he expected to be attended to and picked up when his mother appeared. She wasn't supposed to just *sit* there! Babies start developing anticipations as soon as they are born, and it seems to me that anticipatory schema may be considered a form of abstraction.

HAYEK   This reminds me of another fact of which I have often been told: babies are supposed to recognize only the mother individually while "daddy" remains for a long time the name for any man and only gradually comes to be confined to the actual father.

SMYTHIES   I would like to take up Professor Hayek's point about the sensory order being governed by abstractions. In an adult, the complex, rich sensory order may be to some extent governed by abstractions. It seems to me more helpful to derive the complexity of sensory order, what we perceive, as a function of ordered and specified mechanisms. Now, would it be in accordance with your ideas to suggest that the element of abstraction may operate in the way that these particular mechanisms are constructed—that is in the way they operate. Obviously the mechanisms in the brain which are concerned with ordering perception are extremely complex. And these may well work by extremely abstruse mathematical operations. For example, you suggested that the brain mechanisms could take square roots; it may well do things of this complexity and mathematical feats of this order. Is that what you mean by the abstracting processes at work—that this is what is responsible for sensory order? This is one way this could be interpreted.

HAYEK   I think yes, but I would not like to give too definite an answer. What I want to stress is simply that the complex operations of the mind are composite products of rather simple elements which do not enter consciousness as such but do so only in their composite results. Each such process elicits a preparedness for a class of actions and the overlapping of the different classes finally selecting a particular action.

KOESTLER   I agree entirely with the first half of Hayek's paper. It does not even seem to me as heretical as Hayek modestly pretended. What it seems to boil down to, in a different terminology, is rule-governed behaviour, behaviour with fixed rules which are general or abstract. If I may quote a sentence from my paper—it may almost serve as a summary of part of yours—"it seems that life in all its manifestations, from morphogenesis to symbolic thought, is governed by rules of the game which lend it order and stability but leave sufficient lattitude for more or less flexible strategies, guided by the particular contingencies in the environment; and that these rules, whether innate or acquired, are represented in coded form on various levels of the hierarchy from the genetic code to the structures in the nervous system responsible for symbolic thought." So here we seem to be in complete agreement. Where you talk about abstractions I talk of rules of the game or canons, which take various forms at various levels. In embryology you have Waddington's creodes and homeorhoesis which rule development. You have rules which govern the sensory order in perception, such as the constancy phenomena, you have rules of instinctive behaviour, rules which govern universes or discourse, and so on. You said these rules are abstractions——

HAYEK   May I interrupt? There may be a misunderstanding here. I would avoid saying rules *are* abstractions, I would merely say that the rules are abstract, by which I mean that they merely limit what is going to happen to a fairly wide range. They narrow the range, and even such a limitation by these rules within a wide range of possibilities suffices to specify a particular outcome.

KOESTLER   I agree. The rules impose constraints on behaviour, but leave room for flexibility. But now comes the rub. You spoke about these abstracts or rules as being located in some sort of super-consciousness. Perhaps you just wanted to avoid the term "sub-conscious", or you wanted to designate a different locality. Now this is where we seem to disagree because I believe that there is a continuous scale of degrees of awareness, from focally conscious processes through fringe-conscious ones like tying one's shoelaces absentmindedly, and so on, down to quite unconscious actions and physiological processes. It is a continuum. Now when we learn a new skill like piano-playing or chess, then there is no primacy of the abstract because you have to build up that skill bit by bit from the bottom, you must first learn to hit the right typewriter key and how the knight moves on the chessboard, and then you can gradually build up higher levels of the skill. Now this learning has to be done with great concentration, with focal awareness on those bits, but once you have attained mastery of the skill, then the rules function automatically, unconsciously. No chess player has to think of the rules how to move his knight. So these rules were highly conscious to start with, but now they function unconsciously like the rules of grammar and syntax. The rules occupied first the top floor of my mind and now they have been relegated to the boiler room in the basement, according to the law of parsimony. But before that there was this building-up process and abstractive process, the abstracting of the general rules from particulars. So I think you were a bit hard on empiricism in that limited sense.

HAYEK   I would not deny that there are such *learnt* rules. I do not believe they could explain the process of learning if there were not others. We could not explain the process of learning if there weren't other rules which had not been learned.

KOESTLER   Agreed.

WEISS   This primacy of abstraction conforms perfectly well both with what I said this morning, that is the progress from the more general to the more specific or particular and of course with what I called the other day the primacy of order, all the way up through the universe, way beyond your super-conscious.

INHELDER   I am, of course, in full agreement with two of your statements: first, that every process of knowledge is an active abstraction, "active abstraction" meaning transforming reality and being aware of the outcome of this transformation; and, second, the necessity of going beyond the dichotomy of abstract and concrete. From the developmental point of view we see of course that the small infant has first to act on physical reality before as an adolescent he becomes able to utilize abstract combinatorial operations and hypothetical and deductive strategies. In between these two levels we note the formation of two modes of abstraction and it might perhaps be of value for our discussion to show the distinction between them. On the one hand, we have the logico-mathematical type of abstraction, which plays a part in such activities as classification, seriation, numeration, etc. Here the subject makes the abstraction from his own action; evidently, he acts on objects, but the actual nature of the objects is of no importance since they simply act as a support. On the other hand, we have physical abstraction. Here it is a question of exploring—of determing attributes of objects, such as weight, volume, etc.; here the actual nature of the objects is important. In both cases, there is activity of the subject, but in the first the abstraction bears more directly on the formation of thought structures and in the second it mainly concerns knowledge of the physical world. The development of these two modes of abstraction is not successive but synchronous.

HAYEK That needs to be thought out, I cannot answer it right away.

BERTALANFFY Your phrase, the primacy of the abstract, is, you said largely synonymous with "primacy of the general". This coincides with what biologists, psychologists and system theorists call progressive differentiation, for which any number of examples could be cited. For instance, von Baer's law in embryology says that the developing organism goes from a more "general" (or "abstract") state to an ever more specific one—from blastula to gastrula through a general vertebrate state, to a bird, to a chicken. The particular characteristics of phylum, class, order, family, genus, species arise consecutively. Differentiation is also a central principle in the developmental psychology of Piaget or H. Werner. The same applies to language which starts with a "holophrastic" stage; at first all running things are "wow-wows", later to become dogs and cats, then dachshunds and poodles. This is a general psychological principle starting with the most fundamental differentiation of "I" and "world" from the Piagetian primitive adualism. The same again applies at your "super-conscious" level; only in my vocabulary, I would prefer to call it "symbolic". This symbolic world has its own dynamic, it is self-propelling, if I may use this expression. For this very reason mathematics, to quote a famous dictum, is cleverer than the mathematicians. So I certainly agree that the world becomes richer with progression from the abstract to the concrete, only I would call this progressive differentiation.

WADDINGTON The first point I wanted to make has been largely covered by Bertalanffy. Of course, in the ordinary processes of development in the biological world we do always go from the general to the specific. At the beginning there is a determination of regions of the egg to become some general part of the organism, such as the foreleg, or the head and so on. This is decided long before it is settled whether this particular portion of the egg will become part of the bones or the muscles in the foreleg, or exactly which part of the brain it will turn into.

The rest of what I wanted to say is concerned with the distinction between the abstract and the concrete. The way I use the term may be rather different from yours, but according to my vocabulary the one thing you certainly cannot entertain in your mind is anything concrete (laughter). Everything you can have in your mind is abstract in some sense or other. From this I want to get on to a point that I think I can best approach by referring to a distinction which White-head used to make many years ago—this probably needs a good deal of renovation to bring it in line with modern ideas, but the basic notion he advances seems to me sound. He argued that we perceive things according to two different modes, which he called "causal efficacy" and "presentational immediacy". Causal efficacy is comparable to your abstract logical structures. For instance, if I was a specialized botanist, I would know whether these yellow patches in the field you see out of the window there were buttercups or dandelions, and knowing that, I should really be able to perceive them as that. But I do not know the flora of this part of the world and from here I cannot see them as any sort of flower in particular. As for the other mode, the presentational immediacy—well, if I were a trained painter, I could see all sorts of little purple dashes and flecks of other colours in that field, which I do not actually perceive there at the present time.

It seems to me that this distinction between two modes of perception is relevant to the question of how you proceed from the general towards the more particular. I should like to suggest that there are two ways of doing this, that seem to me rather different. In one way you begin with a confused or chaotic mass of details, jumbled together in disorder so that the totality has no particular characteristic, then within this you gradually discern an outline of a pattern around which you gradually learn to group the whole collection of items, so that eventually they make sense as something with a particular character of its own. I think this is the kind of process that Koestler was mentioning when he spoke of building up a competence in chess, first learning to move the parts correctly and then getting a feeling of how the moves fit together. It is also surely how you learn to ride a bicycle—you learn the movements necessary to correct your balance and then suddenly all these detailed movements fit together and you can do it without thinking. This process seems to be related to perceiving by presentational immediacy, in which you first perceive merely a lot of patches of colour and tone and only gradually discover how to interpret these as a particular scene.

In the other way of proceeding from the general to the particular, you start with something which lacks particularity, not because it is chaotic, but because it is extremely orderly but abstract, for instance, a general logical rule, or a scientific hypothesis; and you then find particular exemplifications of this rule. This is the path that people conventionally suppose most human thought to take.

Recently, however, I have been thinking about how it could operate, not in the realm of thought but in the material world we observe. As I mentioned a few minutes ago, in the development of the embryo, something determines that a group of cells will form some part or other of a complex organ, such as a leg, rather than part of an arm, before it is decided whether those cells shall be bone or muscle. Now I think this is a very astonishing performance. What kind of material substance can there possibly be which characterizes the whole leg with all its bones, muscles, nerves, blood vessels and so on, and differentiates it from an arm with its bones, muscles and so on? One highly speculative but amusing possibility I've been thinking about is this: we have recently started to realize that we have to think of most living cells and similar systems as oscillators. The important processes in cells are controlled by negative feedback loops, and systems of that kind have an inherent tendency to oscillate.

Now, we have a lot of experience of systems that oscillate, or go up and down in a wave-like manner; for instance, sound is an oscillation of the air. Consider what happens in a piece of music—or at least some kinds of music. There is an early statement of something very general—a melodic theme, or perhaps in some forms of jazz only a sequence of chords. The theme or the chord sequence fixes the general character of the subsequent performance, which consists in working out this generality into detailed particulars. In the performance of music, of course, this development is carried out by the intervention of human mentality, but there seems no particular reason why a computer, or even some simpler mechanical system, should not be able to carry out an analogous development of a stated theme. If it did, we should have an example from the non-mental world of the second type of progress from the general to the particular, that corresponding to perception by causal efficacy. It is, I think, an interesting possibility that something of this kind happens normally during embryonic development. This would, I think, be different from the type of building up from chaotic details towards a particular generalization that Koestler was talking about.

HAYEK   On the first point I really completely agree with you and I readily admit that I have been using the terms loosely. Strictly speaking only the real is concrete—it always possesses more properties than we know, and everything mental is abstract compared with the real. But we have to deal with degrees of abstractness and it is inevitable that we should use the word concrete not in the strict sense, but only in the sense of something less abstract. On the other question of how you actually get from the abstract to the concrete. I suggested a superimposition of abstractions, and when you have used a sufficient number of abstractions, you can finally get a combination which in fact defines one unique individual event or action.

WADDINGTON   I would like to make one further point about this. When you go from the general to the particular, the thing that carries you has got to have the character of an instruction. It is a rule of the game, in fact an algorithm. You have got to have something which gives rise to a process. It is that which generates the complexity.

KOESTLER   Somebody has compared the process of generating a sentence with the development of an embryo, and both with the carving out of a figure from a piece of wood. Although one should not carry analogies too far, there are certain basic things they have in common—thus each is a step-wise process. Spelling out step-wise an implicit idea or code in more and more explicit forms, as in the Chomsky schema, or as in your strategy of the genes. There is again a combination of fixed rules and flexible strategies.

FRANKL May I quote what Koestler said, that the mind itself can never know what governs mental operation? In other words, I would say, it is trans-conscious and one could also say that the mind eludes full self-awareness, self-consciousness. Now in this context what comes to mind is Max Scheler's emphasis on the fact that the person or the subject cannot be fully objectified, it eludes objectification by itself or by others. This in turn might remind us of the statement in the Vedanta that that which does the knowing cannot be known, that which does the seeing cannot be seen. In other words that which does the seeing is not perceptible to itself. Now what is mirrored here is a general law we meet again and again, for instance, in clinical practice. Often we are confronted with patients who are so to speak over-conscious of what they are doing, and this interferes with performance, be it the performance of the sexual act (sexual neurosis) or the performance of any artistic work (vocational neurosis). I have published a series of case histories which indicate that artistic creation and performance are impaired to the extent to which this phenomenon is present—I have coined a word, *hyper-reflection*, for it. Hyper-reflection needs to be counteracted by a therapeutic technique that I call *de-reflection*, i.e., turning away the patient's attention from himself, from his own activity.

Thus one could say that full self-awareness and/or self-consciousness are self-defeating. But we need not stick to such a negative formulation, we may as well reformulate it in positive terms by saying the essence of existence is not based on or characterized by self-reflection, but rather by self-transcendence. That is to say human reality is profoundly characterized by its intentionality, its directionality. To be human means being directed beyond oneself, being directed at something other than itself. And human existence falters and collapses unless this self-transcendent quality is lived out. Let me repeat: that which does the seeing is invisible to itself, cannot see itself. But we have to add: except for pathological cases. There are pathological cases in which the eye may well see itself, but to the same extent the function of the eye, the seeing capacity is impaired. Remember the trivial example of the *mouche volante*: this is something within the eye which the eye yet is capable of seeing.

But precisely to this extent the seeing capacity has been impaired all along. It is the exception that proves the rule.

HAYEK   I entirely agree and I am even convinced that the contention can be strictly proved. Yet the proof is very difficult and I have never succeeded in fully working it out. But I can give an example of the sort of proposition it would involve. Though it is a very simple one, I think you will see the analogy with the more general proposition. The example is the thesis that on no adding machine with an upper limit to the sum it can show is it possible to compute the number of different operations this machine can perform (if any combination of different figures to be added is regarded as a different operation). Assume the maximum the machine can show is 999,999. Then there will be 500,000 different additions which give 999,999 as the result, 499,999 different additions which give 999,998 as a result, and so on, and thus clearly a total number much greater than 999,999. It seems to me that this can be extended to show that any apparatus for mechanical classification of objects will be able to sort out such objects only with regard to a number of properties which must be smaller than the relevant properties which it must itself possess; or, expressed differently, that such an apparatus for classifying according to mechanical complexity must always be of greater complexity than the objects it classifies. If, as I believe it to be the case, the mind can be interpreted as a classifying machine, this would imply that the mind can never classify (and therefore never explain) another mind of the same degree of complexity. It seems to me that if one follows up this idea it turns out to be a special case of the famous Goedel theorem about the impossibility of stating, within a formalized mathematical system, all the rules which determine that system.

THORPE   Karl Popper has come to much the same conclusion. I was very interested in some of the things you said about ethology, and the point you made that all of us are capable of a wide variety of actions and then learn to limit ourselves to particular actions. That of course, is true. It is in fact one of the most interesting things about birds; and the reason in fact why I, as an ethologist, took up the study of birds was that some parts of their behaviour are so precisely programmed and so rigid, and other parts so flexible. Now in those contexts where the actions are flexible, birds are capable of a very high degree of learning. There is one particularly interesting case you might like to hear about. If you hand-rear Bee-eaters (*Merops apiaster*) and feed them on artificial food, you soon see the young birds going through shaking movements of the bill; and although these movements serve no function when the animal is being fed on an artificial diet, they continue unabated. Under natural conditions this action serves to stun the bees, or other sting-bearing Hymenoptera, before they are eaten; and there are many other cases of that kind where actions appropriate to dealing with a particular special diet to which the species is adapted, so to speak, force themselves out, whether appropriate or not, under artificial conditions of rearing. At one time, I and some of my pupils became interested in the question whether the choice of food was very highly programmed in different Finch species, each of which has differently shaped bills. So, Miss Janet Kear, a pupil of mine took four species of finches, Chaffinch (*Fringilla coelebs*), Bullfinch (*Pyrrhula pyrrhula*), Greenfinch (*Carduelis chloris*) and Hawfinch (*Coccothraustes coccothraustes*): with different sized and shaped bills, with the idea of finding out whether these birds when they meet different kinds of seeds in nature have their feeding methods all innately programmed; whether in fact the bird was programmed with the correct pattern, with the appropriate behavioural apparatus, to cope with a certain kind of seed most efficiently. What she found was that this certainly was not innately programmed. What the young bird does is to sample all the different kinds of seeds offered, and then concentrate on those seeds which could be most easily exploited by the bill of particular size and shape which the species possessed. And so this behavioural adjustment was learned quite quickly by the bird experimenting on various different kinds and sizes of seed. The bird as a result of this experience then confines itself to one or a few particular kinds of seeds which can be exploited most efficiently. To put it picturesquely, one might almost say the bird is carrying out a time and motion study!

This contribution concerning the choice of food by birds is, perhaps, a very minor one but I was very stimulated by your approach, and the phrase "the primacy of the abstract" will certainly stick with me.

New perspectives in psychopharmacology
Seymour S. Kety

Harvard Medical School

There are more rigorous disciplines in the hierarchy of the neuro-sciences than psychopharmacology, which has as its ultimate goal an understanding of the interaction between chemical substances and the mind. In that very statement I label myself a dualist, an appellation I shall not eschew. I am quite content with reducing the universe to two types of entities—mental and material. A further reduction at the present time I would feel to be an unnecessary oversimplification. I cannot conceive any more than Leibniz was able to do, how a vortex of events in one world can effect a change in the other, yet it is our common experience that they do. Nor, for that matter, can I conceive of how one body of matter attracts another body across empty space, and yet it does.

By means of this dualism psychopharmacology is made a richer science. By recognizing both worlds it does not have to restrict itself to either one and distort its observations or its language to deny the existence of the other.

Having observed in myself and inferred from the reports of others who seem in every observable way to be analogous creatures, that there are chemical substances the ingestion of which produces an undeniable mental change, I adopt a fundamental working assumption—that chemical substances may affect the mind by acting on the brain—and can then proceed to use all of the techniques and findings of neuropharmacology, neurochemistry and the other neurosciences in an effort to elucidate that interaction.

We can classify these chemical substances or drugs which act upon the mind into several phenomenological groups. The first and perhaps the oldest group of psychotropic drugs includes the sedatives, the hypnotics and the anaesthetics. These are probably stages of the same kind of action since one particular agent depending upon its dosage may produce any one of the three effects. Another and very old group is the analgesics like morphine or aspirin, which seem to have a specific effect on pain with minimal hypnotic effect. There are the convulsants like strychnine and the anticonvulsants like Dilantin. Other groups of drugs have more recently been discovered or characterized. There are the psychedelics whose effects are predominantly subjective and in the area of perception, the ataractics or tranquillizing drugs like reserpine or chlorpromazine, the euphoriants like amphetamine, or the antidepressants like iproniazid and imipramine.

One generalization which emerges from the experience of the field is that very few of these drugs were discovered by a rational approach from the desired effect to the development of a new drug. Most were discovered by serendipity and simply improved upon by proper pharmacological techniques. There are other generalizations which are so universally supported by experience that they have almost become dogmas. The first of these is that a drug does not produce a new function but merely modifies existing functions. It is also generally true that drugs more readily correct an abnormal function than improve upon a normal attribute. Nor is there much evidence to challenge the statement that no drug has a single action, all drugs have numerous actions—the desired one and the others which are regarded as side effects. It is the task of applied pharmacology to develop drugs with increasingly specific action, fewer and more benign side effects and the task of the physician to choose the drug and regulate its dosage so that the desired effects are enhanced and the untoward effects minimized.

Since the brain is the most complex structure in the universe we know and its relationship to behaviour only beginning to be understood, it is not surprising that our knowledge of the mechanism of action of psychopharmacological agents is still rather rudimentary. In the case of the sedatives and the anaesthetics it is likely that their action depends upon an ability to block transmission at central synapses, but in the case of the more recently discovered agents— those which more specifically affect mood and awareness—we are still in the stage of developing hypotheses and accumulating evidence with which they may be compatible. Central to most of these hypotheses and the subject of considerable experimental work at the present time are the monoamines which occur in the brain and which have some interesting correlations with behaviour and mental state. Because I believe these relationships to be so compelling, I should like to dwell upon them.

Let us begin with a consideration of the anatomical relationships of the monoamines, especially the catecholamines (dopamine and norepinephrine) and serotonin in the central nervous system. In 1954, Martha Vogt described the gross distribution of norepinephrine in the mammalian central nervous system, reporting highest concentrations in the hypothalamus and brain stem, considerably less in the cortex and low concentrations in white matter. Shortly thereafter, Bogdanski and Udenfriend reported the distribution of serotonin in the central nervous system and this appeared to follow the same kind of pattern except that there appears to be a tendency for serotonin to predominate in lower species and to remain in the primordial parts of the brain giving way, so to speak, to norepinephrine in the most recently acquired portions of the mammalian central nervous system.

A more precise localization of these biogenic amines in the central nervous system has been made possible by a very elegant histo-fluorescence technique (which was developed by Falk and by Hillarp from the original work of Eränko and has been extensively applied by Hillarp's collaborators Dahlström and Fuxe). Despite certain of its drawbacks, this technique has to a considerable extent revolutionized our understanding of the distribution of the monoamines in the central nervous system by permitting their characterization and identification within specific cells. It has revealed that the monoamines in the brain are intracellular and sharply limited to certain cells where they occur in the cytoplasm, in the axon and in highest concentration in the endings of the axons where there appears to be a synaptic terminal. These techniques have indicated more clearly than regional chemical analysis that the monoamines are ubiquitous in the central nervous system of mammals, but that their concentrations are especially high in the regions where chemical studies had previously shown a high concentration. In addition, however, to confirming their presence in dense collections of neuronal endings in the hypothalamus, these techniques have demonstrated many fine monoamine-containing axons, large in number but small in mass, in the cerebral cortex, hippocampus and cerebellum, where chemical studies had not previously suggested their importance. By cutting portions of the brain and examining the increase or decrease of fluorescence proximal or distal to the cut, it has been possible to postulate the existence of certain monoamine-containing tracts. Two of special importance have been described by Anden and his colleagues—a nigro-striatal system of dopamine-containing neurons with bodies in the substantia nigra and with endings in the neostriatum, the caudate and putamen. These workers have also adduced evidence for an interesting new system of norepinephrine-containing neurons which appear to have their cell bodies in the brain stem and which send axons via the medial forebrain bundle to the hypothalamus the hippocampus, the limbic system and throughout the entire cerebral cortex.

In the peripheral sympathetic nervous system the early hypothesis of Cannon that a catecholamine was a chemical transmitter between the nerve ending and the effector cell has been amply substantiated. The evidence is quite clear that norepinephrine is stored in particular vesicles in the presynaptic terminals of sympathetic nerves, released during sympathetic activity and acts upon the effector cell to produce the characteristics of sympathetic nervous activity.

Turning to the physiological evidence for a role of monoamines in the central nervous system, we ask the first question: Is there evidence that these amines are released in association with synaptic activity, nerve function or particular behavioural states? There are few direct demonstrations of the release of any of these putative transmitters within the central nervous system. Although evidence has been reported for the release of acetylcholine from the cerebral cortex or of norepinephrine from limbic structures in conjunction with appropriate stimulation, the observation that urea or even water can be released or rapidly exchanged under similar circumstances raises some question regarding the specificity of this monoamine release.

There is better but less direct evidence for the release of these substances with functional activity. Baldessarini and Kopin have recently demonstrated that brain slices or isolated parts of the brain in vitro will release significant amounts of norepinephrine when their neurons are depolarized with a mild electrical current or with potassium. Even more indirect, but more extensive and physiological, is the evidence for release which is inferred from the demonstration that their concentration in various parts of the brain falls or that their turnover is accelerated in association with certain types of activity. In severe stress, anoxia, exposure to cold, forced swimming, strenuous exercise and electroconvulsive shock various workers have demonstrated a slight or a moderate decrease of one or another amine in total or regional concentration in the brain.

Most of the evidence for a relationship to behaviour, however, depends upon pharmacological studies of which there are a great number.

Probably the first significant entrance of pharmacology into this field occurred in 1943 when Hofman, an organic chemist working in Basle, one April afternoon, suddenly went berserk and had to be taken home from the laboratory. For the next several hours he experienced a number of strange hallucinations, illusions and other psychotic symptoms. He was finally able to get to sleep with the help of a sedative and had recovered by next morning. Had he been a psychiatrist, he might have wondered about the many aspects of his life situation which could have precipitated such an episode. Being a chemist, however, he decided it was something he ate! He remembered that he had been working with a new compound that day, LSD-25, and assumed that he had inadvertently ingested some of it. Next morning he went back into the laboratory and took a very minute dose of the substance—$\frac{1}{4}$ milligram—and promptly reproduced the previous effects. This strange compound which had such potent effects in infinitesimal doses (70 micrograms in a glass of water will produce an acute toxic psychosis), stimulated a great deal of work and it was not long before an objective pharmacological effect of LSD was found. Gaddum in England and Woolley in the United States independently discovered that LSD in very low concentrations would block the effect of serotonin on smooth muscle and both independently speculated that perhaps the fact that serotonin was present in the nervous system and that LSD could block serotonin and produce psychosis suggested that the psychosis resulted from its antagonism to the normal brain serotonin.

Shortly thereafter reserpine came into the picture as an important ataractic agent which produces a marked sedation in animals and, in man, a tranquillization of over-excited states. In some individuals it produces a frank depression which is hard to distinguish from clinical endogenous depression. Shore and his Colleagues at the National Heart Institute, examining the effects of reserpine, discovered that it produced an almost complete depletion of the serotonin in the brain. It seemed reasonable to conjecture that the effects of reserpine on mood were somehow related to its action on brain serotonin. Then by another stroke of serendipity, iproniazid was discovered to have some important side effects. This drug was first introduced as a treatment for tuberculosis, but it was soon realized that not only did it appear to halt the progress of this disease but that it also produced psychic stimulation. The first reports of its success also mentioned that the patients were dancing in the halls at the sanatorium and this was not entirely because they had seen their x-ray films. Iproniazid is a potent excitant in animals and in large doses in man can be a euphoriant and even a psychotomimetic agent. In smaller doses it exerts an antidepressant effect, specifically on certain cases of clinical depression. It was soon learned that iproniazid is a monoamine oxidase inhibitor and that it elevates the levels of brain serotonin by blocking the enzyme which normally degrades the amine and maintains a physiological level. This served as further support for the hypothesis that serotonin was *the* neural hormone involved in affect. Then it was demonstrated that reserpine depleted the brain of norepinephrine and dopamine also, and that iproniazid was capable of increasing the brain concentrations of these amines as well as serotonin so that what looked like a clear and simple hypothesis became, as it inevitably does, much more complicated. I believe it is still fair to conclude that the contrasting effects of reserpine and of monoamine oxidase inhibitors upon mood are related to their opposite effects upon amines in the brain. Which of the specific amines are related to which moods or how they interact in association with a particular state is still an open question.

Let me mention some of the specific findings in connection with particular affective states. In various kinds of physical stress and in stress associated with pain, a significant fall in norepinephrine or increase in its turnover has been found in the brain stem and other regions of the brain which would be compatible with a release of this neurohumor in the brain and an increase in its synthesis. In sham rage produced in cats by stimulating the amygdaloid nucleus of the limbic system or by certain sections of the brain, a significant fall in the norepinephrine level of the telencephalon has been demonstrated. It is possible to breed a strain of so called muricidal rats which will compulsively attack and kill mice, whereas the large majority of laboratory rats tend to ignore them. This genetic strain shows a 25 % increase in norepinephrine levels in its brain with a normal rate of turnover indicating a significant "hypertrophy" of the noradrenergic system in the brains of such animals. There is evidence that norepinephrine is associated with certain forms of appetitive behaviour. When injected in infinitesmal amounts to the hypothalamus, norephinephrine will induce eating in a satiated rat. Stein has adduced considerable evidence suggesting that norephinephrine is involved in the appetitive behaviour of self-stimulation— the phenomenon in which rats given the opportunity by pressing a lever to send a mild current through certain regions of the hypothalamus continue to do so in preference to food or sexual gratification. Drugs which depleted norephinephrine in the brain inhibit this appetitive behaviour, while drugs which enhance norepinephrine effects will accelerate it. Whereas there is some evidence that norepinephrine is associated with arousal, Jouvet has developed considerable evidence that serotonin is necessary for the induction of normal sleep. Animals in which the synthesis of serotonin is blocked by a specific inhibitor show a marked reduction in sleep time which can be brought back to normal by the administration of a precursor which will bring the serotonin levels toward normal.

Although the evidence regarding the relationships of these monoamines to mood in man is highly indirect, it is probably of significance that the important drugs which effect human mood can also be shown to have significant effects upon these putative transmitters in the brain of animals. I have already alluded to the effects of reserpine and of the monoamine oxidase inhibitors. Amphetamine, which is a notorious euphoriant in man, has several actions which would tend to enhance norepinephrine concentrations at central synapses: its tendency to facilitate release of norepinephrine, to oppose its removal from the synapse, and to increase its rate of synthesis. Imipramine, which is perhaps the best of the antidepressant drugs, potentiates norepinephrine peripherally and seems to act by blocking the reuptake of norepinephrine from the synaptic cleft and preserving its concentration at that important site.

Electroconvulsive therapy, however, introduced two decades before modern drug treatments, has proved to be the most effective form of treatment of depressive illness. Although there is general agreement on its efficacy there has been little knowledge with which to formulate an explanation of its mechanism of action. If a deficiency of norepinephrine were involved in depressive illness, one would expect some change induced by electroconvulsive shock on the availability of this amine in the brain compatible with one or another of the effects reported for the antidepressant drugs. Since we had found that stress increased the turnover of norepinephrine in the brain, suggesting an increase in its rate of synthesis, we speculated that the chronic depletion in norepinephrine induced by a series of electrically induced convulsions might induce a persistent increase in the synthesis of brain norepinephrine. This hypothesis was confirmed, our results indicating a 40–50% increase in the rate of norepinephrine synthesis in the brains of rats 24 hours after the last of a series of electroconvulsive shocks. This finding lends some credibility to the concept that clinical depression involves an inadequacy of central norepinephrine mechanisms and that the therapeutic efficacy of electroconvulsive shock may to some extent reside in its ability to stimulate the synthesis and utilization of this amine in the brain.

The correlations, such as those I have reviewed, which suggest some role for the norepinephrine-containing neurons of the brain in affective behaviour, by no means deny the operation of other amines, such as serotonin and dopamine, in these states. In fact, there are some reports which show a better correlation between dopamine and arousal, activity and aggression, than that which is found for norephinephrine. It is, in fact, extremely unlikely that a particular affective state is associated with only one of these amines or can possibly be equated even with an interaction between them. Whereas the function of some organs, such as the liver or skeletal muscle, may be thought of as a simple sum of the functions of its component cells which are small replicas of the whole, the brain is unique, for its neurons, even those with similar morphology and chemistry, differ markedly from each other. The function of a neuron is very largely determined by where it is, with what neurons it is connected, and what has been its past history. It is because chemical processes act not directly on the brain but on the highly specific and complex network of the units which comprise it, that behaviour can never be reduced to neurochemistry.

The apperceptive mass of an individual can be thought of as representing some of those interrelationships incorporating the present situation and the past experience. Neurochemical and pharmacological studies of affect must take into consideration the interactions between chemistry and the apperceptive mass. There is already evidence that the effects of circulating epinephrine depend upon such cognitive factors. Schachter has found that the same dose of epinephrine may produce anxiety, hostility or hilarity, depending upon the individual and the environmental cues at his disposal. Similarly, I would think that neurohumoral and neurochemical processes may determine the volume and the keys in the rendition of human behaviour while the melody is largely provided by these idiosyncratic perceptive factors.

I should like to offer some speculative hypotheses of how neuro-chemical factors and specifically the biogenic amines may operate in the acquisition and consolidation of the apperceptive mass. Let us consider first what are the adaptive purposes which are served by the affective response. We can point out a number that are classical. The emotional state activates muscular, endocrine and autonomic systems to appropriate preparatory and consummatory experiences. The affects of anger or rage are associated with hypothalamic and autonomic stimulation which releases epinephrine and norepine-phrine in the circulation and mobilizes blood sugar, increases the blood supply to muscles and the heart and makes the animal ready for fight or flight. Other affective states do many other things besides preparation for fight and flight; they are involved in the autonomic endocrine and muscular responses necessary for exploring, foraging, eating, courting, mating—in fact, all of the adaptive res-ponses crucial to survival of the individual or the species. But in addition, emotional states play an essential role in the process of learning. Psychologists have known for some time that the intensity of affect associated with an experience determines to a large extent its acquisition and retention. I should like to develop a hypothesis regarding the biological substrates of that association and the adaptive purposes they serve.

From the teleological point of view it is not enough for an animal to remember all of its past experience. The adaptive advantage made possible by behaviour which is learned by individual experience over that which is wired in genetically, is that learned behaviour is more finely tuned to the anticipated vicissitudes of the individual's experience, while that which is genetically endowed can only be an abstract of millions of years of the experience of the race. If learning is to have survival value it must rule out experiences which are irrelevant to that goal, reinforce those which are, and endow them with appropriate avoidance or appetitive qualities which may guide future behaviour.

I should like to think that we are endowed genetically with a system for evaluating current experience in terms of its survival value. Such a system would operate on the basis of a few relatively simple generalizations. Inputs which are familiar and in the past have not been associated with deleterious events are to be ignored. On the other hand, a new input, such as a novel experience, the outcome of which cannot be predicted, should be attended to. Secondly, such a system should be able to discriminate between outcomes which are good and outcomes which are bad, in terms of the general experience of the race. Thus, in its simplest form it would label as good and with appetitive overtones, those experiences which are followed by the ingestion of food and drink, copulation and other outcomes with a positive survival value. It would label as bad and to be avoided, those outcomes which cause destruction of tissue, disturbance of homeostasis, and a host of outcomes deleterious to survival. It would want to assure that those experiences loaded with salutary or noxious components are remembered, and in association with the components.

I should like to think that the diffuse system of monoamine-containing neurons with their cell bodies in the brain stem and their axons ramifying throughout the limbic system and neocortex are, in fact, such a built-in system for discriminating between and facilitating the memory of new experiences significant to survival. Their distribution to the classical sensory-motor systems of the brain and the little we know of the ability of local microinjection of monoamines to inhibit and occasionally to stimulate cortical neurons, would be compatible with their playing a modulating role on the classical systems, mediating arousal and attention.

But more important, I can imagine ways in which the release of biogenic amines at sensory-motor and association synapses could affect learning. Much evidence indicates that experience is first held in a very temporary storage, probably electrical in nature, and only after an elapsed time, between a few minutes and an hour, is the temporary storage converted to a more permanent form of consolidation. It is the latter phase of memory which is blocked by inhibitors of protein synthesis in the brain. These two phases of the storage process may also have an important adaptive function. The first holds the information tentatively awaiting an outcome and the judgement of the evaluative system of whether to discard or retain and what value to associate in the retaining. By regulating protein

M

synthesis at particular synapses the affective state generated by the interaction of outcome on the wired-in judgement system could determine the intensity and the coloration of the memory, and it could do this by releasing at the synapses appropriate biogenic amines.

There are some precedents for suggesting that small molecules like amines may regulate protein synthesis in the brain. The importance of thyroxin in the dendritic proliferation which accompanies maturation of the cerebral cortex or in accelerating the development of a reflex, is well established. From the work of Sokoloff, which demonstrated a stimulant effect of that hormone on protein synthesis in the immature brain, it is possible to explain those cytological and behavioural effects of thyroxin there. In the case of the biogenic amines, Campbell has found some effects on in vitro synaptosomal protein synthesis. But research on the control of protein synthesis in the brain, especially that involved about the synapse, has only recently begun. I have little doubt that as it progresses it will teach us much about the mechanisms involved in memory.

To elaborate further this one hypothesis, however, let us examine some more of the requirements of our postulated system for labelling and storing new experiences in terms of their significance to survival. If the organism is to take full advantage of these mechanisms and profit maximally from idiosyncratic experience, it should not be partial to any sensory modality but receptive to all. Having screened out the trivial and familiar by the processes of habituation and attention, all the novel concomitants of a situation should be eligible for reinforcement and retention when the evaluators indicate a more than passing significance to the outcome. This would undoubtedly involve the temporary reinforcement of numerous irrelevant connections which, in common with those which happen to be relevant, would have an inherent tendency to decay. The repetitive trials which are required for most learning, however, should effectively reinforce the constantly recurring associations while permitting the cancellation of the random and adventitious ones, much as the modern computers of average transients operate. Ideally then, the input which reinforces on the basis of outcome should not be applied in some predesigned manner to selected circuits but indiscriminately to all or at least to all which were novel, which familiarity had not suppressed. The systems of monoamine-containing neurons and their diffuse processes which permeate the telencephalon appear to have some of the characteristics of distribution necessary to fulfil such requirements. Some relationship of these systems to affective learning is suggested by the highly selective loss of the conditioned avoidance response which occurs after drugs like reserpine or alpha-methyltyrosine which deplete them of one or more of their amines.

It might suffice that intensity of affect regardless of its qualitative nature reinforce time-associated connections by directing the synthesis of proteins which in some way favours conductance. This could happen by the sprouting of new connections, by increasing the membrane surface of some already established or increasing the amount of key enzymes to augment the concentration of transmitters. By means of such mechanisms the concomitants of the affect and its appetitive or aversive qualities could be associated with the experience in consolidation and recall by the reinforcement of certain connections to appropriate subcortical or limbic areas involved in the specific affective states. It is, however, also interesting to speculate upon the possibility that the different amines give some characteristics to the newly synthesized protein, imparting a "colour code" of affective qualities to the circuit as it is being formed. Waelsch has adduced evidence for the incorporation of amines into proteins, but other chemical mechanisms can be conceived.

It is not difficult to imagine how a value system, at first very primitive and based entirely on genetically transmitted criteria derived from racial experience, could itself learn from the experience of the individual, becoming more differentiated and discriminating, capable of tolerating greater delays to outcome, susceptible to more subtle grades of affect than crude pleasure and pain, and, in a species capable of symbolic conceptualization, sensitive to imagined, predicted or planned outcomes in addition to those actually perceived.

Now, what is my opinion of the possibility of pharmacological intervention in this process? Strongly believing, as I do, that the unique functions of the brain rest crucially upon the particular connections of its parts in addition to its chemical composition, I see drugs as capable of inhibiting or facilitating the processes involved in those connections but hardly capable of generating them *de novo* and without the genetic or experiential processes which make them adaptive. My belief that specific memories lie in specific patterns of connection in the brain and not in specific molecules, would make me incredulous of the possibility that experiential information can be transmitted from one individual to another by the transfer of a chemical substance however macromolecular. On the other hand, I would quite readily accept the possibility that certain drugs could slow or accelerate the process of making these connections and thus speed or retard the rate of learning. In fact, the hypothesis I have presented, is based upon the possibility that endogenous biogenic amines in the brain are engaged in just that activity. Again, regarding affective state in man as inextricably bound to the apperceptive mass of the individual, I would find it difficult to believe that a drug could establish or correct an affective state out of context with the experience and the intellectual processing which are necessary to generate it. Of course, it is possible today to diminish or enhance the levels of affective states by means of endogenous chemical substances or administered drugs. The aggressiveness of the male animal and its fluctuations during the mating cycle is clearly related to the presence of testosterone, and prehistoric man knew that the ox was more placid than the bull. I have already alluded to the alterations or corrections of mood which can be achieved by drugs which seem to act on biogenic amines—reserpine produces a depression of mood and a loss of aggression, a variety of other agents may produce euphoria or a more specific correction of endogenous depression. It is possible that there is a chemical common path or bottleneck which mediates affective states of varied intellectual and experiential origin, and that simple chemical effects on that process may thus significantly alter the mental state. It should be pointed out, however, that being a common path it affects many neural and cognitive systems and its manipulation lacks the specificity that one would hope for in the most rational ameliorative intervention. A drug may reduce aggression but at the expense of reducing many other affects and, furthermore, without correcting the intellectual and experiential basis of the aggression. For that matter, so will a straitjacket. If a depression is endogenous and really to be ascribed to a simple chemical change in the common pathway, then correcting that change should quite completely correct the abnormal mood. If the genesis of the depression is on the basis of overwhelmingly depressing events in the environment a chemical change produced in the brain by means of drugs may within a certain range help to alleviate it, but is quite incapable of correcting it.

Koestler perceived this and argued it compellingly when he wrote:

". . . it is fundamentally wrong, and naïve, to expect that drugs can present the mind with gratis gifts—put into it something which is not already there. Neither mystic insights, nor philosophic wisdom, nor creative power can be provided by pill or injection. The psychopharmacist cannot *add* to the faculties of the brain—but he can, at best, *eliminate* obstructions and blockages which impede their proper use. He cannot aggrandize us—but he can, within limits, normalize us; he cannot put additional circuits into the brain, but he can, again within limits, improve the coordination between existing ones, attenuate conflicts, prevent the blowing of fuses, and ensure a steady power supply. That is all the help we can ask for—but if we were able to obtain it, the benefits to mankind would be incalculable . . ."

I wish I could be as sanguine as Koestler regarding the potentialities of pharmacology in this area. A rational approach to the treatment of the "paranoid streak in man"—and let us not forget how few drugs have been developed by a rational search for them—would require a more definitive understanding and description of the neurophysiology and especially the neurochemistry which one is seeking to correct. Until we have that, the search for the ameliorating drug will be a hit-and-miss affair.

I am more optimistic about the possibilities of putting additional circuits into the brain, not by means of drugs, but in the manner in which they are normally developed. The fact that there are remarkable differences between individuals and between whole cultures in the degree of true civilization which they have achieved, sustains my hope that it is possible to correct the deficits which are so readily found in most human beings by learning more about the relatively few individuals who do not seem to possess them. I suspect that the differences will be found not in the genes or in the neurohumors but in the cultures and the life experience of these individuals. It will be a laborious process to tease out the crucial factors, and an even more laborious one to have them universally adopted, but no more laborious, I believe, and much more effective than any single biological approach I can suggest.

DISCUSSION

KOESTLER   Before the discussion starts, may I make a factual correction, or the whole discussion will be distorted. I never asked or wished for a drug to block aggression; on the contrary: what I asked for and wish for is a drug which blocks devotion, that fanatical loyalty to a leader, to a flag, to a creed, that hypnotic suggestibility of the mass-mind which caused all the major disasters of history.

KETY   Well, actually that is even harder. We may find a drug that might block aggression—but loyalty, that is an effect that is so subtle and so complex, I don't even begin to see that pharmacology . . .

THORPE   I might say that the three outstanding people that you mentioned were extremely loyal. That was their strength, not their weakness.

KOESTLER   I was talking of misguided loyalty.

HYDEN   I would like to comment on two points, on the possible mechanism which you drew up, the whole mechanism of enzymes and how they could act on synapses first, and then on your very interesting suggestion that the aminergic system could be involved both in attention and in learning. And that discussion should lead up to the question whether our two fields could be drawn together, as you suggested. So let me start with the biochemical view in a very simple way. Some time ago we thought that it would be valuable to see whether the effect of drugs acting upon amines, the content of amines, the rate of synthesis of them, should also bring forth other changes in the nervous system, so we used two types of drugs for that, one is a monamino oxidase inhibitor, a potent one, the other one is tofranil or imipramine. We studied certain parameters. We found for example that the rate of activity of two types of enzyme systems was increased, over a period of time, after administration of these drugs. Of the enzyme systems, one belonged to the Krebs-cycle, the other one belonged to the respiratory chain, succinoxidase and cytochrome oxidase, as a matter of fact. These were increased in neurons and decreased in glia. One could say that this would reflect an increased synthetic activity. Then we studied protein synthesis and RNA synthesis. We found that proteins were synthesized at an increased rate under the effect of these two drugs. Then we went to the RNA and made a quantitative study and a qualitative study and we found differences there between the effect of the two types of drugs. In the case of the monoamino-oxidase inhibitor, the effect was primarily on the neurons with a high increase in RNA. The species of molecules being synthesized were small molecules, but nevertheless it was a surprising increase in the RNA synthesis in the neurons. With respect to the other drug, tofranil, we found that the effect was largely on the glia cells with an increased synthesis of RNA, but the RNA was about the same type as in the other case. And then a little later on, when we had seen the effect on a glia cell for a certain time we could also see it in the neuron, but then there was a clear difference there in the effect. It would be possible to make a correlation between the increase of amines and the increase in the protein synthesis. That was one of your questions. So I think we could answer that question in an affirmative way.

Now to the other question. Could one see a way in which your suggestions could be fitted into both the observations on the synthesis of acidic proteins at the establishment of new behaviour and in the synthesis of specific nuclear RNA species, at such an establishment of a new behaviour. Let's have a look at the limbic system. You pointed out that the amines in this system, the aminergic system would go through the limbic system. It was clearly shown how the state of emotion can be correlated with both localization and the rate of synthesis of amines in this field, as you pointed out. The small pathways leading from the brain stem up into the cortex also contain norepinephrine. If now the amines colour the evaluating system, these pathways so to speak, or the changes brought about by experience, one would suspect that it would do so also in the limbic system. Since protein synthesis can be affected and increased by the amines, here we have a correlation. An effect on the synthesis of protein was shown by two independent methods, an effect on the emotional state and on learning. Now comes the quantitative effect. If one looks at both the amount of protein being synthesized and the amount of 'RNA being synthesized at various physiological types of stimulation and also in learning, the response is a little too great to involve only amines and the effect on protein synthesis we suppose they could bring about. I won't go into the calculations for it, I'd be glad to do that afterwards, but I would say it's perfectly possible that there are two systems here, the protein system involving specific proteins, as you also suggested, and that of the amines and their release. We have one small molecule operating and that is cyclic AMP. In both cases the increased rate of synthesis of AMP will be there. So there is a sort of common link. I think that here they come together—both the Gestalt of specific proteins to which networks of millions of neurons could respond, depending upon the pattern of specific protein, depending on the previous additional activation of gene areas, and also the whole system, the aminergic system colouring the old system in addition to the basic one.

KETY   When you mentioned to me the other night your findings with one of the monamine oxidase inhibitors, I didn't realize immediately the connection. This is a very interesting point. But as you point out, it may be more than enough to be compatible with this hypothesis, because as a matter of fact the amount of protein involved might be very small in terms of establishing new connections. I think you have indicated the course of a strategy, some areas that ought further to be explored, such as the correlation between amines and protein synthesis in specific areas in the brain, and see whether that correlation stands up. But of course there are many non-specific factors which could be involved in this, just as it is possible to demonstrate an increase in protein synthesis in the brain which could be non-specific due to increased cerebral circulation, increased general metabolism in the brain, and so forth, but at least it is comforting to know that your finding would be compatible with that hypothesis.

SMYTHIES   I would like to say how much I am impressed by Seymour's hypothesis of how the emotions and the memory get together in the brain, because as Penfield showed, some memories are laid down in the brain together with the right emotion, that is, each memory is linked with its own emotional state. If you recall Penfield's experiments stimulating the cortex in epileptics, when the memory of the particular event in the person's life was recalled, it came with the emotions that the person had felt at the time. Now the thing that struck me about Seymour's suggestions is that there is one possible way of doing this. At the synaptic level in the cortex, a particular chemical may be fed into the synthesis of protein and then this particular "tagged" protein then marks that memory trace with that emotional state. Now what I would like to do is just to get to the board and describe very briefly some of our own work, and how this idea can be linked with our ideas of what goes wrong in psychosis (Blackboard). The ordinary action of catechol O-methyl transferase (COMT), which is the significant enzyme at the adrenergic synapses, is to put on the methyl group in the 3 position (Figure 1). Now this has always been regarded merely as an inactivation process. Possibly, however, this could be doing something else besides just inactivating the transmitter.

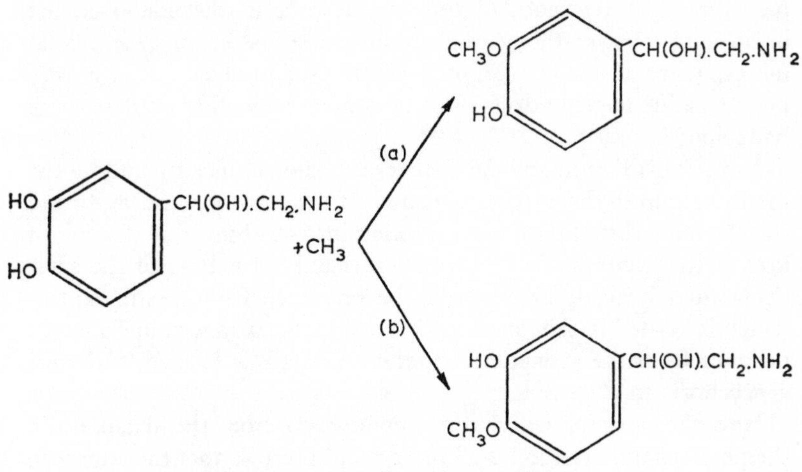

FIGURE 1   (*a*) *normal action of the enzyme COMT*
               (*b*) *suggested aberrant action in schizophrenia.*

We have been trying to find out which elements in the hallucinogenic molecule are necessary for its action and we found that there was only one group that was necessary and that is the 4-methoxy group (Figure 2). This latter compound by itself is inactive, if given

FIGURE 2   (*a*) *Mescaline*
               (*b*) *p-methoxyphenylethylamine.*

by mouth, as it is very quickly destroyed by amine oxidase in the liver. However, if it is protected from amine oxidase by pre-treatment with an inhibitor or by α-methylation (to form p.methoxy amphetamine), it becomes an extremely potent hallucinogenic drug. It has been tested in animal experiments, and has also been tried on man and it is considerably more active as a hallucinogenic drug than mescalin. The relevance of this to our present discussion is that ortho- and theta-methoxy amphetamines are just ordinary amphetamines, whereas the 4- (or para-)methoxy amphetamine is a very potent hallucinogenic drug. And of course, mescalin is a very weak hallucinogenic drug.

Now COMT normally metabolizes catechol amines by putting the methyl group in the 3 (meta) position. If the enzyme were faulty and put the methyl group on the 4 position instead (Figure 1) this might lead to the accumulation of 4-methoxylated derivatives of the catecholamines instead. These would be protected from amine oxidase which is located in the mitochondria inside the axon terminal at some distance from the synaptic cleft where COMT is located, and thus a psychosis might result.

Now the question is, what compound gets from the synapse into the post-synaptic cell to "tag" the protein there so that the latter can exert the coding function Seymour has postulated. And this might well be, not the amine itself, but its O-methylated metabolite. That is to say, the function of COMT may be not merely to "inactivate" the transmitter but to convert it into a form that could enter the cell more readily than nor-epinephrine with its highly-reactive polar groups could. An interesting phenomenon with which this links up is the well-known "flashback". This may happen to many people taking LSD and mescalin. This means that something days or weeks later sparks off the original psychosis and the whole thing can come back again. The most dramatic example of this I have seen was in an experiment I carried out some years ago. I gave 400 mg. mescalin to a big burly fellow, and he had very little response at the time. He had very few hallucinations. He was happy and talkative, and said it was just like taking a couple of Martinis, no more: it was really not any kind of psychosis. About ten days later, he went back to his home abroad. He had a very tiring journey and lost a lot of sleep. He arrived at the airport—and suddenly a full-blown mescalin hallucinosis started up that lasted for eight hours—a full-blown mescalin psychosis in the middle of a foreign airport! Luckily he was rescued. But this indicates this cannot have been merely a learning phenomenon because he never had the mescalin experience in the first place; the "flash-back" was triggered by sleep-deprivation, which is well-known to induce psychosis. But the mescalin itself must have been tucked away somewhere in the body in order to cause this flash-back. But I think that Seymour's idea that what is happening in the brain in normal memory formation links up with this. The nerve endings are getting linked into engrams because they are subject to a dynamic chemical specification. The biochemical mechanism responsible for this specification may be the one that goes wrong in the psychotic brain. I think this is a very interesting and valuable suggestion.

KETY I think it's a very interesting suggestion. As a matter of fact the one thing that all the activating drugs have in common is that they produce an increase in normetanephrine in the brain; and norepine-phrine will potentiate the effects of the parent compound. I tend to favour more the idea that the affect gets hooked into the memory by hooking in an appropriate hippocampal or limbic circuit, so as to let the rudimentary circuits for pain or pleasure get hooked up to different memories through circuitry. Although the other possibility was once suggested, I must say that your idea is an intriguing one, and perhaps makes the possibility more likely that the memory is hooked up as a change in the specific joining protein. Certainly one could speculate about schizophrenia being a disjunction between the memory and the affect associated with the memory. This is an attractive idea.

KOESTLER I would like to make one axe-grinding remark. If we are looking for new ways in which psychopharmacology can help, if you will forgive me for saying so, humanity at large, I do believe it is essential to realize that our main predicament is not individual aggression but individual devotion to totems, flags, creeds and causes; and that the egotism of the group, the aggression of the group feeds on the altruism of the individual and not on the aggression of the individual. I have discussed this at length in the book you mentioned, and I think it is of basic importance. Waddington made a quite similar point when he talked about man, the "ethical animal" being a belief-accepting animal—was that your term?

WADDINGTON Yes, or something like that.

BERTALANFFY I wish to congratulate Seymour on a particularly lucid and penetrating presentation. I was a member of a study group on psychotropic drugs of the World Health Organization in 1957. (*Ataractic and Hallucinogenic Drugs in Psychiatry*, Technical Report No. 152, Geneva, World Health Organization, 1958). At this time, there had been some 6700 investigations on chlorpromazine alone (if I correctly remember the number); the epinephrine story was at its beginnings. Developments since seem to be rather disappointing for the great expectations set into psychopharmacology (apart from symptom therapy). Following Kety's authoritative review, two major advances have emerged, i.e., the elucidation of the epinephrine mechanism and the histofluorescence technique. Not much new light seems to have been shed on clinical psychosis, with two possible exceptions: (*a*) the relation of biogenic amines with emotional components of psychosis which, however, is hardly unexpected; and (*b*) the structural similarities of psychotropics with brain-active substances mentioned by Smythies. These, however, were known and widely discussed 12 years ago and longer, and they leave the thorny problem to what extent hallucinogen effects may be paralleled with clinical psychosis, or hallucinogens produce a toxic psychosis, as Kety himself termed it with commendable caution.

HAYEK   I have just one brief, general remark. If we think in terms of states which affect the whole nervous system, this leads us indeed to think in terms of simple alternatives, such as either *A* or *B*. But it seems to me that at any one moment there may be a great many different dispositions present. It will be the effect of the superimposition of several such dispositions which determine a particular action. It is only thus that we can conceive of the generation of highly complex new actions never taken before in exactly the same manner. I believe that what you describe as "states" is the same as what I described as abstract disposition towards ranges of actions possessing certain properties.

KETY   I heartily agree. I presented this simple two-poled hierarchy, but undoubtedly there is interlocking and interacting between these, just as emotions can interact, and I think it is a lot more complex than would appear.

WADDINGTON   I wanted to make one remark on this question of whether one can imagine a pharmaceutical agent which is specific against aggression and does not have much effect on learning. In the domestication of animals men have succeeded in selecting genetic strains which are not aggressive, but which have not lost their capacity for learning. These domesticated strains have been produced by selecting appropriate genes in the population. Genes, of course, act by affecting molecular processes, presumably in this case processes at the nerve synapses. Now if genes can affect molecular processes to produce this result, there seems no obvious reason why you could not find drugs which would do it too.

KETY Are we talking here about a chemical effect or a neuro-physiological one? I have a lovely English golden retriever who is the gentlest animal in the world, except when he sees another dog. Has this been a change in the biogenic amines or has it been a development of the neocortex which has taught him not to be aggressive in response to human beings?

WADDINGTON I really should have put my remarks to you as a question. Perhaps what the genes do is too complex for any single drug to imitate.

KOESTLER I do want to make an aggressive remark, against this constant insistence that what we want to do away with is aggression. Why? We should all become cabbages. I mean, a minimum of aggression is necessary, but the main danger is fanaticism, that is blind devotion.

THORPE Ah yes, but fanaticism and devotion are different. If we did away with devotion we should all become cabbages and there would be no great scientists, no great artists, no great writers or poets.

KOESTLER Well, fanaticism is misguided devotion. In other words, what is really needed from the psychopharmacologist is to create a balance between emotion and reason. So that they complement each other, instead of being in conflict with each other.

FRANKL It is now 30 years since I published my first paper on psychopharmacology, in Switzerland, and in 1952 I was, I think, the first to introduce, to develop a tranquilizer in continental Europe. Although my experience is purely practical and clinical, I venture to suggest that psychopharmacology can and sometimes must be combined with psychotherapy. There are cases of obsessive-compulsive conditions that respond favourably to certain techniques of psychotherapy. There are cases of obsessive-compulsive conditions that respond favourably to certain techniques of psychotherapy. There are other cases that respond just as favourably to tofranil or to mono-amino-oxidase inhibitors. But there are also cases, that respond to psychotherapy only after one of these drugs, tofranil or MAO, has been administered.

But since we are concerned with psychopharmacology, I would like to conclude with a short remark on a specific issue: the question whether aspirin only affects pain. I deny this, because I and the doctors on my staff found time and again that aspirin has a markedly euphorizing effect, particularly on dreams—apart from the fact that in cases of aged people who react to the conventional sleeping pill by paradoxical excitation, aspirin works excellently as a sleeping drug. In addition to this, my staff is used to prescribing aspirin in all those cases in which a patient is suffering from anxiety dreams by way of so-called chain-dreams; that is to say they wake up and after falling asleep again the same dream returns. This is a very embarrassing situation, and it responds to aspirin. Out of my own experience I would like to offer you the following evidence. Once I remember I dreamed that my brother was wrestling with death, in the process of dying, and I woke up and after falling asleep again, my brother was still dying, and after waking up several times finally I took half an aspirin, and, believe it or not, my brother recovered. Another time I was dreaming that my wife was committing adultery; whenever I woke up and fell asleep, she was again committing adultery. I took one aspirin, and, believe it or not, then *I* was committing adultery. So if what Arthur wants is that loyalty should be weakened, aspirin is the ideal drug for your specific purpose.

KETY   Well, certainly I can support what Frankl said that pharmaco-therapy is synergistic with psycho-therapy, and this would fit in completely with my idea that these moods are not simply chemical, but have a powerful cognitive compound. So that just changing the chemistry without changing the cognitive apperceptive mass is not likely to affect a permanent recovery. I don't have any comments on the aspirin, I can just substitute my own subjective feelings for Dr. Frankl's subjective feelings: in me aspirin doesn't produce infidelity or euphoria, it seems to be a pretty inert substance, but I may not be aware of the other side-effects.

The theory of evolution today
C. H. Waddington
University of Edinburgh

This talk is going to be rather different from most of the previous presentations; they have been in some way or other about man, about human biology or psychology or behaviour. This is about biology in general, with no special emphasis on man. In the first letter I got from Koestler he quoted someone—I think it was Thorpe—who said that there are a large number of biologists who are not satisfied with the so-called synthetic theory of evolution, and who wonder whether the conventional view that evolution simply depends, in the classical phrase, on "random mutation and natural selection" is really sufficient to provide a satisfactory theory; that is the topic that I want to discuss.

The theory of evolution has passed through two main phases, the Darwinian and the Mendelian. The first phase opened when the idea of natural selection was promulgated by Darwin, but originally he had no very definite idea of the hereditary system on which natural selection operates. In the beginning he adopted the conventional view of the time, that inheritance involves some sort of blending of the constitutions of the two parents to produce the heredity of the off-spring. However, it was very soon pointed out that any such system of blending inheritance rapidly reduces the hereditary constitution of all individuals to a uniform intermediate state, so that the differences between individuals vanish, and there is nothing left for natural selection to work on. Darwin was very puzzled about what to do about this situation. He saw that you have to provide some means of renewing the store of variation, and for part of his life at least he was tempted to adopt the Lamarkian theory that this is brought about by changes in the environment, producing changes in the heredity of the indi-

357

viduals subjected to it. This is the idea of the inheritance of an acquired character. Insofar as Darwin was attracted to it, it was mainly because it seemed to provide a way of keeping in being adequate variation which would otherwise disappear through blending.

This problem was, of course, resolved as soon as Mendelian inheritance was rediscovered at the beginning of this century. This ushered in the second, Mendelian, phase of the theory of evolution. The point about Mendelian inheritance is that the hereditary characters are carried by discrete factors, so that factors from the two parents come together in the offspring but do not blend, and they can resegregate again in the next generation, so that there is no tendency for variation to disappear and all individuals to become alike.

The idea that Mendelian heredity provided just what was required to make Darwinian evolution work was fairly generally adopted in most parts of the world at the beginning of this century. In Britain, however, there was an unfortunate controversy—exceedingly stupid it looks now—which has influenced a good deal of subsequent thinking about evolution. There was a violent battle between two groups of prominent characters, Bateson on one side defending Mendelism, and people like Pearson and Wheldon on the other, opposing Mendelism and arguing for some other system of heredity which did not involve discrete hereditary units which segregate from one another. The anti-Mendelians based their claim on the existence of a complete range of variation in such characters as height and weight, with all intermediates between large and small. The question at issue had in reality been completely solved about the year 1902, almost immediately after the discovery of Mendelism. A statistician called Udny Yule, who like Bateson was a fellow of St. John's College, Cambridge, and who presumably discussed Mendelism with Bateson at the high table, published a paper showing that if a population contains a large number of Mendelian factors affecting a character like height, it will show a continuous range of variation of exactly the kind on which the Pearson-Wheldon group laid such stress. However, this paper appeared in an obscure journal and little attention was paid to it. The great battle of Titans went on, with special journals founded to publicize the opposing views, Bateson publishing his results in the "Reports of the Royal Society's Committee on Evolution", while Wheldon edited a special journal called "Questions of the Day and of the Fray". It was one of the large scale academic controversies of the early part of this century; now it is almost totally forgotten, but it was in the aftermath of this battle that the first attempts were made to formulate the Mendelian evolution theory in a more precise and mathematical way. The main purpose of two of the first people who laid the basis of a mathematical theory of evolution was to demonstrate —in the context of this controversy—that Mendelian factors could actually provide a satisfactory basis for natural selection.

One of these two was J. B. S. Haldane. He started the mathematical formulation when he was still an undergraduate, and I am not certain that at first he took it very seriously. At any rate he published the first papers in the transactions of the Cambridge Philosophical Society, which is rather a local journal. A few years later he brought these papers together into a book called "The Causes of Evolution". In that the motto on the title page is "Darwinism is dead! . . . *any sermon*"; and a little later he goes on to say: "I propose to anticipate my future argument to the extent of stating my belief that in spite of the above criticisms, which are all perfectly valid, natural selection is an important cause in evolution". A pretty mild statement surely—"*an* important cause in evolution"; it looks as though Haldane was not really trying to formulate a complete theory of evolution, but simply to demonstrate that natural selection could work on the basis of a Mendelian type of heredity. The same comment I think is true of Fisher.

What Haldane and Fisher did to establish this point theoretically was to express the situation in extremely simple mathematical terms. If you have a single recessive mutation a in a randomly mating diploid population the proportion of the varying types of individuals will come to an equilibrium $u^2AA : uAa : 1aa$. Haldane said that if the recessive is at a disadvantage under natural selection its frequency in the next generation will be reduced, so that the proportions will be converted into $u^2AA : uAa : (1-k)aa$. Fisher's mathematical scheme is logically more complex, since instead of a linear coefficient k he used an exponential coefficient which he called a Malthusian parameter.

Now these formulations employ a number of very obvious simplifications, of which I will mention three which are perhaps the most important. One is that any population containing a number of different genes will normally settle down to an equilibrium after which the proportions of the various genes remains constant. The Fisher and Haldane schemes imply that their populations are not in equilibrium, but they do not say anything about why they have departed from it. Secondly, their formulations are in terms of a very few varying genes—for instance, one single genetic locus with just two alleles. This leaves out of account the extraordinary complexity of most genotypes in actual populations, and it is quite difficult to get this fact back into the mathematics again. Thirdly, and most important of all, the Haldane and Fisher mathematical equations are formulated entirely into terms of the genotypes. They say nothing whatsoever about the actual phenotypes of the animals or organisms concerned. Now natural selection obviously acts on the phenotypes. If, for instance, natural selection demands that a horse can run fast enough to escape from a predatory wolf, what matters is not what genes the horse has got, but how fast it can run. It is irrelevant whether it can run fast because it has been trained by a good race-horse trainer, or because it has got a nice lot of genes. Thus the mathematical formulation commits a basic logical error in attaching the coefficients of selection, not to the phenotypes where they really belong, but to the genotypes. This is a fundamental weakness of the whole scheme.

I should now perhaps say something about the third major figure in the origin of modern mathematical theories of evolution. This was Sewell Wright. He is an American and did not grow up in the atmosphere of the violent, silly English quarrel between Bateson and Wheldon. He put forward a much more sophisticated mathematical treatment of the evolutionary situation. Right from the beginning he considered populations in which large numbers of genes were varying simultaneously, and he considered the fitnesses of whole complex genotypes, not of individual genes. He expressed these fitnesses not by single coefficients attached to the individual genes, but in terms of a "fitness surface", which is quite a complex topological idea. If one considers a multidimensional space with sufficient dimensions to represent all the possible combinations of the large numbers of genes in the system, then you can represent the fitnesses of the various gene combinations by a surface within this space. Expressing the matter in a pictorial way, Wright argued that this fitness surface would be of a hilly character, with the tops of the hills as the fittest points; but each hilltop would be separated from others by valleys of lower fitness. Any actual population left to itself would tend to come to a temporary equilibrium at the top of one of these hills. But there might however be another higher hill, i.e. a genetic constitution of greater fitness, in the near neighbourhood; the question was how did the population come down from one hilltop, cross a valley of lower fitness, and go up another higher hill?

In this way Sewell Wright avoided the first two simplifications of the Haldane Fisher scheme. He pointed out why his populations were not at a state of equilibrium by suggesting that they were at an equilibrium which was only metastable; and he dealt with multiple gene situations and not simple situations. But he too made the third simplification, in that his fitness surface is in a space of genotypes and not a space of phenotypes. But it is when you pay serious attention to the undoubted fact that natural selection impinges on phenotypes, not on genotypes, that you find yourself confronted with the really interesting logical situations which demand very sophisticated mathematics (which nobody has yet succeeded in providing).

I would like to try to describe what seems to me the essentials of an evolutionary situation. What is the paradigm, or ideal model, of an evolutionary process? The Haldane paradigm is simply that of the change in proportion from $u^2 : 2u : 1$ to $u^2 : 2u : 1-k$, from one generation to the next, which we mentioned above. I have discussed my own more complex paradigm at greater length elsewhere (Waddington, 1968, a, b) and I will try to describe it shortly here. It is intended to include all the factors that are necessary to set up a system in which continuous evolution would take place.

We have to start by defining a genetic system. A genetic system involves the transmission from one generation to the next of something or another; let us for the moment call it "information", although in a minute I am going to point out that this is not a very good name, but it is the one most often used at present. This "information" has to be carried by some material structure. So let us start with some material structure P which has some characteristic Q which we are referring to as information. Now a genetic transmission of this takes place if, when you put P into certain conditions, it produces more of Q. This is a very general definition of the basic logical structure of genetic transmission; it can apply to things which are clearly not biological; for instance if you have a crystal which has a dislocation on one of its faces at which the atomic arrays are slightly displaced, then the next layer which forms on this crystal as it grows will usually have the same dislocation. This is an example of the transmission of the dislocation. Moreover the dislocation may be transmitted even if the next layer of material precipitated on to the crystal is different from the earlier material. There can therefore be a genetic transmission into a range of different materials, and the transmission can take place in a variety of different conditions, for instance, of temperature or concentrations of various substances in the solution. There could therefore be a sort of natural selection, in which those types of dislocation which were most able to transmit their characteristics under a wider range of conditions would be favoured (for a fuller discussion of this, see Cairns Smith, 1966).

This example shows that you could find processes with the logical characteristics of genetic transmission, and even of natural selection, in systems which nobody would be tempted to call living. That is why it is inadequate to try to define life, as many biologists have been tempted to do, by saying that its essential characteristics are genetic transmission and natural selection. But one might ask, why do we deny the title of living to such a system as a crystal which transmits its various dislocations to the next layers of molecules that settle on its surface? I think the answer is that this sort of "information" is purely genotypic. It is simply transmitted, and itself does nothing to its surroundings so as to produce from them anything that can be regarded as a phenotype. The purely genetically transmissible aspects are simply not interesting enough to be called living, and compared with the biological systems in whose evolution biologists are interested. In order to have an evolving system comparable to the biological ones it has to include not only genetic transmissible "information", but this "information" has to do something to its surroundings so as to form out of them a phenotype which can exhibit properties which are as interesting as those we see in the living things of this world.

It is for this reason that the word "information" is not adequate in this connection. Information is something which is simply transmitted; what we need is something which is not only transmitted, but which is also active in changing its surroundings. We need transmissible *instructions*, or programmes, or to use another technical word algorithms.

The biological systems in this world have settled on a rather complicated way of solving these problems, using several different but closely related substances to fulfil the various necessary roles. At the core of the system is DNA, a substance which is sufficiently inert and unreactive to serve as a reliable memory store. By itself it can do almost nothing; it is very inefficient (even if it can operate at all) at producing a copy of itself which could be transmitted to the next generation, even when it is provided with all the necessary building blocks. However, it occurs in association with protein enzymes, which in the first place enable it to be replicated, so that there are copies to pass on to the daughter cells, and which also use the information stored in the DNA to produce a corresponding RNA and that in turn is used to guide the synthesis of a corresponding protein. This whole system works extremely efficiently, but as you see it involves separating the function of reliability in storing information from that of actually using the information as instructions to change the surroundings. It is, I suppose, theoretically possible to imagine a substance which was both inactive enough to be reliable and active enough to have effects in producing a phenotype. Possibly some evolving systems on Mars or elsewhere in the universe have discovered such a perfect answer to the evolutionary problem, but life on this earth has not. It is stuck with a system in which there is an inescapable difference between the genotype—what is transmitted, the DNA—and the phenotype—what is produced when the genotype is used as instructions.

It is now necessary to say something about phenotypes. At the very beginning of evolution, and in the most lowly organisms, such as viruses or bacteria, the phenotype can be relatively simple and not so very different from the genotype. However, we have to have a system of ideas which can also deal with the highly complex phenotypes which have been produced during the course of evolution. The genotypes of higher organisms contain enormous numbers of genes. There are a few tens of thousands even in a bacterium, and at least some hundreds of thousands, possibly up to a million, in such highly evolved organisms as ourselves. Now if we consider each gene as an instruction, and think of the number of ways these instructions can be combined with one another and interact with the surroundings, the possible number of combinations is truly astronomical. If one wants to make a diagram of the situation, one cannot really do it on a blackboard of two dimensions; but topologists nowadays have made us get used to thinking in terms of spaces with an almost or quite infinite number of dimensions. In these terms we can imagine a multidimensional space with one dimension for each type of gene, so that a particular genotype can be represented as a single point within it; and we can even indicate this by a two-dimensional space on the blackboard. Now from any particular genotype there eventually develops a corresponding phenotype, which again we could locate as a point within a multidimensional phenotype space. However, between the genotype space and the phenotype space we must remember that there is a whole series of processes in which the various genetic instructions interact with one another and interact also with the conditions of the environment in which the organism is developing. The system therefore moves from the genetic space into the phenotype space through what we may call an "epigenetic space", i.e. a space of developmental processes, which we may represent by vectors, or diagrammatically by arrows, which are tending to push the developing processes in one direction or another. Now, not all these arrows or "epigenetic operators" as we may call them, arise from the instructions in the genotype; some of them originate from the environment. The same genotype can therefore produce a number of phenotypes according to what the environment of the developing system has been. This means that if you start with a phenotype, as natural selection does, there is an essential indeterminacy in the relation between that phenotype and the genotype; the relation only becomes determinate if you take into account the environment also (see diagram).

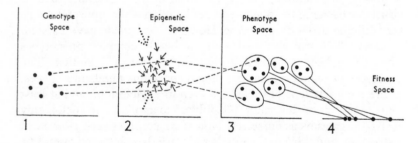

And in fact in evolution the indeterminacy is even worse than this, because natural selection does not pay attention to all the characters of the phenotype, but only to its fitness, that is to say the number of offspring it leaves. Natural selection operates in a one dimensional fitness space. To get back from there to the genotypes you have first to go from the one-dimensional fitness to a many dimensional phenotype space (a one to many mapping) and then from the phenotype space through the epigenetic space with its environmental components, to the genotype space (and in this transition one can have examples both of one phenotype corresponding to many genotypes or of many phenotypes corresponding to one genotype). This sort of indeterminacy leads to a really interesting logical situation, which is not at all amenable to the kind of simple mathematical treatment that Haldane and Fisher tried to give it. It calls for mathematics much more like the Theory of Games.

And finally there is another factor in the evolutionary process which also seems to call for handling by the Theory of Games. The environments in which organisms live, develop, and are selected, are neither wholly dependent nor wholly independent of the genotypes of the organisms involved. To some extent an organism's phenotype (particularly in its more elaborate development, the behaviour of an animal), can influence the type of environment in which the organism lives; putting it crudely, many organisms if they do not like the situation where they are have the ability to move away and try somewhere else. On the other hand there are aspects of the environment which the organism cannot influence and cannot in fact foretell; there may be a new ice-age round the corner, or a new predator or a new disease may suddenly appear. An evolving population is, as it were, playing a game in which it has some choice as to which card it puts down for any given trick. It hasn't much choice as to which cards it is dealt.

The trouble about trying to discuss such situations in terms of the Theory of Games is that although that branch of mathematics has got such an interesting title, which seems to promise such a lot, in practice as far as I can make out it contains almost no theorems. There is one, a Theorem of Minimax Strategy, which essentially teaches you how to play safe. But beyond that the content of the Theory of Games seems to be almost non-existent, except for discussions of elaborate ways of formulating problems, which you then find you can't solve.

HAYEK   John von Neumann died too early.

WADDINGTON   Possibly he did. Anyhow I wish someone else would now put some meat into the Theory of Games.

Before going on to other aspects of evolution I want to make some remarks about this epigenetic space. During the development of an organism there are a large number of genetic factors, each producing enzymes, and many of these enzymes may be synthetized out of the same raw materials, or be acting on the same substrates or inhibiting or facilitating each other's activity in all sorts of complex ways. We cannot yet describe all these individual processes in detail. One can however, point to some general characteristics of the type of result which emerges from this whole complex system. The point I want to emphasize is that such systems usually exhibit a kind of stability. I have used the words canalization or homeorhesis to describe this. The latter is a new word. It is related to the well-known expression homeostasis, which is used in connection with systems which keep some variable at a stable value as time passes. A thermostat, for instance, is a device for producing homeostasis of temperature; or in the biological world it is well known that there is a homeostasis of the level of carbon dioxide in the blood, which is adjusted by varying the rate of breathing, for instance, so that the concentration of dissolved gas remains constant. We use the word homeorhesis when what is stabilized is not a constant value but is a particular course of change in time. If something happens to alter a homeorhetic system the control mechanisms do not bring it back to where it was at the time the alteration occurred, but bring it back to where it would normally have got to at some later time. The "rhesis" part of the word is derived from the Greek word Rheo, to flow, and one can think of a homeorhetic system as rather like a stream running down the bottom of a valley, if a landslide occurs and pushes the stream off the valley bottom, it does not come back to the stream bed at the place where the diversion occurred, but some way farther down the slope.

There seems to be no recognized word for a stabilized time trajectory of this kind. Since they are the most important features of developing biological systems, I have invented a name for them—the word chreod, derived from the Greek Chre, it is fated or necessary, and Hodos, a path.

The general characteristics of the epigenetic space can be described by saying that it contains a number of chreods, each of which is defined by the instructions in the genotype which interact together to produce a system which moves along a stabilized time trajectory. Environmental influences may operate in such a way as to tend to push the system off the trajectory, but the canalization of the chreod, or, otherwise expressed, its tendency towards homeorhesis, will tend to bring the system back on to the normal path again. This system of ideas, which I have been discussing for quite a long time, has recently been taken up and formulated in much more precise mathematical language by the French topologist René Thom. He is an extremely profound mathematician, who a few years ago was awarded the Lieper Prize, which is often known as the mathematician's Nobel

Prize. He is in the process of publishing a large book called "Structural Stability in Biology", in which he discusses chreods in detail, and in particular makes some very important advances in our understanding of the way one chreod may branch into two. This is, of course, an essential feature of embryonic development. What is, at an early stage, a single undivided mass of tissue may later differentiate into two quite different and distinct organs. This can be expressed by saying that the chreod which the tissue was following in its early stages has later branched into two.

WEISS Do you know that Witowski is working this into biological systems now? He is a Polish student of Landau in Russia.

WADDINGTON I know that Thom talked about his work at the last International Mathematics meeting in Leningrad a year or so ago, and he may well by this time have some followers in Russia.

Thom has been particularly interested in the spontaneous breakdown of structural stability in various systems; he calls them "catastrophies". Some physical examples are a wave breaking or a jet of liquid breaking up into drops. Thom claims to have demonstrated that in real four-dimensional space-time there are only seven possible distinct type of catastrophy. He has described each of them; and given them somewhat funny names such as "the parabolic umbilic" to give one example, but at least we are getting a really sophisticated mathematical description of the kinds of epigenetic processes by which the phenotype is produced.

One of the most important problems for theoretical biology at the present time is to try to understand how these chreods come into being. Are they purely the result of natural selection which has built up genotypes in which the various genes interact with one another to form a harmoniously operating control system, which ensures the canalization of the chreod? Certainly it can be easily shown that natural selection has played some part in producing the stability; but it is also at first sight possible that some degree of stability, though perhaps not a very high one, might be an inevitable consequence of any system in which very many genes are interacting with one another. I have always felt intuitively—and I have once been brave enough to put the idea down on paper (Waddington, 1968)—that some sort of stability would arise spontaneously in any complex interacting system, and I felt that the greater the number of genes any one gene interacted with, the greater the chance of a stable system.

There is now beginning to be some evidence on the point and it looks as though my hunch about increasing stability with an increase in a number of interactions was probably wrong. There is a very young man in America, a pre-doctoral student, who has really excited most of the theoreticians concerned with this subject by an experiment he has recently done on a computer. He set up a system of a fairly large number—a few hundred—of simple binary on-off switches; he then coupled these together at random in such a way that each switch has two inputs coming into it from two other switches which are chosen at random; then to each of these inputs he assigns, at random, one of the Boolean functions, which controls the way the switch will respond to that input—it will compare its own state with that of the input and respond either yes, no, and, or, and so on. Having set up this system with two randomized factors—the random connection of the inputs and the random assignment of Boolean functions—he starts the whole thing off in an arbitrary state, with randomly chosen switches at the on or off position, and lets the computer follow what will happen. The really surprising thing is that in quite a large number of cases the computer shows that the system quite soon gets into a steady cyclic condition, and proceeds to go round and round a certain sequence of changes without ever deviating from it. It reaches in fact a stable "limit cycle". This is an example of a certain degree of stability emerging in a system which had been set up purely at random, so that the stability is not dependent on natural selection, or anything else, carefully choosing just those values which together produce stability . . .

WEISS   Do you know the name of this man?

WADDINGTON   Yes, he is Stuart Kaufman. I am not sure where he is now. He was at California Medical School, but so far as I can make out half the best Universities in America are now saying to him "Would you come and be our graduate student?" I think he is now in Alex Fraser's Department of Biology at Cincinatti.

Well now let's get back to evolution. I have by now described a system that is beginning to work as an evolutionary system, but there are still a number of factors to add to it.

First of all, if the system is not to come to an equilibrium and stop, one has to arrange for there being an ever-increasing number of environmental niches or "life situations" into which new living beings can evolve. Now this requirement can be met very naturally. The environment of any organism is partly composed of the other organisms in its surroundings—those it eats or which eat it for instance, and there can be many subtler relations of parasitism, commensalism and so on. So as soon as any organism evolves, this changes all the environments of other organisms with which it interacts. Therefore, if one imagines a static non-evolving world, as soon as any evolution started in any of its constituent species this would in effect engender an infinite number of new environments, as everything else had to react to the one which had started changing first.

If the living system is going to be able to exploit all these new environments which come into being in this way, there have to be mechanisms which will, in the first place, produce new hereditary variations and, secondly, ensure that the organisms are not fixed in one place, but dispersed sufficiently to discover the new environments as they arise. There is no particular difficulty about this dispersal—all organisms have in fact evolved some means of spreading their offspring around the scenery sufficiently to give them a reasonable chance of running into new evolutionary opportunities. The first requirement, of producing new hereditary variation, has also been adequately dealt with. The DNA of the genes, although pretty inert and stable, does change from time to time and thus give rise to new hereditary properties. Theoretically, it would be possible to imagine a system in which these gene mutations were produced by the direct effects of the varying environments in which the animals were living. This was, of course, the hypothesis of one of the earliest writers about evolution, namely Lamarck. There is nothing wrong with the idea, except that it does not actually happen. It is only very rare, special and untypical environmental influences which can directly cause gene mutations—even then effects of the altered genes on the phenotype are rarely, if ever, at all appropriate to the natural selective demands of the environment which brought them into being. It would perhaps be an efficient system if it could be realized. The difficulty I suppose is that which we discussed earlier, namely that the material in which the genetic information is stored must be very inert and unreactive if it is to be a reliable store. If it was capable of being changed by all sorts of environmental influences, of the kind which exert natural selection on the organisms, it would soon be reduced to a jibbering nonsense. In fact the earth's living creatures—and let's wait and see what we find on Mars—have settlep for the rather inelegant process of relying on purely random mishaps

to produce changes in their genes. The vast majority of the alterations are, of course, harmful, and have to be weeded out by natural
selection, but at least the supply is inexhaustible. The genetic structure of populations of highly evolved organisms, which are mostly
diploid, having two of each kind of chromosome, has the effect that
a great many of the altered genes, even the harmful ones, hang around
in the population at a low frequency for many, many generations—
and eventually natural selection may find a use for some of them.

The whole question of randomness in evolution needs careful
consideration. Many people seem to feel that the most difficult point
to accept about modern theories of evolution is their insistence that
the complex and orderly world of living things has been produced
from a basis of merely random gene mutations. I think that this
appears less surprising if we consider carefully just what is meant by
random in this connection.

A gene mutation consists in some sort of alteration in the sequence
of nucleotides in the DNA. From a chemical point of view these
alterations in nucleotides are presumably not wholly at random. There
may well be quite considerable regularities in the processes by
which the alterations come about; however, we know very little
about them as yet. But even if we fully understood the physical and
chemical processes involved, this would not be very relevant to
evolution. From that point of view, the important thing is that the
occurrence of a mutation and the nature of the mutational change is
not directly connected with the environmental circumstances which
will exert natural selection on the result. As far as natural selection
and evolution is concerned, mutations can therefore be considered as
effectively at random.

At very early stages in evolution, when there is only a minimal
distinction between genotype and phenotype, the randomness of the
mutational process is the controlling factor in the situation. In
particular, during the processes by which the first living systems came
into being it is difficult to see any other way of discovering proteins
which had some enzymatic activity, except waiting for random mutations to produce them. The situation gets very different as the
organisms concerned become more complex. There are two major
factors which reduce the importance of the randomness of mutations.
In the first place, evolution produces systems which allow for genetic
material to be changed not only by random mutation, but also by
processes of recombination between existing genotypes. In recom-

bination whole blocks of DNA become shuffled around, and this is a very different process from changing one or a few nucleotides as occurs in mutation. Such recombination can go on to some extent even in lowly organisms such as bacteria which have only one set of genes. Most higher organisms are, of course, diploids with two sets of genes, which makes recombination very much easier. It leads to a startling increase in the efficiency with which the system can alight on a favourable result. Many people have tried to simulate the natural selective process on computers. They have on the whole come to the conclusion that random mutation, with natural selection of any favourable results that turn up by chance, is an exceptionally inefficient process. However, most of these investigators are mathematicians or physicists who seem to have known little biology, and therefore did not bother to simulate the diploid condition and to allow for recombination. When this has been done by more biologically-minded computer experts, such as Fraser and Lewontin, it turns out that random mutation, recombination and natural selection is actually remarkably efficient.

The second factor which mitigates the effects of randomness in mutation is the increasing complexity of the epigenetic processes by which the genotype is developed into the phenotype. This means that the effects of any given gene mutation will be affected by interactions with a great many other genes. As we have seen, epigenetic processes tend to be chreodic, that is to say to have certain stability characteristics. These may be sufficient to buffer out the effects of a good many gene mutations, which will then bring about no noticeable change in the end products of development; or if the genes do produce an effect this will be as much dependent on the stability characteristics of the epigenetic system as on the nature of the gene mutation. Many different mutations may in fact cause the same alteration to the phenotype, and that is another way of saying that, if natural selection is pressing for some particular alteration of a phenotype, there may be many different genetic ways of producing it. In such highly evolved complex organisms the randomness of the basic gene mutations is, as it were, buried deep in the complexity. It is rather like the randomness of the shape of the pebbles which form the aggregate of the concrete out of which a bridge has been built. It remains true enough to say that the ultimate units, the pebbles in the concrete, or the genes in the organisms, have been produced by random processes, but this is almost irrelevant to the engineering of the bridge, and in many cases not much more relevant to the anatomical or physiological construction of the organism.

Consider for a minute a well-known example of evolution, that of the horse family during the Tertiary period. There were a series of ancestral horses, of which we have rather large collections of fossils so that we can follow their evolution in some detail. They underwent a fairly complex set of changes—the evolution was not one simple continuous process of change in one or two directions, as people used to think before we had as much fossil evidence as we do now. The changes concerned such things as modifications of the legs, lungs and so on to enable the animals to run very fast for long distances; alterations in the mouth, teeth and digestive apparatus to enable them to live on grass rather than herbage; and other such complex alterations related to the activities of living. Now, as I pointed out earlier, in adopting the particular strategies of evolution, which they did—for instance to run away from enemies rather than to stand up and fight them—the horses were playing a game, taking a gamble on one particular tactic rather than another. But when a given population of ancestral horses adopted a particular style of living, which subjected them to certain natural selective pressures (such as running away, when the pressure was to run faster and farther) success in this is not likely to have depended on altering some particular sequence of nucleotides by chance mutation. Very large numbers of genes are concerned in the development of limbs efficient at running. What has to be discovered by the evolving population is not one particular sequence of bases in DNA, but rather one out of a number of combinations of genes which produces an effective complex organ. In a diploid population there are always a great many unusual alleles present in low frequencies. When natural selection calls for the evolution of a modified phenotype this can usually be produced by selection of appropriate recombinations of genes which are already present in the population. It is rare for a population to have to wait for a mutation to throw up just the one gene it is looking for. The importance of random mutation is only to replenish the stores of variation contained in the population which would, of course, eventually run out if selection operated for a very long time and there was nothing to keep on producing new variations.

There is another process which helps an evolving population to find a way of meeting the demands of natural selection more directly than merely by waiting for an appropriate change in some nucleotide sequence. Natural selection has built into all the more highly evolved organisms some capacity for reacting to a stress in ways which tend to make the organism more effective in dealing with it. Such responses can be considered as a very generalized form of learning. It is clear enough that responding to a stress in this way would be useful to the organism and would therefore be favoured in natural selection.

The capacity to response to the stress, and the character of the response, are genetic characters dependent on the genotype, just like any other physiological characteristics of the organism. The genes concerned with determining such characters will normally have no noticeable effect on the organism unless the stress is applied. If the stress is one which members of a population often encounter, suitable genes which control an appropriate response will be kept in being in the population by the natural selective advantage they confer when the stress happens.

There is a mechanism by which such genes can be used in evolution to bring about permanent alterations of the phenotype in later generations. One has to remember that developmental processes are normally chreodic, that is to say, they exhibit some resistance to change. When an organism responds to a stress by an appropriate modification, the stress has had to overcome the resistance of its developmental chreods. The genes which control the response to this stress do so by controlling the stability of some chreod and the way in which it can be altered. Now, consider what will happen if some novel stress is applied to a population. There will be natural selection for those individuals which respond to it in the most appropriate way. This selection, continuing over the generations, will gradually bring together and concentrate in the population the genes which bring about the most appropriate response to stress, that is to say genes which allow the environment to modify developmental chreods so that it finishes at an appropriately altered phenotype. Now it seemed to me likely that these genes would actually modify the chreod, so that less and less influence of the environmental stress was necessary to produce the altered phenotype. If this were so, we would find that after some generations of selection the chreod had been effectively modified, and it might in fact prove quite unnecessary for there to be any environmental stress at all to condition the development of the altered phenotype which had been selected for. The phenotypic alteration produced by the stress would then have been assimilated by the genotype.

I have made a number of laboratory experiments to test this, and have shown that this process of genetic assimilation actually does occur. If you take a population of animals and subject them to a stress, some of them will respond by some sort of phenotypic modification. Such a modification is what is usually referred to as "an acquired character". If in the next generation the stress does not occur the alteration does not happen either. The acquired characters are *not* inherited. However, the *capacity* to acquire the character—to respond to the stress in this particular way—is inherited. If in the first generation you select and breed from these individuals which respond most strongly in the appropriate way, in their offspring there will be a higher proportion which respond to the stress in the same manner. As selection goes on from generation to generation, the proportion showing the appropriate response increases. If after a number of generations of selection you stop applying the stress it is often found—in all the cases which I have tested—that quite a large proportion of the population still shows the phenotypic alteration which was originally produced only in response to the stress, but which has now been genetically assimilated, by the building up of a stabilized chreod which leads to it. This process, therefore, gives exactly the same effects as the Lamarckian inheritance of acquired characters, but it does not operate by any direct influence of the environment on the genetic constitution, but rather by long continued selection of appropriate responses. It not only gives a satisfactory explanation for all the examples which in the past have led people to invoke the Lamarckian inheritance of acquired characters, but more generally provides a mechanism for evolving the coordinated alterations which are usually required to make a phenotype more efficient. Physiological responses to stresses are usually more or less co-ordinated—certainly more coordinated than random changes in genes would be—and the selection is, of course, for those individuals in which the response is best organized and most appropriate. Thus genetic assimilation makes it possible for evolution to exploit what one might call the cleverness of physiological reactions to stressful situations.

DISCUSSION

MCNEILL   May I point out a very impressive similarity between the course you described here as homeorhesis and the course followed in the generation of language.

WADDINGTON   Yes, I am sure there are many contexts in which the idea of chreods is appropriate in linguistics. In fact one of the chapters in René Thom's book "Structural Stability in Biology" argues that a word should be considered as a chreod, fundamentally because it is a name used for a number of related concepts which all fall within what he calls "a basin of attractors".

HYDEN   You commented on the sequences of the DNA and its mutations and the relation of the phenotype to the genotype. I would like to take up an observation which may be subject to the influence of the epigenetic space, an observation made lately by Briten and Roberts, namely that if we consider DNA hybridization, around 15% of the DNA in higher organisms re-associates many times faster than lower organism DNA. What does that mean? It means that large populations of simple nucleotide sequences do exist in the DNA component, so there may be 100,000 copies of rather similar but slightly different nucleotide sequences existing in the DNA. They have also shown that there seems to be a repetition of such nucleotide sequences and that they are interspersed in various places. So what comes out of this observation is that—and there I would like to hear your opinion—if a group of genes shows such similar but slightly different nucleotide sequences in the DNA, let's say 10,000 of them, coming up suddenly at one stage in an organism, these may be utilized in evolution. These may be the gene areas floating around in a pool, because if we consider the intercellular relationships in an organism, they must depend both on similarities and differences of the surplus proteins. That means different products of a set of genes.

WADDINGTON   This is a very interesting question. I have quite a large group of people in my laboratory working on DNA reiteration, amplification and so on. The subject is still not at all well understood. As to the genes which are floating around in the gene pool of a population, I think that these are not only genes from the reiterated stretches. In the human species, for instance, we know at least 20 genes controlling different forms of haemoglobin. These are straight structural genes, and it is very unlikely that any of them are reiterated.

Of the reiterated stretches, the only one that is at all well understood is that which is concerned with the formation of ribosomal RNA. As you know proteins are synthesized on ribosomes which contain both RNA and protein. Both in Drosophila and in the toad Xenopus, which is the species we work on mainly in our laboratory, there is considerable reiteration of the genes which code for the ribosomal RNA. In the toad there are about 2,000 for each of the two main components of the ribosomal RNA (the so-called 28S and 18S components). They are all collected together in one place in a particular chromosome, and a great advantage of the toad is that we possess a mutant in which the whole of this section of chromosome has been removed, so that it has a chromosome which has no DNA capable of making tRNA (or ribosomal RNA). One interesting point is that not only are there a lot of ribosomal genes incorporated into one chromosome, but at a particular point in the toad's development, when there is a necessity to make a great quantity of ribosomes in a hurry, this piece of chromosome is still further amplified. About 1,500 copies of it are made and these are shed from the chromosome and float around in the sap of the nucleus, where they appear as little nucleolar bodies at which the ribosomal RNA is synthesized. It is very interesting to find that this process of amplification which occurs during the ripening of the egg cells in the ovary, is under some sort of general control by the rest of the cell. If one has a female which is heterozygous for the lack of the ribosomal genes she has one complete chromosome and one chromosome from which these genes are missing, whereas normal individuals of course have two complete chromosomes. Nevertheless the heterozygous female produces just as many amplified stretches of the ribosomal DNA as a normal one. This must mean that there is some general cellular control which ensures that the amplification reaches a certain level even if it has to start with only one set of genes instead of two.

There seems always to be, in higher organisms at least, a good deal more reiterated DNA than that concerned with the ribosomes. Just what the function of the rest of it is nobody has any clue as yet. For instance, in the mouse there is a considerable fraction of DNA which appears to consist of a single gene reiterated about a million times (it is detected by the fact that if you take mouse DNA and denature it by heat, and then study the kinetics of the renaturing, this fraction renatures exceptionally fast—but we need not go into these technical matters now). Then there is a lot more which seems to consist of genes reiterated around 10,000 times. All this reiterated material is far too big to pack into a single chromosome, and it almost certainly must be scattered widely throughout the whole set of chromosomes. Nobody knows what its function is—my own guess is that it is something to do with the system for controlling the action of whole batteries of genes simultaneously rather than only single genes one at a time—but that is purely guesswork. Another possibility is that it is concerned with making the nuclear envelope. Perhaps it is more important to mention that there seems to be something very odd about the evolution of such genes. Another worker in our laboratory, Dr. Forbes Robertson, has compared the reiterated DNA in two extremely closely allied species of Drosophila—D. melanogaster and D. simulans, which are so similar that they are referred to as "sibling species". He found that their DNA was astonishingly different in these reiterated stretches. It is very difficult to see how or why natural selection can have brought about these differences. There is probably going to be some very interesting point about evolution emerging from this work in the fairly near future.

HYDEN May I make a last, short comment. You were speaking about the organisms selecting genes to respond to stress, the capacity of responding to it. And there I cannot help thinking of the various types of cells in the brain. In the brain, in a stress situation, it's easy to show that high amounts of largely ribosomal RNA are being formed in a very short time, in an amazing amount, it seems to be a sort of an unspecific reaction; so that may reflect—since RNA is a sort of a mediator between the genotype, and the phenotype—it may reflect such an evolutionary thread.

INHELDER You must already know how much we appreciate your work and how helpful it was in our studies on cognitive development, and particularly on learning. May I ask for more information about the "time tally" problem?

We found that learning, in the strict sense of the term, may within certain limits accelerate development. Such acceleration apparently obeys limiting conditions of assimilation which in turn are subject to time regulation, reminiscent of your notion of the "time tally" and of the chronological succession of competence in embryology. We made the hypothesis that there are some optimal time conditions. Harmonious development must be neither too fast nor too slow, but as yet we do not know the nature of this time regulator. I would be glad to know if such an analogy makes sense to you and if you can tell us more about the biological mechanisms governing the time tally.

WADDINGTON   I think this is an important analogy between learning and development, but I do not think we can now give much explanation of the time conditions in embryology. It is certainly the case that developmental systems pass through a series of stages. For instance, there is a first phase when they are competent with respect to a particular choice between alternative future powers of differentiation. There is only this particular phase when such a choice can be made; and if nothing is done to make it, the system will pass out of the phase. It is then too late to make this choice, but some other one will become relevant. We cannot yet give any explanation of how these changes come about. Molecular biology, deriving its ideas largely from work with bacteria, usually suggests that the fundamental process determining developmental changes is the attachment or release of particular proteins from certain sections of the chromosomes. This does not seem to me to give one sufficient flexibility of control to account for all the phenomena in higher organisms. In higher animals we have to find explanations not only for the activation of a gene so that it actually starts controlling the production of a protein, but also for earlier phases, such as that in which a cell becomes "determined" as to the future path of development it will follow, which may happen long before it actually starts to follow this path; and then there is the still earlier phase of competence with respect to a particular type of determination, which I have just mentioned. We can regard these as stages through which a cell approaches a bifurcation in the chreod it is following, but this is of course only a description, not an explanation of what is happening.

SMYTHIES   Could we go back to the horses with different options open to them? You say that the genes that are selected are those with a capacity to respond to stress. And I can see how this would increase the capacity of fight or flight. How do you, however, from the stress mechanism account for such things as increased bone growth, for example? Would you say that the physiological mechanisms, which are being potentiated as by-products of this, act on the growth of the bones and thus it happens that the legs become longer?

WADDINGTON   Yes, I think so; if you take an ordinary individual, and let him grow up in a very static situation where he does not take much exercise, his muscles will be rather poorly developed. Certainly improved nutrition obviously affects the growth of the muscles and skeleton—it has put six inches on to the average stature of the Japanese in a generation.

SMYTHIES   So what is being selected is a direct physiological mechanism?

WADDINGTON   Yes, you are selecting a physiological mechanism which is based on genetic potentiality.

SMYTHIES   The other point I wanted to raise is the question about the spontaneous emission of behaviour which has adaptive value—as in the case of the tits who started drinking the cream in the milk bottles. This I presume is also another factor, assuming that it happened to have survival value.

WADDINGTON   I think I gave an example of this when I said that horses had several strategies available, either to stand and fight or to run fast and escape. Of course in using terms like "strategy" one is being rather anthropomorphic. I think there are both dangers and advantages in being anthropomorphic in thinking about evolution. If you are not at all anthropomorphic you run a great danger of being confined to very simple-minded mechanical models; if you sometimes ask yourself what happens in human populations, you are likely to be stimulated to see more of the possibilities. On the other hand, of course, there is the danger of being too anthropomorphic and implying that the evolving horses made a conscious choice whether to run away or stand and fight. I think the actual mechanism that would really operate in evolution is that certain groups or populations of horses began acting in one way and other groups in other ways, and that there was then a form of selection between the groups as to which was more successful and survived and left further offspring.

WEISS   I want to underline the isomorphism of basic principles in Waddington's presentation and mine. Perhaps in mine there has been a deliberately heavier emphasis on realistic physics rather than on symbolic mathematics, but these are just opposite ends of a continuous spectrum. As for the various questions that have been raised, I would like to discuss only one: the macro-determinacy and micro-indeterminacy which we encounter on each level. And by indeterminate I mean not only that the thing is indeterminable, but that it would be scientifically utterly irrelevant to determine, because it is unique for a particular case in a particular individuum, at a particular instance. Take two parts of a joint. You separate, for instance, the femur from the socket—their rudiments—in the embryo. The head of the femur is convex, the socket is concave. You separate them in an early stage, and grow them separately in tissue culture. Then each part will form its specific shape independently of the corresponding part. But if you try to put them together, they won't fit. In other words there is that much latitude for microadaptation. Phenomena of this kind are quite common, because you are not dealing with a single-tracked linear event in time, but with a network, or reticulation, or whatever you want to call it. And so there are critical phases with some degrees of latitude.

BERTALANFFY   I am sorry that I had to present my paper before Waddington's arrival. As I had more than my share in the discussion of evolution, I believe that I should pass at this late hour, and mention only one item. It seems to me that the concepts of "homeorhesis" and "chreod" are closely related to "equifinality" in open systems. I do not have at present a precise formulation, but believe the point should be looked into.

HAYEK   I am puzzled about a problem in the theory of environmental factors. Sometimes the most important part of the environment is the other members of the same species. Now adaptation may be adaptation to characteristics of other members of the species which the adapting individual doesn't possess. What is the consequence of this? There may be a temporary advantage for one or the other members of the species which still have that characteristic and when it disappears this is an advantage which is due to a passing phase of evolution. It may be that some are more intelligent than others in a particular respect, but if that lack of intelligence in the others has disappeared, it may no longer be an advantage, but it may still be retained. I cannot clearly see where this leads to. Does it lead to a dominance of the characteristic which was an advantage only so long as there were other members of the species who did not possess it? Or does it lead to a curious kind of alternation in the development?

WADDINGTON   There is quite an elaborate mathematical theory on the effects of selection operating on characters whose selective value depends on their frequency. They may be highly advantageous when they are rare, but become less advantageous or even disadvantageous when common. The results arrived at by the mathematics are too complex to be summarized in a couple of sentences. Like all this type of mathematics it is, of course, based on assigning the selective coefficients to the genotypes and not to the phenotypes.

It is possible it might have to be modified if one took the phenotypes fully into account, but there are many contexts in which the omission of the phenotype, although incorrect in basic logic, does not in practice make so much difference; and I am not certain it would make very much difference in this particular connection. The general type of result arrived at is that a gene of this kind settles down to a frequency at which its selective effects are more or less neutral— but that summary is a drastic oversimplification of the case.

THORPE Could this cope with what to my mind is one of the great problems, the preservation and continuance of actions which are individually deleterious, but valuable to the group. One sees so much of this in animal behaviour. Have you got any sort of scheme whereby you can see the way through that? Is this partly the same argument?

WADDINGTON No, I think that question has in general to be dealt with in terms of selection between groups rather than between individuals. I think the only general solution is to suppose that this "evolution of altruism", as J. B. S. Haldane called it, depends on selection between semi-isolated groups within a population.

KOESTLER You mean things like the warning cry of the bird, warning the flock against a predator, and thereby exposing itself to danger. Or the surrender gesture.

THORPE Yes, that would come in. The mobbing action of birds, which for the group is undoubtedly valuable, for the individual certainly dangerous. The bird usually gets away with it, but it's a dangerous thing to do.

KOESTLER For the surrendering wolf to sprawl on its back and expose its jugular vein does not seem to have very great survival value.

WADDINGTON There is an extremely interesting type of selection which people are now beginning to study, although they have not yet got very far. That is selection for the stability of a system; the character selected for is not what is going to happen in the next generation or two, but for the production of a system which is stable enough to deal with difficulties that may arise many generations in the future. For instance, consider ecological systems, in which many species of plants and animals are related by interactions of predator with prey, or parasitism, commensalism and so on, into a complex ecological organization. It is often found that if there are only a relatively small number of different species involved, as for instance in arctic or desert ecosystems, the whole thing may be extremely unstable and suffer from violent fluctuations of numbers. It may indeed fluctuate so largely that at a low ebb some species may disappear entirely, and thus disrupt the entire system of relations on which the ecosystem depends. Ecologists seem to think that there is some selection all the time for increasing the complexity of the ecosystems, at least sufficiently to make them adequately stable.

Some sort of selection for long term stability may also operate on genetic systems. One of the controversies which has been going on in evolutionary genetics in the last few decades was the question of whether evolution produces groups of genes in which positive and negative selection values are linked together to form balanced blocks in the chromosomes, each block consisting of many alternating positive and negative genes, so that the overall effect is pretty neutral. So long as geneticists were confined to analytical algebraical methods of considering such matters, the subject was extremely difficult to deal with, but on the whole the balance of opinion was against this idea. However, recently people have been able to simulate such systems on computers and it now appears to be emerging that selection *will* set up such balanced blocks, not for any immediate advantage they may convey, but rather for the stability they impart to the system— but it may take many hundreds of generations before this stability begins to be effective in determining the way selection acts.

KOESTLER I would like to ask Waddington a few questions; the first relates to a classic experiment of his. As some of you might not know about it, let me explain it briefly, speaking under correction. He exposed Drosophila pupae to intense heat and a certain quite specific change in wing structure occurred as a result. Now after a number of generations, this change, this acquired characteristic in the wing structure, becomes hereditary . . .

WADDINGTON It only becomes hereditary after a number of generations of *selection*: after a number of generations in which you have selected those animals in which the change in wing structure does occur. All the animals in these experiments have had this life experience, but its effects only become hereditary if in each generation you select and use for breeding those individuals that respond most strongly. If you did not make this selection in each generation you can repeat the application of a stressful situation indefinitely, and nothing will happen, and it will not become hereditary.

KOESTLER That narrows it down and qualifies it, but the ambiguity is still there. Your own comment was that it *looks* like a Lamarckian process, but it could be imitation of Lamarckian inheritance by means of a Darwinian mechanism. How would that Darwinian mechanism work?

WADDINGTON The Darwinian mechanism here is that you have

normal Mendelian genes controlling the ability to respond to the environment. You then select those genes which produce the best response to the environment. You then combine this with the fact, to which I referred before, namely that developmental processes have some structural stability, so that once you have got a developmental process going in a certain direction it tends to go on there independently of changes in the environment.

KOESTLER Your own comment, at the time, was: "it seems wisest to keep an open mind on the subject"—on the subject whether this is a Darwinian mechanism, camouflaged as Lamarckian inheritance, or whether it is really a Lamarckian phenomenon. And you further commented that it would be unwise to throw the latter possibility out of court.

WADDINGTON I think that what I said, or at least what I meant to say, was that the mechanism involved in my experiment *is* a Darwinian mechanism. I said in my "Strategy of the Genes" book that it might be wise to keep one's mind open to the possibility that there *could be* Lamarckian mechanisms in other situations. However, I now think this is even less likely than I did then. Our new knowledge about the molecular basis of genetics makes it extremely difficult to see how strictly Lamarckian mechanisms could operate.

KOESTLER I just wanted to clarify this point. My next question concerns the old puzzle of the ostrich's callosities—or the warthog's. The ostrich has a certain ungainly posture of resting on two callosities fore and aft on its undercarriage, and these callosities are just there where the ostrich needs them. But they appear already in the embryo. In other words, the callosities are not caused by friction with the earth, but are inherited. They are inherited and placed exactly where the animal needs them. Now your comment was that if natural selection works primarily in favour of plastic, adaptable behaviour, then the process of canalization during development will become so flexible in itself that it no longer requires any particular gene mutation to endorse the new feature—such as the acquired callosities of the ostrich—but merely "some random mutation to take over the switching function of the original environmental stimulus". Now here I am at the end of my tether, because there seem to be a lot of missing links. How could any random mutation act as a trigger, stabilizing a specific acquired characteristic so that it becomes inherited?

WADDINGTON If you have a pistol with a hair trigger, just chuck it on the floor and it will go off. What I am saying is that you build up a system which has certain possible structurally stable types of development open to it. There are several, but only a few things that it can do (a pistol can only do one thing, it can explode, but a developing system can do one of several things, it has a number of developmental performances all of which it can easily do). Once you have got a system set up like this, then all you need is some little thing to pull the trigger. If you have got the system all ready to produce callosities at the right places, it only needs something to, as it were, pull a trigger and lo and behold it will produce the callosities.

KOESTLER   Oh, that part is clear, but how do you get it into a state where it is ready to develop the callosities just in the right places?

WADDINGTON   Because you selected it for many generations for developing the callosities in the right places when a specific stimulus is given. Natural selection has therefore built up a developmental system which is all set to produce callosities just on the places where the adult ostrich sits down in contact with the hot sand. The action of the environment has been like using your whole hand to cock and fire a pistol—your thumb to pull back the hammer and your forefinger to release the trigger, but selection has eventually produced a pistol which is always cocked, and does not need any action of the thumb to pull back its hammer. All it needs is something to jog the trigger to set it off.

Perhaps I should put that in a rather less figurative way. Let me draw on the blackboard a diagram of the developmental systems of some of the individuals at the beginning of such a process of selection (see figure on p. 364). Their developmental processes can be represented by a series of arrows which will push the developing cells in various directions. In the individuals at the beginning of the selection these arrows will not be much organized, at least in relation to the particular phenotype that the environmental stress is going to call for, and which will be favoured by natural selection. But the environmental stress also acts during development. We can represent this by some dotted arrows. Owing to the general tendency of organisms to respond adaptively to stresses, the epigenetic effects of these environmental factors tend to produce the appropriate end result; but they will do so more frequently and to better effect when they operate on an individual whose own developmental arrows tend to push the cells in the same direction as the environmental factors do. Thus there will be a weeding out of those individuals whose own developmental arrows point in the wrong directions, and a favouring of those in which they point in much the same direction as the environmental arrows. As the favoured individuals of one generation breed with one another to produce the next generation, more and more of these favourably directed developmental arrows will come together, and eventually there will be enough of them to guide development into the right path without any assistance from the environmental arrows.

KOESTLER I'm sorry if I seem stubborn, but I simply cannot see, and a lot of my biologist friends with whom I discussed it are equally unable to see, the following point. The development in the epigenetic landscape is pre-natal; it is completed before the animal is born. Now the callosities were originally acquired by the mature animal. In the environment of the developing embryo, there seems to be nothing that has any relation to the future locus of that callosity. And if this is so, how can the acquisition of the callosity in the right places, by generation after generation *of adults*—how could this influence embryonic canalization?

WADDINGTON The reason I brought this question of the callosities into my discussion of genetic assimilation was because it was one of the most puzzling of the situations which older biologists used as arguments calling for a Lamarckian explanation. I brought it in because I wanted to discuss the difficult cases as well as the simpler ones. Actually it is not a very good example to use as a basis for explaining how genetic assimilation works, because obviously, as you point out, the embryo itself never actually encounters the environmental influences. Presumably in the earlier stages of their evolution the callosities were formed only as a response to the stimulus of sitting on the ground by the young chick soon after it emerged from the egg. The natural selection would be for the response to this stimulus in the early post-egg life of the creature; but once you had got the developmental system set to carry out this particular performance, there might be a lot of latitude in the period of life history in which it actually did so. After all, the basic determinants of the mechanism—namely the selected genes—are present from the time of fertilization onwards. There is nothing to prevent them operating and playing their tune, as it were, at any time after the embryo has developed a skin which can be thickened. It is, I think, a fairly general observation that embryos tend to carry out a developmental process long before the organ concerned will actually come into use. For instance, mammals form kidneys, lungs and so on, at a much earlier period in life than they start using them. I suppose there is a certain factor of safety involved in this, and that is the essential reason why developmental performances tend to get pushed back into earlier stages of life.

KOESTLER   So let me ask my third and last question. It concerns the Baldwin effect, which, I think, was rediscovered independently by you and Alister Hardy . . .

WADDINGTON   Well, to be brutally frank I think that the revival of the Baldwin effect is nothing more than a piece of American nationalism. Actually it is not very clear that the effect was really discovered by the American Baldwin. According to Hardy, in "The Living Stream", p. 164, Baldwin (and Fairfield Osborn) first spoke about it in a discussion following a lecture of the Welshman Lloyd Morgan in 1896. This was, of course, before the rediscovery of Mendelism, and the ideas were discussed in pre-Mendelian terms which makes them nowadays extremely difficult to interpret. The whole idea was almost totally forgotten from about 1902 onwards, until it was revived by G. G. Simpson in 1952, after I had started working on genetic assimilation.

As I have said it is difficult to make out just what Lloyd Morgan and Baldwin did mean. The usual interpretation—for instance by Huxley in his "Evolution, the Modern Synthesis" and Mayr in his "Animal Species and Evolution"—is that their idea was that animals which could make a physiological adaptation to a stressful situation could persist and breed in that environment until such time as mutation had produced an appropriate gene which would bring about the required phenotypic modifications. This is, of course, not the same thing as my genetic assimilation, which involves selection within a population for genes in response to environment. Mayr in fact concludes (his p. 611) "that the conceptual assumptions underlying the hypothesis of the Baldwin effect make it desirable to discard this concept altogether". Hardy in "The Living Stream" pp. 161–170 interprets Baldwin more broadly in a way rather closer to genetic assimilation, but with the main emphasis on the tendency of organisms to select particular habitats to inhabit. I had some correspondence with Hardy about his interpretation, and, if anyone is interested, it has been published in "Theoria to Theory" Vol. 2, p. 226 (Pergamon Press, 1968).

As a matter of fact, if you are looking for forerunners of the idea of genetic assimilation, there is a much more closely related one, the

Russian Schmalhausen. His book "Factors of Evolution" was published in Russian at just about the same time as I first mentioned genetic assimilation in 1942. Indeed in his introduction he refers to my letter and regrets that he did not come across it early enough to deal with it in his text.

KOESTLER What I am driving at is the interpretation which Alister Hardy put upon the Baldwin effect. His point is that instead of passively responding to environmental stresses, the animal takes the initiative, develops a new habit, and this habit spreads at first by imitation. He gives a very pretty example of this—the tits and the milk bottle. One genius blue tit once discovered that you can pierce the cap of the bottle which the milkman left in front of your door and drink the milk and apparently all the tits in Europe soon learnt this trick. Thus, Hardy argues, the first step towards an evolutionary novelty is the functional initiative of the animal. Then sooner or later a mutation may turn up which so to speak gives its blessing, its genetic sanction to this new function. So here function precedes structure in the genetics of behaviour. And that problem goes all the way back to the question of the Galapagos finches. Could you comment . . . ?

WADDINGTON The general idea that the first step towards a new evolutionary change is for the animal to acquire a new habit, or a new mode of behaviour, is one of the fundamental ideas advanced by Lamarck. It goes back long before the Baldwin effect. I am sure it is a very important point, though not an absolutely general one, since it is obviously difficult to apply it to plants. I tried to make the point this morning when I spoke about horses having a number of different strategies open to them, to run away or to stand and fight. The introduction of considerations about behaviour into evolutionary theory is, I think, very important, but it is very unexplored logically. It raises difficult problems because it introduces another sort of indeterminism, in the sense that the behaviour of the animal quite largely determines what kind of natural selective pressure it is going to be subjected to, so that you have a circular system in which the environment through natural selection determines the genotype, and the genotype by its influence on the behaviour determines what the environment shall be. This is a very important consideration, but it is rather independent of the genetic assimilation work: in my Drosophila experiments for instance I simply assumed that the flies were living in a very high temperature, but did not go into the question of whether in nature they would actually choose to do so or not.

KOESTLER So in your Drosophila what we have is still a passive adaptation, as it were, to an imposed stress, whereas in Hardy's example of the tit the animal takes the initiative. And that of course, from a philosophical point of view is a very attractive idea. And it has an additional advantage. When the animal discovers a new habit, a useful new habit, then the proverbial monkey is no longer hammering wildly away at individual keys on the typewriter, producing random mutations, but the monkey is now working on a pre-programmed typewriter. In other words, since the whole habit is functionally already there, it is sufficient for the monkey to hit just one key to trigger off its hereditary stabilization, or genetic assimilation as you call it. So that you have got here a whole hierarchy of processes which reduce the role of the random mutation to that of a mere trigger. The monkey's typewriter is programmed.

WADDINGTON Yes, I think that in the evolution of higher organisms new mutations are usually only of importance when they act as triggers. Of course mutation also provides the mixed bag of enzymes which the organism has at its disposal and which are, to use my previous analogy, like the pebbles in the stream bed out of which you are going to make concrete.

KOESTLER If this point is agreed on, then it is sheer nonsense to say that evolution is "nothing but" random mutation plus natural selection. That means to confuse the simple trigger with the infinitely complex mechanism on which it acts.

WADDINGTON Yes.

WEISS The idea that "anything goes", that a mutation can do anything, grow wings on a monkey, is absolutely unrealistic. We must realize that there is an extremely limited number of choices, of evolutionary repertories available. That's what bothers me.

KETY One afterthought . . . Reflecting on the original invitation that called this group together, I remember that Arthur quoted Thorpe as saying that there was a great deal of current dissatisfaction with the Darwinian concepts of evolution. And that there was a feeling that random mutation and survival of the fittest did not explain all of life. Reflecting on that I felt a little disappointed that a case for that critical position had not been presented. Arthur tried to do it . . .

BERTALANFFY I did, surely . . .

KETY Well, I thought you began to do it, but unfortunately there was not enough confrontation between you and Professor Waddington. Because I didn't find that Dr. Waddington's position was really seriously challenged. Not that I don't agree with his position, but I would have liked to have a thoughtful good case made for the opposition which Dr. Thorpe indicated was currently rampant. Anyway, after this exchange between Arthur and Dr. Waddington, I have only a very trivial question. I was fascinated by the truism you enunciated that it is the phenotype which competes with the environment, and that it is the genotype which is transmitted. It is one of those self-evident things that becomes self-evident only when someone has pointed it out. And I wanted to ask you whether you would comment upon the impact of environmental variability in the phase of epigenesis as contrasted to the phase of survival, of competition, of the phenotype. Is it possible that in consequence of that consideration, one could make an evolutionary case for a homeostasis in the epigenetic period in contrast to a very wide-spread change in environment during the other period?

WADDINGTON The environment comes into the story of evolution in two ways which are logically distinct, though they may be occurring at the same time. On the one hand the environment is important as partially determining the phenotype, in cooperation with the genotype. Secondly the environment is important in exerting natural selection. These two operations may be quite different. For instance an annual plant grows up in the weather of one year, but most of the selection on it may go on next year when its seeds germinate in quite different conditions. On the other hand, in other circumstances the epigenetic and selective effects of the environment may not be separated but very much mixed up together.

When you have heterogeneous environments, varying both in space and in time, there are two major strategies in evolution. One of them is to canalize, to make your epigenetic processes very highly buffered and homeorhetic, with a great deal of structural stability, so that you finish up with much the same end-result whatever the circumstances. A good example of this is the mouse. If you catch mice which live in cold storage depots, and compare them with mice living in the high tropics, the mice from the cold storage have fractionally shorter tails and fractionally thicker fur, but that is about the only modification made by this enormous temperature difference. On the other hand, the opposite strategy is to have an extremely flexible development, although of course it is necessary to ensure that the flexibility produces always a rather well adapted phenotype. As an example of this you can take many little crustacea of the kind that live in shallow water pools. With sufficiently refined measurements of body proportions and so on, the population of almost every pond can be distinguished from the population of any other pond. Their development is extremely flexible in relation to temperature, the saline content of the water and so on. These are the two extreme poles of strategy. On the whole the stabilization one seems to be the most frequent, since most animals vary phenotypically remarkably little in relation to quite large differences in the environment.

KETY　Can you demonstrate that mathematically? Logically? It seems to me that with that very nice paradigm you ought to be able to come to an operational demonstration that homeostasis in the epigenetic phase and a maximum environmental heterogeneity in the survival phase would be most advantageous from an evolutionary point of view.

WADDINGTON　You can certainly describe the situation mathematically. I have done a little mathematics of a very simple kind, in which I separated the environment's action on development and its action in exerting natural selection. I produced a paradigm case that is almost as simple as Haldane's, but mine does involve two environments operating in these two different ways. I showed that under some circumstances it would pay better to canalize, in others it would pay better to be flexible. This could certainly be enormously elaborated into much more impressive algebra, and I dare say it will be, but I doubt if I shall do it myself.

KOESTLER　You have said, I think in *The Strategy of the Genes*—that although Lamarckism is a highly unlikely hypothesis, an unlikely mechanism, we must nevertheless keep an open mind about it, and you have even designed a tentative model of how Lamarckism could work. Could you repeat it for our benefit?

WADDINGTON　That was a very speculative and schematic idea, and was written some time ago. To discuss it nowadays one would have to think of it in terms of the sequences of nucleotide bases in DNA. When I wrote it, it was still possible to think that the sequences of amino acids in proteins might be almost as important as the DNA in determining the nature of the gene. My suggestion was that, since we know that the shape of enzymes may be altered, by interaction with their substrates, for instance, one could conceive of the possibility that an enzyme might distort the template on which it is made in such a way as to produce a new template which would later form an appropriate enzyme. However, such an idea is much more difficult to reconcile with what we now know about the constitution of genes and the DNA which makes the essential templates, We still think that the structure of proteins is to some extent modifiable. However, it is just possible that the structure of DNA is not quite so inflexible as we now think. When I wrote that passage I said that I did not want to shut my mind to the possibility. What I am saying now is that I should shut it a good deal more firmly, although even now I would leave just a tiny crack open.

KETY Isn't there at least one theory about specific antibody formation which relies on processes of this kind?

WADDINGTON Yes, of course specific antibody formation is a notoriously difficult case.

WEISS I want to come back to the question how a change in a single gene, could lead to such a tremendously complex coordinated end result . . .

WADDINGTON I agree in general, but there *are* some processes in which single genes are important, particularly in the prebiotic stages of evolution, when life was originating and the challenge was to find some sort of protein that could combine with oxygen or act as an enzyme. The sort of process I was envisaging—if it could occur at all—would have been particularly useful at those stages.

HYDEN In 1958 I wrote a rather lengthy paper for the Congress of Biochemistry in Vienna about an instructional mechanism which is a slight modification of what you said, that electrical fields in the enzyme system could replace one base for another one and lead to a slightly different protein, which then could effect a change in the synapse and serve a memory mechanism. I wrote this hypothesis as an instructional mechanism. But I think this question must be left open because in the discussion here on the first day it was brought out that hysteresis phenomena, for example, could work in RNA, so it still is a possible thing.

The other thing I wanted to say is that geneticists are presently living in a hazardous time, because the definition of mutant, mutation, how to define a mutant is becoming a difficult problem. Two observations lead to such a difficulty in defining a mutation in the short-term range: one is that only a certain part of the genome is active, and the other is that there can be gene activation by external factors. Well-known studies deal with hormone effects on cells and on the end products. So now you have a link between a change in behaviour which leads to an additional activation of gene areas, you have a phenotype change and this goes on for as long a time as the new behaviour stays there, but it's not like a mutation where you can take away the environment—the change goes back again. If you are dealing with an organism with a limited life-time and you stop the experiment and you do not perform control experiments with respect to the environment, then it is easy as far as I can see to confuse mutation and inactivation. Is that so?

WADDINGTON   I think so. I think we are going to see extraordinary changes in our ideas about evolution quite soon. I should like to mention two things. One arises from what Hydén was talking about earlier this morning, namely gene iteration. I think I mentioned then that Forbes Robertson in our laboratory has been studying the iterated sequences of DNA in two very closely related species of Drosophila; Drosophila melanogaster and Drosophila simulans, which look so alike that it takes a very clever taxonomist to tell them apart by just looking at them. Their comparative genetics have been studied for a long time, and people thought they differed by just a few gene mutations and a certain shuffling of the order of the sequences on the chromosome. However, when you look at their DNA, particularly their reiterated DNA, you find a really extraordinarily large difference. Now what this means, why it has happened, how it has happened, we just have not yet the beginning of an idea.

THORPE   So isn't this in effect saying that the gene pool is changing while the species remains essentially the same?

WADDINGTON   Yes, and probably quite a lot of this goes on. For instance, the rodents have been studied and we know there are considerable differences in reiterated sequences between the mouse and the rat, and some species of woodmice and the guinea pig; but those are fairly widely separated species which evolved apart quite a long time ago. This study of Drosophila is the first case of two very closely related organisms.

THORPE   But isn't this going to raise great difficulty for the concept of homology?

WADDINGTON   Homology is a difficult enough concept anyway to talk about in genetical terms. Personally I rather doubt whether it retains much value. The really difficult question here is to see what sort of selection process could possibly be causing these enormous differences in the reiterated DNA.

I wanted to mention a second thing which it seems to me may cause a revolution in our evolutionary ideas. If we take ordinary tissue cells, for instance of a mouse, and grow them in culture they usually only grow well for a certain limited number of generations, perhaps 80–100, and then gradually peter out, *unless* they undergo some sort of crisis which is spoken of by cell biologists as "becoming an established strain". If they do this, then they are capable of indefinite propagation in tissue culture for ever. Now becoming an established strain seems to happen quite quickly, certainly over a few cell generations. It often involves very large genetic changes, possibly the loss of a few chromosomes, so that cells have an abnormal chromosome number, or again some of the chromosomes may be enlarged or reduced or have abnormal shape. There seems to have been very considerable shuffling around of the genetic material taking place in a few cell generations.

Nobody knows what happens genetically when the cells turn into an established strain, but it looks like a different sort of mutational event from the ones we normally come across. It is, of course, a question whether the results of this sort of mutation could be propagated in evolution, but if we do have here a new type of mutation, as appearances suggest, this may turn out to have some relevance to large-scale evolutionary changes.

THORPE Has this any bearing at all on the problem I just mentioned to you—namely the extraordinary constancy of some groups and species? It seems to me as being outside the existing system of genetic theory. It appears to be more difficult to explain the constancy of certain groups than it is to account for their evolution. Every time one reads about the discovery of some new fossil evidence (and one seems to do so about every other week nowadays), one finds that the probable origin of the group has been put back by a step of about a million years or so! To take a homely example: that Wagtail (*Motacilla*) there in the garden was here before the Himalayas were lifted up! This constancy is so extraordinary, that it seems to demand a special mechanism to account not for the evolution but for the fixity of some groups.,

WADDINGTON I think that to some extent the answer to this problem is to be found in selection for developmental canalization, so that the group has a pretty standard phenotype and does not show much variability. If a species has been selected for a very well canalized developmental system, then there will be few gene mutations that can produce modifications of the phenotype. If you had such a species and it lived in some sort of environment, such as grassland or forest, which will always occur somewhere on the earth's surface, then there seems to be no particular reason why it should change over very long periods of time.

THORPE Another question arose the day before yesterday which I wished you had been here to discuss. It was the problem of this tendency to greater complexity which seems to be operating widely in the animal world and is relatively lacking in plants. I suggested that this must have something to do with the fact that animals are mobile and plants sedentary. Would you care to say anything about that?

WADDINGTON I should agree that the greater complexity of animals as compared with plants is probably related to their greater mobility. I think it is also derived from the fact that animals more often utilize other animals as part of their living space; that is to say foxes eat rabbits, but one plant does not usually live off another plant, except for relatively few parasitic species. Plants such as epiphytes do of course live on other plants, utilizing them as a platform on which to grow, but a really active interaction between two species is more usual in the animal world.

THORPE To make a mere off-the-cuff suggestion—could you envisage a genetic mechanism which would account for a general tendency towards increasing complexity? And I wonder whether such a mechanism which you may have in mind would operate equally in plants and animals or not?

WADDINGTON No, much more in animals I would say, because if you have an ecosystem involving several different species it is always possible to imagine evolving a new species of animal which utilize several of them. Once you have snails in the streams and men walking into streams to bathe you can get trematode parasites that live half their lifetimes in the snail and the other half in man. An animal which exploits other species can always go one degree better than the existing system. I think that is the general explanation for the increase in complexity in the animal world. It does not mean, of course, that this tendency is quite general and everything has to get more complex. It only means that as evolution goes on the most complex animals existing are likely to be more complex at later stages than they were at earlier stages; but it is certainly not a universal tendency.

REFERENCES

CAIRNS SMITH, A. G. (1966) "The Origin of Life and the Nature of the Primitive Gene". *J. Theoret. Biol. 10*, p. 53 (1948) "An Approach to a Blue Print for a Primitive Organism" in *Towards a Theoretical Biology* (ed. Waddington), Vol. 1, p. 57. Edinburgh University Press and Aldine Press, Chicago.

WADDINGTON, C. H. (1969) *Towards a Theoretical Biology*. Edinburgh University Press and Aldine Press, Chicago.

———— (1968a) *Symposium on Population Biology*, Syracuse University Press, N.Y.

———— (1968b) in *Towards a Theoretical Biology*, 1, p. 13.

Reductionism and nihilism
Viktor E. Frankl
University of Vienna

The overall title chosen by Arthur Koestler for this symposium reads
"New Perspectives in the Life Sciences",[1] using sciences in the plural.
We are confronted with this pluralism and challenged by the question
how to maintain or to restore a concept of man that does justice to the
humanness of man and more specifically to the one-ness of the human
person. This question in turn boils down to the problem how to
maintain, or to restore, a unified concept of man in the face of the
scattered data, facts and findings as they are furnished by a thoroughly
compartmentalized science. The pictures we obtain today from the
various individual sciences are very disparate. These pictures differ
from each other so much that it is becoming more and more difficult
to arrive at a unified world-view. But it should be pointed out that
such differences, *per se*, need by no means constitute a loss in knowledge. On the contrary, such differences may well make for a gain in
knowledge—just consider stereoscopic vision. There is a difference
between the right and the left pictures that are offered to you. But
it is precisely this difference that mediates the acquisition of a new
wholeness, of an additional dimension, the third dimension of space.
To be sure, the precondition is that we achieve a fusion between the
picture on the right and on the left. And what holds for vision is
also true of cognition: unless we obtain a fusion, confusion may be
the result.

Now the pluralism of science is reflected in the individual scientist
by an increasing trend towards specialization. We could say that ours
is an age of specialists; and alas all too often the specialist could be

[1] This was the original title suggested for the symposium. Eds.

defined as a man who no longer sees the forest of truth for the trees of facts. But the wheel of history cannot be turned back, and so society today cannot do without specialists, because too much research is based on teamwork, and in teamwork the specialist is simply indispensable. However, I for one think that the present danger does not really lie in the loss of universality on the part of the scientist, but rather in his pretence and claim of totality. That is to say when a scientist who is an expert in the field of biology attempts to understand the phenomena of human existence in exclusively biological terms, he has fallen prey to biologism. And at the moment biology becomes biologism, science is turned into an ideology. What we have to deplore therefore is not so much the fact that *scientists are specializing*, but rather the fact that *specialists are generalizing*. That is to say indulging in over-generalized statements. We have for long been familiar with the *"terrible simplificateur"*; but now we meet more and more frequently another type, the *"terrible généralisateur"*. He tends to turn biology into biologism, sociology into sociologism and psychology into psychologism.

But let me confine myself to the phenomenon of psychologism. Psychologism is frequently combined with something I once called pathologism. That is the tendency to go all out to detect neurotic flaws and to discover sexual symbols and so forth. Let me offer two quotations from Sigmund Freud—I am quoting from the book by Fabry, *Logotherapy applied to Life* (1968). On page 160 he is quoting Freud to the effect that parental love is narcissism born again; and friendship is a sublimation of homosexual attitudes. Now Freud himself had been wise and cautious enough to remark once that sometimes a cigar may be a cigar, and nothing but a cigar. However, his epigones are less cautious and feel less inhibited in this respect. One of them, a famous Freudian analyst, has recently published a book on Goethe. Let me quote from the review of the book by J. Heuscher:[2] "In the 1,538 pages the author portrays to us a genius with the earmarks of a manic-depressive, paranoid and epileptoid disorder, of homosexuality, incest, voyeurism, exhibitionism, fetishism, impotence, narcissism, obsessive compulsive neurosis, hysteria, megalomania, and so forth. The author seems to focus almost exclusively upon the instinctual dynamic forces that underlie the artistic products. We are led to believe that Goethe's work is but the result of pre-genital fixations. Goethe's struggle does not really aim for an

[2] In the *Journal of Existential Psychiatry* 5, p. 229, 1964.

ideal, for beauty, for values, but for the overcoming of an embarrass-ing problem of premature ejaculation. These volumes show us again" the author of the review concludes, "that the basic position of psycho-analysis has not really changed."

Small wonder if this state of affairs takes its toll; only recently Laurin John Hatterer, a Manhattan psycho-analyst, pointed out in a paper that "many an artist has left a psychiatrist's office enraged by interpretations which suggest that he writes because he is an 'injustice collector', or a sadomasochist; acts because he is an exhibitionist; dances because he wants to seduce the audience sexually, or paints to overcome a strict bowel training by free smearing". I could offer you a collection of such examples. I would suggest that the unmasking of motives is justified, perfectly justified, but it must stop where the man who does the unmasking is finally confronted with what is genuine and authentic within a man's psyche. If he does not stop there, what this man is really unmasking is his own cynical attitude, his own nihilistic tendency to devaluate and depreciate that which is human in man.

It is an inherent tendency in man to reach out for meanings to fulfil, and for values to actualize. But, alas, we are offered by two out-standing American scholars in the field of value-psychology the following definitions: "Values and meanings are nothing but defence mechanisms and reaction formations." Well, as for myself, I am not willing to live for the sake of my reaction formations, even less to die for the sake of my own defence mechanisms. And I would say that reductionism today is a mask for nihilism. Contemporary nihilism no longer brandishes the word nothingness; today nihilism is camou-flaged as *nothing-but-ness*. Human phenomena are thus turned into mere epiphenomena. If you allow me a brief digression, I would say that, contrary to a widely held opinion, even existentialism is not nihilism; the true nihilism of today is reductionism. Although Jean-Paul Sartre has put the word "néant" into the title of his main philo-sophical work, the true message of existentialism is not *nothingness*, but the *no-thingness* of man—that is to say *a human being is no thing*, a person is not one thing among other things.

And that brings up the problem of the impact that reductionism might have, particularly on the young generation. I well remember when I was a Junior High-School Student, how our science teacher used to walk up and down the class explaining to us that life in its final analysis is nothing but combustion, an oxidation process. In this case reductionism took actually the form of oxidationism (laughter). On one occasion I jumped to my feet and asked him: "Dr. Fritz, if this is true, what meaning, then, does life have?" At that time I was twelve. But now imagine what it means that thousands and thousands of young students are exposed to indoctrination along such lines, taught a reductionist concept of a man and a reductionist view of life.

This situation accentuates a world-wide phenomenon that I consider a major challenge to psychiatry—I have called it the existential vacuum. More and more patients are crowding our clinics and consulting rooms complaining of an inner emptiness, a sense of total and ultimate meaninglessness of their lives. One should not assume that this state of affairs is confined to our western civilization; several publications have come out in recent years from behind the iron curtain. Thus, for instance, Stanislav Kratochvil, a Czech psychiatrist, has published several papers on the existential vacuum. His contention is that this experience is by no means restricted to the capitalist countries, but is more and more noticeable today in Eastern Europe as well. I was invited to lecture at the Karl Marx University in Leipzig, in Czechoslovakia and in Poland. During the question and answer period it transpired again and again that really the same problems are confronting our psychiatric colleagues in communist countries.

To attempt a short explanation of the causes of this phenomenon, I would say that, in contrast to animals, man is not told by drives and instincts what he *must* do. Nor, in contrast to man in former times, is he any longer told by traditions and values what he *should* do. Sometimes he does not even know what he basically *wants* to do, but instead he just wants to do what other people are doing—which is conformism. Or else he just does what other people want him to do—which is totalitarianism. But in addition to conformism and totalitarianism we can observe a third side-effect of the existential vacuum, and that is neuroticism. I am referring to a new type of neurosis that I have termed noogenic neurosis, in heuristic contrast to the conventional type of psychogenic neurosis. And this new type of neurosis, this noogenic neurosis mainly results from the existential vacuum, from existential frustration.

The director of the psychology laboratory at the Veterans Administration Hospital in Gulfport, Miss., James C. Crumbaugh, has taken some pains in developing a special test, which he calls the PIL-test, or purpose-in-life test. He has tried it to date on 1,152 subjects. The data have been computerized, and the conclusion that seems to have emerged was that noogenic neurosis is indeed a new type of neurosis that can be differentiated by tests from the conventional types of neurotic illness. There is also agreement between various authors in various countries, that about 20% of neuroses today are noogenic by nature and origin.

We may define the existential vacuum as the frustration of what we may consider to be the most basic motivational force in man, and what we may call, by a deliberate over-simplification, *the will to meaning*—in contrast to the Adlerians' will to power and to the Freudians' will to pleasure. Two days ago Dr. Hyman, a brain surgeon from California, delivered a paper to the Austrian Society for Medical Psychotherapy, of which I am chairman. He said that he was again and again confronted with patients whom he had completely relieved from intractable pain by stereotactic brain surgery, and who then said to him: "Doctor, I am free from pain—but now more than ever I ask myself what the meaning of my life is, because I know that life is transitory, particularly in my situation"—and so on. So people do not (as most current motivational theories seem to assume) care so much for pleasure and avoidance of pain, but they do care for meaning. And that is why I am speaking of the will to meaning.

Let me just add that if you attack my theoretical position by saying that psycho-analysis also recognizes something called the reality principle, this doesn't alter the scenery, because according to Freud's own statements, the reality principle is in the indirect service of the pleasure principle. But we could even go a step further by saying that the pleasure principle itself serves a more general principle and this is the homeostasis principle—the principle of tension-reduction on which most of the current motivational theories are still based. Now we have learned long ago from Ludwig von Bertalanffy that the homeostasis principle is no longer tenable as an over-all law within biology; and even less is this the case in psychology. We are in this respect indebted to Kurt Goldstein's brain pathology, because Goldstein arrived at the conclusion that the homeostasis principle is actually an indication of disease, but not of a normal state. And within psychology we are in this respect indebted mainly to Gordon W. Allport, Charlotte Bühler, and Abraham Maslow. Now actually the motivational theories that still stick to the homeostasis principle are, by what they imply, true monadologies. They only know a closed system. Man is depicted as a being primarily and basically concerned with his inner equilibrium or something within himself, be it pleasure or anything else. But actually I dare say that being human—I anticipated this yesterday—being human is always pointing beyond itself, is always directed at something, or someone, other than itself. Be it a meaning to fulfil or another human being to encounter. And actually man is not concerned, not primarily concerned with pleasure or with the so-called pursuit of happiness. Actually—due to his will to meaning—man is reaching out for a meaning to fulfil or another human being to encounter; and once he has fulfilled a meaning or has encountered another human being, this constitutes a reason to be happy. And once there is a reason for it, happiness ensues automatically. Whereas, if this normal reaching out for meaning and beings is discarded and replaced by the will to pleasure or the "pursuit of happiness", happiness falters and collapses; in other words, happiness must ensue as a side-effect of meaning-fulfilment. And that is why it cannot be pursued, because the more we pay attention to happiness, the more we make pleasure the target of our intentions, by way of what I call hyper-intention, to the same extent we become victims of hyper-reflection; that is to say the more attention we pay to happiness or pleasure, the more we are losing sight of the primary reason for happiness, and blocking its attainment; happiness vanishes, because we are intending it, we are pursuing it. This makes it impossible for fulfilment to ensue and as I also indicated yesterday, we observe this phenomenon in about 95% of sexual neuroses. Whenever a male patient is trying deliberately to manifest his potency, or a female patient to demonstrate her ability to experience orgasm, the very attempt is doomed to fail. (In therapeutic practice we are counteracting this hyper-reflection and hyper-intention by the technique of de-reflection particularly in cases of sexual neuroses).

Now what holds for pleasure and happiness also holds for self-actualization. Self-actualization is a good thing; however, we can actualize ourselves only to the extent to which we have fulfilled a meaning, or encountered another human being. But we have no longer any basis for self-actualization at the moment we are striving directly for it. We could epitomize this state of affairs by quoting first Pindar and then Karl Jaspers. Pindar said you should "become what you are". And Jaspers once said "what a man is he becomes only through that cause which he has made his own". In other words we may obtain self-actualization by living out the self-transcendent quality of human existence. You remember yesterday I pointed to one aspect of this self-transcendent quality that permeates human existence: the capacity of seeing is dependent on the incapacity of the eye to see itself.

Reductionism is more than just saying time and again that something is nothing but something else. It is an approach and procedure that deprives the human phenomena of their very humanness by reducing a human phenomenon in dynamic terms to some sub-human phenomenon, or deducing human phenomena, in genetic terms, from sub-human phenomena. In other words we may well be justified in defining reductionism as sub-humanism as it were. Take the two most definitely human phenomena: love and conscience. They are among the most important and significant human phenomena in so far as love can be defined as the capacity to grasp another human being in its very uniqueness, while conscience is the capacity to seize the unique meaning of a situation, and each situation implies unique meaning. Now love, to the reductionist, is derived from sex; it is conceived as a sublimation of sexual instincts or, as Freud has put it, "goal-inhibited" sexuality. And conscience is reduced to the mere super-ego. Both views are erroneous. For if sublimation is to take place, it presupposes the capacity for loving, because ultimately only for the sake of a person whom I love am I capable of integrating my own sexuality into my personality as a whole. If the ego is to integrate its own *id*, it must in the first place be lovingly directed to a thou. So what is the pre-condition of sublimation cannot be the result of sublimation. As to conscience: if one attempts to reduce conscience to the super-ego, one disregards the fact that conscience often opposes, if need be, the conventions and standards, traditions and values, which are transmitted and channelled by the super-ego. So, if conscience is able to contradict and oppose the super-ego, it cannot be identical with the super-ego.

Some years ago, Konrad Lorenz, in a lecture he gave to the same Austrian Society for Medical Psychotherapy, spoke about *Moralanaloges Verhalten bei Tieren*, animal behaviour that is analogous to moral behaviour. But he was stressing the limitations of the analogy, because if one says that a dog that has wetted a floor and slinks under the couch with its tail between its legs, obviously manifests a bad conscience, this is a naïve error. I would say the dog is manifesting anticipatory anxiety, the fearful expectation of punishment. However, true conscience has nothing to do with the mere fear of punishment or longing for reward. Those people, those behaviourist psychologists or whatever they call themselves, who insist that what is observable in a man must also be observable in animals, and who wish to derive from this attitude a justification for reductionism or subhumanism, these people remind me very much of the rabbi in a Viennese joke. There were two neighbours; one of them contended that the other's cat had stolen and eaten five pounds of his butter; there was a bitter argument and finally they agreed to seek the advice of the rabbi. They went to the rabbi and the owner of the cat said: "It cannot be, my cat doesn't care for butter at all", but the other insisted that it was his cat and so the rabbi decided: "Bring me the cat." They brought him the cat and the rabbi said: "Bring me the scales." And they brought the scales and he asked: "How many pounds of butter?" "Five pounds." And believe it or not, the weight of the cat was exactly five pounds. So the rabbi said: "Now I have the butter, but where is the cat?"

This is what these people remind me who say in fact: we have some type of conscience in the animal as well; so whatever we find in humans may be found in animals as well. And then they ask: we have this behaviour in man, we have these "innate releasing mechanisms" and so forth, but where is man? The humanness of the human phenomena has necessarily disappeared. Or if you turn from the rat model to the machine model, you may come across a recent book in which man is defined as "nothing but a complex biochemical mechanism powered by a combustion system which energizes computers with prodigious storage facilities for retaining encoded information".

Now I hope you won't misunderstand me. As a professor of neurology I declare that it is perfectly legitimate to use the computer as a model for certain activities of the central nervous system. The mistake is exclusively in the phrase "nothing but". In a sense, in a way, man *is* a computer, but at the same time he is infinitely more, or let me say dimensionally more than a computer. If you imagine a cube that is constructed on the basis of a square, you are justified in saying that in a way the cube is also a square. The square is contained and included in the cube. It serves as its foundation and basis. However, if we say the cube is nothing but a square we have then been reducing, removing, shutting out a whole dimension, the third dimension.

In other words, reductionism is a kind of projectionism. It projects human phenomena into a lower dimension. It must be counteracted by what one might call dimensionalism (I have termed it dimensional anthropology) in order to preserve the one-ness and humanness of man in the face of the pluralism of the sciences. Pluralism after all is the nourishing soil on which reductionism flourishes. Dimensional anthropology uses the geometrical concept of dimensions as a model, as an analogy. In other words dimensional anthropology is an *imago hominis*—to allude to Spinoza's ethics, "*ordine geometrico demonstrata*". Dimensional anthropology rests on two laws (Blackboard). The first law as I conceive it, reads as follows:

A phenomenon, say, a cylinder, projected out of its own dimension—i.e., three-dimensional space—into different dimensions lower than its own, for example into horizontal and vertical planes, yields pictures that are *contradictory* to one another. Here we get a circle

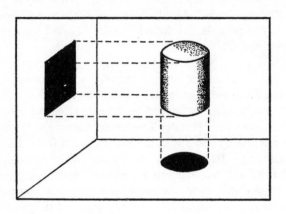

FIGURE I

and here a rectangle. And this is an evident contradiction. What is more, if you imagine this cylinder is to represent a tumbler, an open vessel, the openness of this vessel completely disappears in the projections into the lower dimensions. The circle as well as the rectangle are closed figures rather than open vessels.

Second law of dimensional anthropology: different phenomena, e.g., a cylinder, a cone and a sphere, projected out of their own dimension into a dimension lower than their own, result in pictures that are *ambiguous*. The shadows of these different spatial figures, are equal, interchangeable, you can never infer what it is that has cast the shadow (see Figure 2).

Now, how can we benefit from these analogies, from these two laws of dimensional anthropology, in the science of man? If you project a human being into a purely biological frame of reference, and/or into the frame of psychology, then in the first case you obtain somatic data, while in the second you obtain psychic data. There is again a contradiction. What seems to be even more important: there was an open vessel and this is depicted as a closed system. Now we know that human

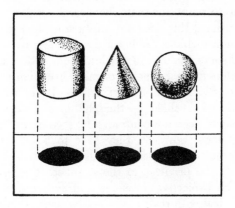

FIGURE 2

existence also is characterized by its intrinsic openness: this has been evidenced by Max Scheler in his anthropology, by Arnold Gehlen, the sociologist, by Adolf Portmann, the biologist. I referred to this intrinsic openness of human existence by the term "self-transcendent quality": it is pointing beyond itself rather than being a closed system. Now we know, thanks to the teachings of Ludwig von Bertalanffy, that man, even in purely biological terms, is not a closed system either; but in biology, as in psychology, he is of necessity represented as if he were a closed system of physiological reflexes or psychological responses to various stimuli. The openness of human existence, once we have projected man out of his own dimension into a dimension lower than his own, necessarily disappears.

Now we are still far from solving the body-mind problem, we are far from bridging this gap between somatic and psychic data. But at least, in the light of dimensional anthropology, these apparent contradictions no longer contradict the oneness of man. The apparent contradiction between the circle and the rectangle does not invalidate the fact that they are projections from the same cylinder. We cannot solve the body-mind problem, but we can at least explain why it is unsolvable. You remember, art has been definied as unity in diversity; I would say man could be defined as unity in spite of diversity. But the *coincidentia oppositorum*, as Nicolas Cusanus called it, is lost when we project man out of his own dimension. His unity cannot be found within the lower dimensions; it must be sought in the human dimension. And the same holds for the openness of human existence.

I would like to point out that whenever I speak of "higher" and "lower" dimensions, this does not imply any value judgement. The higher dimension includes the lower dimensions; by the same token the sound and sober findings in any lower dimension, are not invalidated by the dimensional concept of man. Freudian psychoanalysis, Watsonian or Skinnerian behaviourism, Pavlovian reflexology, Adlerian individual psychology—though they may neglect the human dimension as such, need not contradict it; seen in the light of dimensional anthropology, however, they have to be re-interpreted, re-evaluated. In other words, these findings have to be re-humanized.

It is not only the privilege of the scientist, it is also his duty to embark on such projective measures and procedures. It is his responsibility to deal with reality as if reality were unidimensional; to neglect the multidimensionality inherent in reality. However, the scientist should also remain aware of the pitfalls of his method; he should be aware of the second law of dimensional anthropology. Let me give you an example out of my own field of research. Replace the three shadows by neuroses. Yesterday I briefly referred to the evidence that there are not only conventional psychogenic neuroses, but also somatogenic neuroses. I described agoraphobias that are due to hyperthyroidism. And I could also show that there are cases of claustrophobia that could be traced back to tetanoid disturbances of metabolism. There are depersonalization states that could be understood and therapeutically dealt with as hypofunctions of the adreno-cortical glands. But there are not only somatogenic and psychogenic neuroses; there are also those noogenic neuroses I mentioned before: neuroses which cannot be traced to oedipal situations, or maladjustments and so forth, but which derive from spiritual problems, from moral conflicts, from conflicts between one's true conscience and the mere super-ego; and lastly, from existential frustration, from the despair of man over the apparent meaninglessness of his life. The etiology of a neurosis is multidimensional. A neurosis might be somatogenic, psychogenic or noogenic. And to the extent to which the causation of a neurosis is multidimensional, to the same extent its pathology is ambiguous—much the same way as the shadow, of which you could not tell whether it was cast by a cylinder or a cone or a sphere. Likewise, as long as you confine yourself to the plane of psychiatry, pure psychiatry that is, you are unable to distinguish diagnostically, between what is ultimately either hyperthyroidism, or castration fear, or else the existential despair of an individual over the seeming meaninglessness of his life. Pathology is ambiguous in that the *logos of pathos*, the meaning of suffering may hide in a dimension different from the dimension in which symptomatology dwells. If suffering is to become transparent as to its meaning, we have to transcend the level of mere symptoms, the plane of pure psychiatry.

And what holds for theory also holds for therapy. A few years ago, at the Montreal Meeting on psychopharmacology, several speakers expressed their apprehension that by introducing drug treatment in addition to shock treatment, psychiatry might become dehumanized and the patient himself depersonalized. Now in my department at the Polyclinic, we are using drugs, and using electroconvulsive treatment; I signed authorizations for some lobotomies without having cause to regret it; in a few cases, I even carried out transorbital lobotomy. However, I can promise you that the human dignity of our patients is not violated in this way. On the other hand I do know psychotherapists who would abhor giving a prescription for a drug, or giving shots or shocks, and yet by their depersonalizing approach to the patient, based on a reductionist picture of man, they violate the human dignity of their patients.

What matters is never a technique or therapeutic approach as such, be it drug treatment, be it shock treatment, but the spirit in which it is being carried out. Professor Petrilowitch (at present head of the Psychiatric University Clinic in Mainz) has pointed out recently, that logotherapy, in contrast to all other schools of psychotherapy, transcends the dimension of neurosis itself. If you consider psychoanalytic therapy, you find a pathogenic kind of psychodynamics which the therapist tries to counteract, by means of, say, a transference relationship. And if you think of so-called behaviour therapy— reflexological therapy—think of Eysenck, Wolpe and others—there the neurosis is considered to be the outcome of pathogenic conditioning processes and accordingly it is dealt with by introducing re-conditioning processes—progressive desensitization and the like; but all these therapies remain in the plane of the neurosis itself. Logotherapy, on the other hand, is transcending this plane, is following man into the specific dimension of his humanness, and is tapping resources available in that dimension such as intrinsically human potentials of self-transcendence and self-detachment.[3]

I have mentioned before that suffering is ambiguous in so far as its meaning may be localized in a dimension higher than the suffering itself. This leads to the question whether the human dimension is the ultimate, the final dimension. Instead of theorizing on this problem, let me report a little episode in my hospital when I once stepped into the room where my first assistant was conducting a group therapeutic session. He had to deal, among other people, with a woman who had tried to commit suicide and was under therapeutic treatment in my department. She had lost a son of 11 years of age, and she was rebelling against her fate. I decided to participate in the group session and asked a question directed at the whole group: "Imagine an ape that is punctured, again and again, in order to manufacture an antipoliomyelitis serum. Is it possible that this ape should ever understand the purpose of its pain?" The group of simple women responded by saying that this was sheer impossibility, because the purpose of inflicting this pain was located in a higher dimension, in the world of man, and an ape is not capable of reaching out into this dimension. Then I asked: "What about the human world? Are you absolutely sure that it is the last dimension, a terminal in the cosmos as it were, that there is no dimension beyond the human world, a dimension in which the ultimate meaning of our human sufferings would be understandable? But then we too have no access to this higher dimension". There was general agreement.

This question of course cannot be answered by either doctor or scientist. It must be left open. As a psychiatrist I must leave it to the patient to answer the ultimate question as to the ultimate meaning of suffering. The same holds for us here: I raise a question, but I must leave it open. This is all the more indicated as the answer to the question after the ultimate meaning of human existence can never be given intellectually, but only existentially: not in words, but by our life, by our whole existence.

---

[3] *Re* "self-transcendence", cf. Frankl (1960); *re* "self-detachment", cf. Frankl (1965).

DISCUSSION

KETY Well, I would certainly heartily agree with Dr. Frankl regarding his comment on reductionism. In fact he might be interested in a paper that I wrote in 1961 called "A biologist examines the mind and behaviour" in which I took issue with reductionism of mind to matter and reductionism of behaviour to biological events, without paying regard to the informational storage of the human being. Reductionism is an unsound hypothesis. The techniques upon which reductionism is based are not necessarily unsound, it is only the extrapolation from them which certain people indulge in that is unsound. And I thought that Dr. Frankl came dangerously close to a denial of appropriate scientific techniques in his comments on reductionism. I was glad that in his discussion of the projection of the cylinder he finally came round to stating that the projections in different planes were not contradictory, they were only contradictory to the most naïve kind of examination. They were complementary, if one took into account the plane of projection, the manner in which it was obtained, and so on, but I think Dr. Frankl came to the same conclusion himself at the very end. Certainly science has analytical techniques which are reliable, provided we take into account that these give us only one segment of the picture. We must put these together, we must try to fill in the gaps which the analytical technique necessarily introduces—indirect techniques—and of course we must never draw inappropriate conclusions from the analytical techniques to the extent or to the conclusion that the cylinder is a circle or is a rectangle.

FRANKL What you have said sums up what I have tried to convey. As I said, the pictures are contradictory to the naïve mind since it is the naïve mind that is in need of the re-evaluating, the re-interpreting. I explicitly said that it is not only the privilege, but even the responsibility of the scientist to embark on reductions, that is to say projections, but he must be aware, remain aware of what he is doing. For some years I have given three lectures a day, 4 p.m. to 7 p.m., the first on neurology, the second on psychiatry, the third on psychotherapy. In the first lecture I sometimes discussed a patient in a reductionist manner: if it was a case of brain tumour, I had to check and examine the reflexes, I was treating him as if he were a closed system of reactions and responses. In the second lecture I discussed psychogenic neuroses, and in the third lecture the first patient was discussed

again, but as a human being—a person whose human dimension I
had to shut out while examining him neurologically.

INHELDER   I know that in many countries psychiatrists are alarmed
about the feeling of existential emptiness, mostly in the young generation; I wonder, however, if you have found out whether this same
feeling is shared by science students and research workers in the
exact sciences, such as mathematicians, physicians, biologists and
experimental psychologists; in other words by all those who are seeking a small part of the truth concerning the laws which govern the
universe or our human conduct. If students today experience this
feeling of emptiness, is it not because we who teach them do not
sufficiently convey our enthusiasm and our faith in scientific research?
Is this in your opinion a widespread sociological phenomenon or is it
more common in specific sections of society?

FRANKL   I have a feeling that you are right, and I would go even one
step further by contending that the existential vacuum is in a way a
contagious disease, because the existential vacuum in the youngsters
is reinforced by the existential vacuum they sense in their teachers.
But you must distinguish between two generations of professors. The
older professors, they still have their idealism and enthusiasm, although it is some old-fashioned type of idealism, but to the youngsters this is preferable—as compared to the emptiness of the younger
professors, those between 30 and 40. This is a remarkable fact. I have
been lecturing at over a hundred universities within the United
States alone, and also at universities in Australia, South America,
Africa, Japan and Israel. I can only speak of impressions, but what
I said is my impression. Something else might be of interest to you:
some time ago I made a statistical research among my students.
Forty per cent of the Swiss, West German and Austrian students
confessed that they knew from their own experiences the existential
vacuum, this inner void and emptiness. Among my American stud-
ents, however, the percentage was not 40, but 81 %. At least partially,
I think, this might be traced to reductionism, which is more prevalant,
I find, on the campuses of American universities than in central
European universities.

HAYEK   Would you agree then that the students revolt because the
professors are too young? (laughter).

FRANKL   No, the student rebellions are evidently connected with
the fact that they are offered no meanings, and this is again partially
due to the subjectivism that permeates most of the theories on mean-

ing and values. I believe that meanings are objective, but this is not merely a personal philosophical conviction on my part, it is the result of psychological experimentation; may I remind you that Max Wertheimer, one of the founders of gestalt psychology, once said that each situation implies a certain quality of "requiredness", and this, he explicitly said, is an objective quality. I would say that, in contrast to those values and traditions that today are evidently on the wane, we should not forget the meanings—life can still be meaningful, in spite of the crumbling and vanishing of traditions, due to the simple fact that this crumbling only affects the values, but spares the meanings. The values are universals, I would say, while the meanings are always unique, pertaining to unique life situations, and to the unique person who is engaged and entangled in such a life situation.

What we call the phenomenological approach seems to me to be a methodological attempt to analyse the immediate data of experience of the man in the street, and to translate these experiences into scientific terms. And then we find that there are three principal ways of finding meaning in life, in any condition, even the most adverse conditions. The man in the street will teach you, if you analyse him adequately, that life can be meaningful by a deed we are doing, or by experiencing what is good and beautiful and true in the world; but if need be also by the way in which a man is shouldering his unavoidable, unchangeable fate in a heroic way, thereby transmuting and turning tragedy into triumph. The man in the street is fully aware, although on a non-verbal level, that this is possible, and if time would allow, I could offer you evidence, drawn not from philosophers, but evidence drawn from utterances of prisoners in California's ill-famed San Quentin Prison—prisoners who were confronted with the gas chamber in which only recently a man had to die. I have visited them, I had to address them. Well, these people may teach us what's going on in a man who is setting out on a valueing process. Man does not originally interpret himself, say, as a battleground of civil war between the id, ego and super-ego. But the man in the street has a basic self-understanding and interprets his own existence as being involved in situations that constitute a challenge, situations that "mean" something to him; anyhow he feels that he has to try hard to do his best, to seek out, to smell out, to sort out the meanings, as it were. And if you systematize this knowledge drawn from the man in the street, you arrive at a phenomenological analysis of the valuing experience in the sense of "finding meaning".

WADDINGTON   I want to make two remarks which are both foot-notes to what has been said already by Dr. Frankl and Dr. Kety.

Firstly, the reductionism, by which you lose dimensionality when you project into a lower dimension, operates throughout the whole of science, not only when you go from the human to the sub-human world, but when you go from any level of organization to any lower one. It operates most drastically when you go from the existent world to any sort of conceptualization. The existent world is obviously of infinite dimensions. You can never exhaust its content by concepts, otherwise you would only have to think of those concepts to re-create it. There is therefore a very extensive loss of dimensionality at that point when you pass from existence to concept, but whenever you go from one level of discourse to a more abstract one you lose dimensionality.

For instance, until a short time ago, chemists operating within the world of chemistry—which you might say is a fairly reductionist one—used a concept of valency. Now this concept could not be included in the still more reductionist world of discourse of physics until London and some others worked out some appropriate quantum-mechanical equations which made it possible to give a satisfactory physical explanation of valency. Before that, however, chemists were employing a concept which was absolutely essential to deal with relations within their field of discourse, but which became lost when one made a further step into the still more reductionist field of physics. This is an example which shows that the loss of content on reducing to a lower level is general throughout the world of science, and not only characteristic of biology.

I should like to develop this theme a little more, into two further points about possible recipes we might follow in dealing with the situation. Consider first the situation when you remain with your concrete multidimensional world, say within your human world, without descending to any lower level. Now it does not at all follow that you necessarily realize what that human world is. You have to develop capacities to discover the full richness that exists within that world. Consider the example of the cup which was projected on to a surface and appeared as a rectangle. What I am saying is that it is a problem to discover that it is a cup. This calls for a very definite effort to explore the full complexities and subtleties of the non-reduced world. Certainly that is one of the major tasks of any science—the task which one could describe in general terms as the attempt to discover what are the interesting questions. Discovering the questions depends on becoming aware of the things that you can ask questions about.

Complementary to this is the attempt to improve your methods of projection into a lower dimensionality. Take again the analogy of the projection of the cup. When projected by light it showed a rectangular shadow or a circular shadow; if, however, you had projected it by X-rays it would have shown a rectangular shadow with a darkened edge, or a circular shadow with a darkened edge. You would have got much nearer to expressing its cup-ness by improving the methods of projection. When the chemists' valencies were adequately translated into physics, this could be regarded as an improvement in the method of projecting from the chemists' dimensions into the physicist's dimensions.

The other point I wanted to make was quite a different one, dealing, from the point of view of a genetic biologist, with this feeling—what did you call it—this "will to meaning". One of the most important defining characteristics of man is that he is involved in the cultural transmission of information, that is to say that he can pass on things from generation to generation, not only through the DNA in his chromosomes, but also by symbolic communication. I think that to the genetic biologist the major thing that distinguishes mankind from other animals is that man has developed symbolic communication so much further than even the cleverest bird. If this is so, then we may say that the very definition of man involves the idea of meaningful symbols. I don't think you could apply the concept of meaning to any sub-human species. It seems to me therefore that a real drive for meaning is something you would have to expect as a fundamental aspect of human nature. Of course the sort of meaning Frankl is talking about, the meaning of life in general, is a very highly developed form of meaning compared to the meaning of symbols. You can have people who can communicate symbolically through speech or writing, without necessarily having any idea of the meaning of life in general. However, I wanted to make the point that it seems to me that the idea of meaning is an almost necessary part of the definition of the human species.

FRANKL   I think we agree that we may call a man a reductionist only if he makes projections, in a scientific way, but is not aware of what he is doing. This is what I explicitly said. But if you asked me for a remedy, I would again say that we should take the lesson of stereoscopic vision, i.e. we should not mind apparent contradictions, but embark on a stereoscopic style of research, as it were. I mean multi-dimensional approaches, as they are carried out in interdisciplinary research. But in addition I would say that we should also learn from Edmund Husserl and Max Scheler, from the message offered by phenomenology. Phenomenology interprets itself as the *Verzicht*, the renouncing of any preconceived patterns of interpretation of given phenomena. For instance, a preconceived pattern of interpretation is established the moment you lie down on the couch of a Freudian analyst and he is setting out to interpret in "psychodynamic" terms, whatever you are saying. Even if you are inhibited or silent this is the expression of a certain "psychodynamics". But this is precisely the same *a priorism* as the preconceived conviction on the part of the rabbi: "If there are five pounds it must be butter—so where is the cat?" That's the same, absolutely the same! Aaron Ungersma, a professor at San Anselmo, California, said in his book on Logotherapy "Freud's motivational theory is valid for the child, that is out to get pleasure. Adler's will to power concept is valid for the adolescent who strives for power, while Frankl's will to meaning concept is valid for the adult, the mature individual." However, the will to meaning might still be the basic human motivation, although in the first decade of life it does not as a rule manifest itself.

SMYTHIES   I would like to reply as a clinical psychiatrist to Dr. Frankl's point of view, and I certainly confirm his clinical statement about people coming to the clinic suffering from the vacuum of meaning. But I think we should think about the other causes of neurosis which one sees today and the curious way in which these are changing. Now you can read all the famous psychiatrists—Freud, Jung, Adler, and so on, and you get the impression that they are describing natural laws, i.e. that neuroses are due to certain causes, and these causes operate at all times and at all places for all men. But clearly the causes of neurosis which bring the patient to our clinics are in a very real sense culturally dependent. If you apply Freud's basic law of repression: what is repressed causes neurosis afterwards—I think no one would dispute that. But what is repressed is what the culture disapproves of. Now different cultures disapprove of different things, so in different cultures you get different things repressed. In Freud's Vienna of 1890, for example, sexual matters were taboo, therefore sexual matters were repressed and so Freudian psychology came to be based on a whole gamut of repressed sexual conflicts, which are presented as the cause of neurosis. As for Arthur's paranoid streak—I don't mean *Arthur's* paranoid streak (laughter)—he speaks of a universal paranoia; now somebody might suggest that in all cases and all types and all lands in every possible world, paranoia is caused by repressed homosexuality. But nowadays, seventy years after Freud, the whole cultural climate *re* sexual matters has changed, sex is no longer taboo; and this came about largely as a result of Freud's own teachings. Thus one could imagine a society in which homosexuality was not in any way something bad, and therefore did not have to be repressed. If you go back to ancient Greece, to Theban culture, you remember homosexuality was perfectly accepted and some of the leading people—for example the cream of the Theban army—were homosexuals, and this was regarded as perfectly normal. Therefore if it did not have to be repressed, it could not be the cause of paranoia. But unfortunately, human psychology being what it is, once one taboo is overthrown, another one grows up in its place. And we have other taboos at the moment.

Sex has been taken away as a taboo, but now one of the main ones which has taken its place and which fulfils the same neurosis-generating function is the concept of death. Every culture has a particular complex cultural machinery for dealing with death and dying. I am not so much talking about the concern of one's own death, but about the effect of death on the person who has been left behind. The forms that bereavement take vary from the Irish wake, where death is regarded as being a transition to a happier world, to our taboos on death. In the wake the party is a great event, a two-day-long party, with plenty of singing and dancing—a very effective way of dealing with grief and bereavement.

It is extraordinary that the first serious scientific paper on the subject did not appear until 1941 when Eric Lindemann at Harvard wrote his famous paper on the analysis of the grief reaction. He studied the victims of the Coconut Grove fire in Boston. Two hundred people were burnt to death in a nightclub fire—and he went round to interview the relatives of the people who had been killed, and so he wrote the first scientific account in Western science of what the grief reaction is and what aberrations it could form. Now as our Judeo-Christian basis for ideology progressively decays, the support it was able to give to the people left behind in bereavement likewise vanishes. If the dead person is thought to have gone to a happier world, then this is a support to the patient. But nowadays this belief has very largely decayed and one commonly sees more and more of the patients coming to the clinic suffering from Lindemann's grief reaction. We have no culturally valid method for dealing with the death of the loved ones, and this leads them to their state of loss of meaning. The meaning of their life has been lost, the meaning that was invested in somebody else, and no other course of the meaning has arisen. These abnormal grief reactions can last for months or years and are very crippling—the person is often psychologically completely crippled. This is an example of another cause of neurosis which supports in general terms your particular view; it expresses your idea of "loss of meaning" looked at from another aspect, a very personal one.

WEISS   I want to say something about the university unrests. I have seen some of these incidents in Milan; we had them in Columbia, I know something about Frankfurt and we've made some investigations. They are not yet completed, but it was quite characteristic that the shock troops among the student activists came almost exclusively from the departments of the humanities and sociology, and there were hardly any scientists among them.

HAYEK   Paul Weiss confirms the half-joking suggestion I made before. But it is not the humanities in general—for instance the classics —which provide the revolutionarist: it's a few subjects like sociology and political science where the professors *are* very young (laughter).

REFERENCES

CRUMBAUGH, J. C. (1968) "Cross-validation of purpose-in-life test based on Frankl's concepts", *Journal of Individual Psychology*, 24.
FABRY, J. B. (1968) *Logotherapy Applied to Life*. Beacon Press, Boston.
FRANKL, V. E. (1960) "Beyond self-actualization and self-expression", *Journal of Existential Psychiatry*, 1, p. 5.
———— (1962) *Man's Search for Meaning: An Introduction to Logotherapy*. Washington Square Press, New York.
———— (1965) *The Doctor and the Soul*. Knopf, New York.
———— (1968) *Psychotherapy and Existentialism*. Washington Square Press, New York.
———— (1969) *The Will to Meaning: Foundations and Applications of Logotherapy*. World Publishing Co., New York.

GENERAL DISCUSSION

KOESTLER   I am very glad that Frankl's was the last presentation on our programme, because the real, the underlying subject of the conference was the rejection of this crude type of reductionism and, frankly, the selection of the participants was guided by this consideration. Now Frankl has stated what was implied throughout our proceedings, even in the most technical papers, and stated it in a very explicit and forceful manner. It was very useful, I think, to cut closer to the bone of what naïve or gross reductionism means, and how far on the other hand reductionism is legitimate as an analytical technique in the solution of limited technical problems. But as you have pointed out, when we speak of "naïve" reductionism, then we have to include among the naïves even men like Sigmund Freud; and quite a number of heavyweights in the various sciences still adhere to this naïve type of reductionism.

When you ask, however, for an alternative view, for instance in the field of psychology, then I have the feeling that one is caught between Scylla and Charybdis: one rejects the Behaviourists and the Freudians, but where is the alternative? There is in America a new movement in psychology—I do not want to go into names—they have a journal, and I am even a member of the editorial board of it, but I am unable to read much of it because it seems to me all wool—very pretty wool, cotton-wool—(laughter).

SMYTHIES   Some of it is even worse.

KOESTLER   So here we are between Scylla and Charybdis, and if one tries to define the frontiers . . .

BERTALANFFY   Did you say between Skinner and Charybdis? (laughter)

KOESTLER   And of course you have the same situation in biology. If you reject the gross mechanism derived from nineteenth century classical physics, classical vitalism is certainly no alternative.

WADDINGTON   I want to bring together a statement by Dr. Frankl and another by Dr. Smythies, and in this connection raise the point whether we are not perhaps treating this subject on a too professional level. Dr. Frankl said that he comes across neuroses based on the frustrated search for meaning and the inability to find meaning. Smythies says that you cannot have a neurosis unless a culture actively represses some natural tendency and prevents you from getting satisfaction for a particular drive or motivation. Do not these two statements taken together mean that our present society is actively repressing the search for meaning? It is not merely that there are naïve reductionists who are not clever enough to see that theirs is a too simple reductionism, but there are actually some forms of social activity which are, as it were, positively frustrating the search for meaning—making meaning into a dirty word.

BERTALANFFY   That is what I call the return to the conditioned reflex by way of the mass media.

THORPE   But Waddington's is, I think, quite a new, important point.

WADDINGTON   Personally I am quite ready to be persuaded that the present social intellectual system of the West actually does repress the search for meaning and tends to regard meaning as a dirty word. Certainly in the last couple of decades people have hardly dared to use the word "progress", which implies positive values. You can hardly talk in intellectual company of the meaning of progress, it has become a dirtier word than sex. Now why should this be so? And if it is so—and I think that the existence of neurosis on this basis is ample evidence that it is—then it seems to me that this is something much more difficult and important to deal with than the possibility that there are some silly reductionists about who don't understand.

KOESTLER   I agree with what you say, but it goes even further. You might say that logical positivism, which calls all questions meaningless which refer to God, and man, and the universe—which calls all these questions meaningless, you might think this is a philosopher's dispute and does not affect the public at large. In reality it affects the public enormously, the semi-literate public through the mass-media, through popularization, indirect indoctrination, and so on; and although Freddie Ayer has now mellowed and admits that some of these questions may have meaning, which he previously denied, this kind of positivism is still going very strong.

BERTALANFFY   I would say it is the principle of replacing human symbolic decisions by Pavlovian-Skinnerian conditioning. Which seems not to work too well.

WADDINGTON   There is something oppressive in the thought that such implausible systems as either logical positivism or Skinnerism could become fashionable.

SMYTHIES   Could I suggest that Professor Waddington's question might provide the subject for another symposium—it is a vitally important question. Which are the organized forces in society that are suppressing meaning by devious and often unconscious means?

WEISS There is a tendency to suppress not only meaning, but also the freedom of expression—these two problems are linked. And then you get these outbursts of protest.

FRANKL On the other hand freedom is too much stressed in a one-sided, in a negative way. All the protesters are actually anti-testers, they have no "pro", no positive alternative to offer, but they are fighting against, rather than struggling for something. In a recent editorial in the campus paper of the University of Georgia, Becky Leet asks: "For today's younger generation, how relevant is Freud or Adler? We've got The Pill to free us from the repercussions of sexual fulfilment—today there is no need to be frustrated and inhibited. And we've got power, witness the sensitivity of American politicians to the 25-and-under crowd, or look at China's Red Guards. On the other hand, Frankl says that people today live in an existential vacuum and this existential vacuum manifests itself mainly in a state of boredom. Boredom—sounds familiar? How many people do you know who complain of being bored—even though we've got everything at our fingertips, including Freud's sex and Adler's power? It makes you wonder why. Frankl may have the answer." Of course I have no answer. Logotherapy is a catalyst, as it has been described in a letter I recently received from a young American soldier fighting in South Vietnam—he says: "I have not yet found an answer to my questions, but your books have started my own wheels of self-analysis turning once again." This is the catalyst function. Now, Koestler said that Freud belongs, in a way, among the naïve reductionists. Listen to what Freud has to say: "I have confined myself to the ground-floor and basement of the edifice called man. But in this basement I have already found a place for religion by putting it under the category of the neuroses of mankind." And now we understand how come that so many young people, and even middle-aged people, are repressing their existential despair, are repressing their longing for an ultimate meaning—their religion has been devaluated and de-preciated, and this of course takes its toll, because such a judgement on religion again is an over-generalized statement. There is a religion in the widest sense, Einstein has conceived of it by saying "if a man finds a satisfying answer to the question what is the meaning of his life, this man I would call religious". Unfortunately religious longings—in this wide sense of religion—have been repressed, and this leads to existential frustration, because no positive alternative has been offered.

HAYEK   May I make just one observation: it is very characteristic of these "nothing but . . ." reductions that the enumeration of factors which follow the "nothing but" always ends with "etcetera" (laughter). I call this "etcetera reductionism" which of course does not really achieve what it claims to do.

THORPE   It seems to me particularly significant that our discussion has come back to this basic problem of reductionism *versus* organicism. For that seems to me what the conference is really about. So I will continue with some thoughts and questions on that general topic.

Now it seems to me that you can attack, or consider the problems of reductionism at some particular points in the story much better than you can at other points; and for various reasons it also seems to me that the points at which reductionism can be looked at most closely and most usefully is probably at the level of the nervous system and especially the neuron. For that reason I must say that I was particularly impressed by Paul Weiss's contribution, because it seemed to me that this was getting down to the fundamental basis of understanding what the problems are when you speak about reductionism in relation to the nervous system.

I think it is very easy to suppose, or to have a logical faith, that everything can be reduced to a simpler unit or element. But I don't think there is much real evidence for this at present; although from the point of view of scientific method it seems overwhelmingly attractive. On the other hand you can take the opposite view and say that organicism is everything and that we cannot construct the more complex organism out of the bits and pieces, cells or other units, whatever they may be, which you know exist. Now I don't think we shall really obtain any evidence to decide between those two until we can be virtually certain what the individual bits can and cannot do by themselves. It's no good saying that when you put two cells together you get some new property until you are sure that neither of those cells had that property when isolated and before you put them together. And so to be precise, you must know, as completely as it is possible to know, the properties of the individual components before you investigate the resulting properties when they are acting together.

That is why I look upon the nervous system as one of the most promising subjects for investigation. For it seems to me that the chances of unravelling the unitary and distinguishing it from the multiple are better there than in almost any other tissues. A particularly good example of this is the investigation of the behaviour of very small ganglia, such as the cardiac-ganglion in the lobster, where, as I remember, you have nine neurons. I don't know whether we are as yet anywhere near explaining the properties of the nine-neuron ganglion in terms of what we know of these nine cells by themselves—perhaps Paul Weiss has some up-to-date information. But this is the kind of complex which must be analysed as fully as possible before one is in a position to say much about either reductionism or organicism in that particular type of situation.

Another level where a similar problem arises and which happens particularly to interest me, is the level at which animals begin to show evidence of being able to make some sort of a "map" of their environment; that is to find their way about in it. I assume there is no evidence that any *Coelenterate* can do this—indeed it is still not quite certain that they can learn. You cannot think without a nerve net, and there is no evidence that Protozoa can "make a map". But, although the evidence is in many respects conflicting, it now seems fairly plausible to suppose that some Protozoa, e.g. *Stentor* display some true learning—i.e., a daytime change in behavioural organization as a result of individual experience—and sea-anemones too can learn. However, the fact remains that when you come to insects, and molluscs, you can detect a surprisingly high ability to, so to speak, make an internal representation of the environment so that the animal can find its way around, and can find the shortest way back to its nest, or home-range, or whatever may be the objective for its return. And you get these astonishing performances not only of bees, but also of wasps (*Vespidae*), hunting wasps (*Ammophila*), etc., where a "locality study" or "survey flight" of a few seconds will enable the animal to go right away and then return perhaps an hour or more later to its nest, with only a very small chance of getting lost.

I have some estimates here of the number of cells involved in insect and other invertebrate brains. Our knowledge of the size and number of nerve cells in the brains of animals is exceedingly scanty and mostly rests on estimates of long ago. Even the common statement of ten thousand million as the number of nerve cells in the human brain rests on an estimate of over sixty years ago (H. B. Thompson 1899). J. Z. Young gave us in 1963 an estimate of 5×10^6 nerve cells for the nervous system of *Octopus*. But for insects the position is much less satisfactory. For the estimated size of insect brains we go back more than one hundred years to Dujardin (1850), who gives the volume of the brain of the honeybee, *Apis sp.*, as 0.62

cu. mm. and that of the ant, *Formica sp.*, as 0.065 cu. mm. The brain of man has a volume of about 1,600 ml. Accepting Dujardin's figures, the brain of the bee is about 4×10^{-7} the volume of that of a man, and the brain of an ant is about 4×10^{-8} of that of man. Even though the size of a nerve cell body is very far from being proportional to the size of the organism, it would seem unacceptable to be forced to conclude from the above figures that a bee-brain contains only a few thousand ganglion cells and that an ant only a few hundred: the larger the brain the larger is the proportion which must be assigned to connecting axons and the smaller the number of cells per unit volume.

A satisfactory estimate of the number of nerve-cell bodies in a bee's brain is urgently needed in view of the enormous complexity of behaviour of these animals which has become apparent following the work of von Frisch. But the point is of such interest that even a rough guess may prove useful. Rockstein (1950) gives figures for the number of cells in two arbitrarily placed 5μ transverse sections across the brain of *Apis*. The numbers obtained are about 800 cells per section; since he states that the smallest neuron cells have a nuclear diameter of 5μ, this number is likely to exceed the actual mean number of cells per 5μ section. But if we take a mean value of 800 cells per section, and take Dujardin's value for the size of the brain and allow for its shape, we arrive at a value of the order of 10^5 nerve-cells. The number in the brain of *Formica* is likely to be very much smaller.

Such very rough estimates point to the great importance of getting accurate ones. Of course the individual nerve cells are not simple standard units, the same in every organism and the same in all its nervous parts. The pattern of its axons and dendrites is of essential importance to the significance of a nerve cell as a unit. All the same, a configuration of 10^5 structures is by no means beyond our power of apprehension when we bear in mind that a computing machine such as "Titan" is built up from units of the order of 10^6 in number. Now a bee brain is, of course, infinitesimally small compared to the cerebral equipment a vertebrate has with which to organize its behaviour. Yet I would hesitate to say that the direction-finding capabilities of one of the hunting wasps, or of the honeybee or the bumble bee are strikingly inferior to that of the homing salmon or the homing bird. Some of the achievements of the hunting wasps are striking enough to puzzle us human beings, and wonder whether we should be able to do as well in similar circumstances. Thus the hunting wasp, *Ammophila pubescens*, digs burrows for its young. It hunts over a considerable distance for spiders and caterpillars, which it paralyses and puts in its burrows. Later the eggs hatch and the young feed upon the paralysed prey. If the prey is too large to be carried by flight, the wasp will drag its prey around obstacles along the ground towards the burrow. If wasp and prey are transferred in a closed box to a new place some distance away, the prey will nevertheless be dragged towards the burrow. The wasp behaves in fact as though it had an internal model of the district round its burrow, just as K. J. W. Craik suggests (*The Nature of Explanation* 1952) that we, ourselves, have an internal model of the external world which we use to control and predict action. The wasp is behaving in relation, not to stimuli, but to a world of objects, and to a world of objects identical with that accepted by our own everyday naïve realism. (Pantin, 1968, *loc. cit.*). So it seems to me that the problem of how many cells an animal requires in order to do a particular thing again becomes very crucial.

Finally, I would like to come to an entirely different point. I was particularly interested in the mention this morning of Gödel and Tarski. As I understand it, Gödel's theorem amounts to saying that no mathematical system that is completely closed and consistent can serve as a precise language which is universal. Now I'm sure that some of those here can say much more about this than I; but would you say that it follows from this that one cannot have an account or a universal description of nature in a single closed consistent language? If that is so, then it seems to me that we need not be surprised, however much we may regret it, if we require different languages to describe effectively different parts of nature. And in fact the dualistic approach—whether a general dualism or a sort of double-aspect dualism—may be essential and inevitable when we are trying to describe various stages in biological evolution and development.

Weiss made a remark in his presentation about the changing view of physicists, in particular he mentioned cosmologists, as to the question of a basic randomness of the natural process. If in fact randomness is going out of the physical picture, then the whole situation has been fundamentally changed. Perhaps we are seeing the beginning of what Whitehead foresaw—that biology would take over physics, and not physics take over biology. Certainly there has been an emphatic change in basic physical concepts over the last 20 years;

how far this has gone, and how far it is likely to go in the near future I couldn't guess; but any information about this, and particularly Paul Weiss's ideas on the subject, could be extremely valuable.

WEISS   Well, I guess that biologists could make a legitimate case on the basis of their postulates from hard and fast actual observations, from experiences with the living system, and that is perhaps what is going to tempt the physicists to try harder to expand their concepts. All this leads up to the problem of randomness—whether we should start from chaos, or whether it is an artifact of the human mind.

WADDINGTON   The same sort of thing is also going on among quantum physicists. I have recently been running a series of symposia on Theoretical Biology—the first volume is just out—"Towards a Theoretical Biology: I Prologomena" (Edinburgh University Press). In the second volume, which is to appear shortly, there is a considerable discussion by David Bohm—one of the theoretical physicists who are dissatisfied with the present outlook on quantum indeterminacy. He wants to go beyond this, to what he calls "orders of order"; his basic point being that the fundamental notion in any logical system is the notion of order.

KOESTLER   You have on the same side de Broglie and Vigier.

SMYTHIES   You say the universe started out of either chaos or order. One normally assumes, if one accepts the big bang theory, that the primary mass of matter in the universe before the big bang was entirely homogeneous. But there could have been, I suppose, a high degree of specificity in that initial mass of matter. It is possible that everything was not entirely random to start with, but highly organized, and that this order has been progressively unfolding ever since. It seems to me an entirely arbitrary assumption that—going backwards—primary matter was random: although the derivation of the complexity of the universe as it is today from a homogeneous creation of highly dense particles is certainly an aesthetically more pleasing theory.

WEISS   Of course, that's exactly it. I mean the alternative is open and this has been authorized by the physicists, so to speak. So the biologists may have the courage of their own convictions.

WADDINGTON   It was, of course, biology in the person of Darwin which first introduced randomness into the deterministic world and made it respectable.

WEISS   Sociology is the experimental play pen where they demonstrate the existence of randomness, isn't it?

THORPE I don't know whether this is going to affect our ideas of the probability of an accidental production of enzymes. This seems to me at present one of the really tremendous blocks to any attempt to connect the biological and physical world through the evolutionary point of view. Thus with regard to the improbability of complex machines M. Dixon and E. C. Webb (1958 *Enzymes*, London) have very relevant remarks on the crucial question of information transfer and storage in the intracellular mechanism for protein synthesis.

Thus, they say:

The code is built into the biosynthesis mechanism in two places: (*a*) it is embodied in the DNA of the genes, which use it to represent the corresponding proteins, and (*b*) it is built into the specificity of the ligases. One of the most interesting and fundamental questions in biology is how it comes about that the genes and the ligases use the same code. The answer that no life is possible unless they do is unsatisfying; the chances against it coming about without some controlling mechanism to relate the two are enormous, but it is extremely difficult to picture such a mechanism.

And again:

The structures of both the specific centres in an enzyme which is subject to feedback inhibition are determined by its structural gene. The genes indeed display an astonishing amount of "knowledge" about the sequence of chemical processes in metabolism. One may well ask how the gene-forming enzyme 2.4.4.14 "knows" that phosphoribosyl pyrophosphate will be converted by the consecutive action of ten or more different enzymes into a purine nucleotide, or how the gene for the first enzyme of histidine bisynthesis, which acts on the same compound, "knows" that its product will be converted into histidine by a different series of enzymes. Even with this information, how do these genes "know" what amino acid sequences in their enzymes will act as specific centres combining with purine nucleotides or histidine respectively? Evidently there must be some mechanism whereby information derived from the metabolic processes themselves is transmitted back to the genes and there incorporated in the form of polynucleotide sequences. The manner in which control was established in the first place, and the nature and mode of action of this mechanism, is one of the most fascinating and fundamental questions of biology.

WEISS   May I say one thing still, not because I have to have the last word, but because I'm the oldest it seems here. I think that the last day's discussions have indicated the real success of this symposium, and that certain general viewpoints have emerged, that there has been a remarkable opening of new vistas. They are ser—what did you call it?—serendipitous, in the sense that we didn't look for them. They came from all disciplines. We didn't know that you can get so much conformity provided you get the right kind of people together.

THORPE   Was this random or selection?

KOESTLER   Cunning selection.

WEISS   There has been to my mind a great amount of really valuable information, and integration which made one recognize that there are so many things that really fall into place, if one is careful to avoid doctrinaire prejudices, and avoids acting the way one usually does in one's own guild, where nobody in modern society dares to step out, and to speak his mind quite as freely as we have done here in this inter-disciplinary group, where none of us gets jobs from the other, and nobody has to worry. I think that what has emerged is a confirmation of Arthur's idea of self-regulating open hierarchic order, the capacity for self-organization of a random group, slightly pre-selected of course (laughter); the emergence of a higher level of socialization, may we say, and this is something for which we owe Arthur Koestler a tremendous debt of gratitude. And so I would like, I hope I speak on behalf of everyone, to thank him for this "act of creation". But we must thank Cynthia Koestler at the same time, who has shown such a remarkable ability on the tape recorder to control the ghost in that machine (Laughter). May I also wish that Arthur would extend our thanks to all the members of his staff for the help they have given to all of us.

KOESTLER   Thank you. I considered my function as that of an Honorary Catalyser.

INHELDER   Then let's have some auto-critique.

KOESTLER   Well, it is difficult to talk of intangibles. I am glad that Paul felt that in these few days a few bridges of understanding have been built, and a certain integration has taken place. This, of course, could not have ocurred without a certain pre-selection—this point has already been made. But the criteria of this particular Alpbach selection are difficult to define. It was to be a meeting of people in the mainstream of research, of specialists in their fields and in their experimental laboratories, who nevertheless feel a discontent with the *Zeitgeist* and its reductionist "nothing but" attitude. That was to be the common denominator, admittedly a rather negative one, but it seems to have worked out quite well. Everybody has of course his own formulation for his holy discontent. I have tried elsewhere to give my own tentative formulation—the rejection of what I called "the four pillars of unwisdom". In a simplified form, the four pillars to me are the doctrines:

(a) that biological evolution is the result of nothing but random mutations preserved by natural selection;

(b) that mental evolution is the result of nothing but random tries preserved by reinforcements;

(c) that all organisms, including man, are nothing but passive automata controlled by the environment, whose sole purpose in life is the reduction of tensions by adaptive responses;

(d) that the only scientific method worth that name is quantitative measurement; and, consequently, that complex phenomena must be reduced to simple elements accessible to such treatment, without undue worry whether the specific characteristics of a complex phenomenon, for instance man, may be lost in the process.

These are my four pillars of unwisdom, and the Symposium has shown that we all seem to agree that the pillars are hollow and cracking. We cannot go much further than that, but even a negative agreement is something. To expect that in five days a positive, new philosophy would emerge would have been silly. So let me close with my favourite motto—a sentence I read in a science fiction magazine:

"I have yet to see any problem, however complicated, which, when you looked at it the right way, did not become still more complicated" (Poul Anderson).

Finally, let me thank, in the name of us all, Bill Thorpe for having been such a model moderator.

THORPE Thank you—the conference is closed.

Retrospect
by W. H. Thorpe

On reading the typescript of our Symposium after a lapse of nine months, I found it gratifying that a conference which ranged so widely over so many fields in the life sciences, nevertheless managed to stick to the essential points so consistently in its attempt to answer the challenge of both the title and sub-title of this volume. We were given an impressive series of essays on new perspectives in the life sciences; yet almost every contribution and discussion touched upon the main problem, namely, *reductionism versus organicism*. In this postscript I wish to run briefly through some of the main points of importance which our meeting raised; referring especially to those where we were able to suggest some further steps of promise along the path towards a solution.

First, let us consider reductionism. About this, there was on the whole a remarkable unanimity. I think none of us felt, and certainly none of us said, that reductionism as a technique or a way of scientific approach to the understanding of nature, is either avoidable, or indeed in any way reprehensible. But many of us hold that because in the last three hundred years the scientific technique of reductionism has been so successful in gaining control over the forces of nature, our present society is far more receptive to rationalistic-mechanistic philosophies than to others simply because it considers such views as more "scientific" than other alternatives. Much of society's acceptance of this is, in fact, based on a misunderstanding of the nature of science itself. Reductionism is only one aspect and not the most fundamental aspect of the activity of scientists and the progress of science. The scientist who has a real understanding of his work and is not simply a glorified laboratory technician finds, as Paul Weiss said, that controversies, on an abstract level, on reductionism versus holism vanish in the light of realistic studies of actual phenomena. In all our studies of nature we find hierarchic order a demonstrable fact. To paraphrase again Paul Weiss' keynote presentation, we can only focus our attention on objects or systems which persist long enough, or recur often enough to deserve a name. This determines the nature of the objects and events of biology. But unless the process of synthesis to the organismic level is carried on at the same time, biology will eventually grind to a halt. When we shift, as we must, from analysis to synthesis, we do so by raising our sights from single objects to their interrelations with others. Arthur Koestler's concept of holons and his description of them as "janus-faced entities" put the essential facts in a very clear and dramatic way. We see that holons are not simple conglomerates but are systems; or ordered conglomerates. And it now appears that hierarchy is a conceptually inevitable idea, and that the part of it which we perceive at any one time depends upon the level at which our attention is focused. But the hierarchical idea is more than this, and I advise readers of this book to look again at Paul Weiss's calculations (p. 13) concerning the organization of the brain and also his *Canon for Determinacy* (pp. 32–33).

Although there are many parts and aspects of experimental biology where, for the time being, it does not matter very much whether or not the experimenter understands this relationship between determinism and organicism, yet as we go up the scale towards the sciences of man the dangers inherent in the acceptance of this complete reductionist view become increasingly serious. For it leads inevitably to a view of the universe as a great system of physical forces, and the mind with all its powers of imagination and creative insight as a mere by-product of those forces. It is most instructive in reading this book to see the deficiencies and dangers of this view exposed on various levels in ascending order: from the basic neuro-physiological approaches of Hydén, Kety, MacLean, through the complementary aspects on the cognitive level discussed by Bruner, Hayek, McNeill, Piaget, Inhelder and Smythies; to the system-theoretical considerations of v. Bertalanffy and Koestler; and, as a climax, Victor Frankl who, in his study of "Reductionism and Nihilism", shows most impressively how false beliefs of this kind can lead to some of the major psychiatric disorders current in the world today. And I believe he is correct when he says that reductionism is the nihilism of today. Finally, let me conclude this section of my retrospect by returning to the micro-level and agreeing with Paul Weiss, that there is a flaw in the scientific approach which leads to a belief in micro-precise causality (micro-determinism) since such a view of a physically determined micro order, or for that matter, of a physically determined brain, has been basically untenable since the coming of Heisenberg.

Since it seems to have been to some extent a remark of mine which sparked off this conference (c.f. p. 1) I would like to come now to the question of the general validity of the modern genetical-selectionist argument as a complete and satisfactory "explanation" of the evolution of the living world. A partial basis for this dissatisfaction was touched upon in the discussion of Ludwig von Bertalanffy's paper (pp. 77–82). The selectionist's argument is one that can be expanded or elaborated to cover anything that may conceivably have happened during the evolution of animals and plants. If selection is taken as an axiomatic and *a priori* principle it is always possible to imagine auxiliary hypotheses—unproved and by nature unprovable—to make it work in any special case. But as von Bertalanffy points out, this procedure corresponds exactly to that of epicycles in the Ptolemaic system: if planetary motion is *a priori* cyclic, then any orbit, however seemingly irregular, can be explained by introducing more epicycles. Similarly, some adaptive value (and consequent selective advantage, survival due to differential reproduction, etc.) can always be construed or imagined. It follows, I think, from this that no amount of theoretical argument as to how the evolutionary results which we now see may have been brought about can carry complete conviction. What can and must carry conviction is the results of experimental genetics, which show how a change can actually be brought about in the laboratory in experimental populations; or, alternatively, in wild populations under close genetical analysis. I personally derived great satisfaction from Waddington's wonderfully complete, yet cogent and condensed, account of the "Theory of Evolution Today". I think we all found his concept of the "chreod" and its elaboration to be valuable and illuminating. And I think I interpret Waddington correctly when I say that he is no more "a reductionist" than the rest of us.

One essential point to note is, I think, that we have now good evidence for the biological transmission of information by systems lying outside the chromosome. Waddington himself has developed this point very interestingly—see his paper *The basic ideas of Biology* (*Towards a theoretical Biology*. 1. Prolegomena. I.U.B.S. Symposium, Ed: C. H. Waddington, Edinburgh Univ. Press 1968, see pp. 16 and 17). It is not even certain that it is nucleic acid which transmits the non-nuclear information, and some biologists think that the cortical information in an egg cell can be transmitted over many generations. As Waddington says, "the majority of biologists doubt this, but there seems no good reason why such information should not be transmitted through the formation of at least a certain number of replicates, as, for instance, in the growth of mitochondrion or nuclear envelope".

I personally found Waddington a little "cagey" concerning the significance of biological organization at a relatively low molecular level. J. B. S. Haldane (1959 in *Darwin's Biological Work*, Ed: P. R. Bell, Cambridge Univ. Press) was himself forced to the conclusion that it is almost certain that facts will be discovered which will show that the theory of natural selection is not fully adequate to account for evolution. Waddington does not seem to have much to say for the idea of "internal" factors in evolution as set forth by L. L. Whyte in his book of that title (London, 1965), but I still think that this book contains some important points which geneticists as a whole have failed to stress sufficiently—and I was glad to see that Paul Weiss also thinks that Whyte makes a significant contribution. The essence of Whyte's book is the conclusion (shared among others by Haldane, Spurway and von Bertalanffy) that the conditions of biological organization restrict, to a finite discrete spectrum, the possible avenues of evolutionary change from a given starting point. The nature of life limits its variation and is a basic factor directing phylogeny. Thus the mutations of which the consequences reach the Darwihian test have already been sifted by an internal selection process. This internal selection restricts the directions of evolutionary change by internal organizational factors. Whyte says, "There may be no mutations which can be fully ascribed to chance." This, I think, amounts to saying that there is a selection going on not merely at the phenotypic level, but also at the molecular and chromosomal levels. And he quotes Medawar (P.B., *The Future of Man*, London, 1960) as having suggested that there is in systems of polymers a repetitiousness or tendency towards elaboration, so that ever more complex sets of genetical instructions are offered for trial and that this may provide a basis for advancing complexity. I think we are all agreed that it is the development of complexity, which in the animal series as a whole, seems to show a continuing trend; and this trend is one of the major problems of present-day evolution theory. Also, G. Bachelard (1953 *Le Matérialisme Rationel*, Paris), quoted by Whyte, argued that there is implicit in quantum mechanics a structuring tendency for complex systems to form more complex forms of ordering. And this again relates to the fascinating problems which are here and there ventilated in this book concerning the possibility that there is a non-random feature, perhaps at the very basis of the natural order, which may well have to be taken ultimately into account by biological theorists.

With Waddington's approval, as I interpret it, for the view that the animal has an active "strategy" in seeking new environments, I feel that another big difficulty that I previously had with orthodox genetics is fading away. But there are still one or two problems which strike me as formidable. The first is that of homology. Geneticists tend nowadays to dismiss as unimportant, or else ignore altogether, the concept of homology which most zoologists, at any rate until recently, regard as absolutely fundamental to the idea of evolution. Sir Alister Hardy in his recent book *The Living Stream* (Collins, 1965) says "in the 1930s it all seemed so obvious, the same homologous structures must clearly be due to the same hereditary factors handed on generation after generation from the early ancestor with occasional changes by mutation; the wide variety of form seen in different animal groups being due to natural selection acting upon these factors or genes which were handed on, with mutational changes, from the original ancestral form". This same view of homology was championed by J. B. S. Haldane in 1932, by Sir Gavin de Beer in 1938, and I suggest, in much more recent writings by zoologists. It seems, however, that recent developments in genetical theory can no longer envisage the fairly static gene-pool or "gene-cluster" which is maintained intact over long periods of evolutionary time and is responsible for the slowly changing development of organs such as the wings, limbs and mouth parts of insects, or the fore and hind limbs of mammals and birds, through all their marvellous structural adaptations, of which we have fossil and recent evidence. Now it is suggested that these constant systems may exist only in our imagination and that the genetic control of the development of such homologous organs may shift relatively rapidly while the organ remains the same! This seems to me to raise a quite fantastic difficulty, and if the orthodox genetical argument leaves us with this riddle, it becomes increasingly hard to my mind, to see how it is that life has progressed beyond its simplest forms—since increase in complexity or each major change in an organ makes life that much more precarious. And coming to a group I happen to know particularly well—I find the fantastic development of the beautiful and complex patterning of male birds very hard to account for as merely part of the isolation mechanisms which keep one species separate from another. They seem to have gone too far and developed too big a momentum of their own to be explicable on that basis (see John Thoday "Chance and Purpose" in *Theoria to Theory*, Vol. 2, 1967, pp. 29–38; and together with this Sir Alister Hardy's reply *Theoria to Theory*, Vol. 2, 1968, pp. 312–323).

Not being a geneticist myself I might feel more hesitation in putting forward difficulties of this kind had it not been for the apparently extraordinary state of flux in which genetics finds itself at the present moment. As we may gather from the final pages of discussion in this book, geneticists are presently living in hazardous times—not least because the definition of so many of their basic concepts is becoming more and more difficult. Let me remind you of Waddington's words: "I think we are going to see extraordinary changes in our ideas about evolution quite soon." And I also draw your attention to the extraordinary genetical happenings which seem to be involved, and to take place very rapidly, in the production of what is called "an established strain". Also in this connection, I would refer to the remarkable conclusions (quoted by Waddington) concerning the work of Forbes Robertson in his laboratory.

Lastly, a concluding word on reductionism: I would call your attention to a remarkable dialogue on this subject between Carl Frederick von Weizsäcker, Professor of Philosophy at the University of Hamburg, and Martin Garstins, University of Maryland and Office of Naval Research, Washington. (*Theoria to Theory*, Vol. 3, October 1968, pp. 7–18.) Von Weizsäcker says that "those reductionists who try to reduce life to physics usually try to reduce it to primitive physics—not to good physics. Good physics is broad enough to contain life, to encompass life in its description since good physics allows a vast field of possible descriptions. There is no reason why living beings should be compared to primitive machines which don't make use of feedback". In other words, if we had a real understanding of the nature of physics today there might be very little danger left in reductionism even as an article of belief. But unfortunately, the ordinary man's physics is still that of 100 or more years ago! And von Weizsäcker in the article just quoted says "the concept of the particle is itself just a description of a connection which exists between phenomena, and if I may jump from a very cautious and skilled language into strict metaphysical expression, I see no reason why what we call matter should not be 'spirit'. If I put it in terms of traditional metaphysics, matter is spirit as far as spirit is not known to be spirit."

Finally—if reductionism were right in the sense that the mental, spiritual, artistic, and ethical values which we experience really *are* in the electrons and other primary components of which the world is made—then all one can say is they don't *appear* to be there. It follows that a great and unjustified leap of faith is required, a leap without any scientific evidence, to believe it. Thus reductionism requires at least as great a faith, if not much greater faith, than the organismic and hierarchic approach combined with Weizsäcker's open-mindedness

By the first we are required to believe what we can in no way detect. By the second we are required to believe in a source of value added to or injected into, the natural process as complexity develops, which we are totally unable to understand.

INDEX OF NAMES

4/73
2406 2.

438 *Index of names*